J. Zittoun   B. A. Cooper (Eds.)

Folates and Cobalamins

With 31 Figures and 39 Tables

Springer-Verlag
Berlin Heidelberg New York
London Paris Tokyo Hong Kong

Prof. Dr. Bernard A. Cooper
Division d'Hématologie et d'Oncologie médicale,
Hôpital Royal Victoria et Département de Médecine interne et de Physiologie,
Université McGill, Montréal, Québec, Canada

Dr. Jacqueline A. Zittoun
Laboratoire d'Hématologie,
Hôpital Henri Mondor et Faculté de Médecine,
94010 Créteil, France

ISBN-13:978-3-540-50653-9   e-ISBN-13:978-3-642-74364-1
DOI: 10.1007/978-3-642-74364-1

Library of Congress Cataloging-in-Publication Data
Folates and cobalamins / J. Zittoun, B. A. Cooper (eds.). p. cm. Includes index.
ISBN-13:978-3-540-50653-9(U.S.)   1. Megaloblastic anemia-Etiology. 2. Folic acid deficiency. 3. Vitamin B12 deficiency. I. Zittoun, J. (Jacqueline). II. Cooper, B. A. (Bernard A.). [DNLM: 1. Folic Acid.
2. Folic Acid Deficiency. 3. Vitamin B12. 4. Vitamin B12 Deficiency. WD 120 F663]   RC 641.7.M4F65
1989 616.3'99-dc20 DNLM/DLC for Library of Congress

This work is subject to copyright. All rights are reserved, whether the whole or part of the material is concerned, specifically the rights of translation, reprinting, re-use of illustrations, recitation, broadcasting, reproduction on microfilms or in other ways, and storage in data banks. Duplication of this publication or parts thereof is only permitted under the provisions of the German Copyright Law of September 9, 1965, in its version of June 24, 1985, and a copyright fee must always be paid. Violations fall under the prosecution act of the German Copyright Law.

© Springer-Verlag Berlin Heidelberg 1989

The use of general descriptive names, registered names, trademarks, etc. in this publication does not imply, even in the absence of a specific statement, that such names are exempt from the relevant protective laws and regulations and therefore free for general use.

Product Liability: The publisher can give no guarantee for information about drug dosage and application thereof contained in this book. In every individual case the respective user must check its accuracy by consulting other pharmaceutical literature.

2127/3145-543210   Printed on acid-free paper

Preface

It is usual to associate megaloblastic anemia with folate or cobalamin deficiencies. However, this notion, even if true in most cases, is too restricted. Megaloblastosis in blood may also be observed in blood diseases without vitamin deficiency, and also after treatment with certain antineoplastic agents; in these conditions, the mechanisms vary with the etiology. On the other hand, folate or cobalamin deficiency may induce various clinical or biochemical disturbances without – as yet – macrocytic megaloblastic anemia.

That the biochemical basis of megaloblastosis is the same in folate and cobalamin deficiencies is due to the close metabolic interrelationships between these two vitamins. However, the role of cobalamin deficiency in folate metabolism is still a matter of debate.

Morphological abnormalities such as macrocytosis in peripheral blood and megablastosis in bone marrow, long considered to be the best indices of vitamin deficiency, are not always constant. Indeed, the improved diagnostic methods often lead to an early diagnosis of deficiency before the appearance of the usual hematological abnormalities.

Actual vitamin deficiencies may be detected without anemia and without macrocytosis or megaloblastosis. Conversely, reduced levels of folate may be found without true vitamin deficiencies, for example, in myelodysplastic syndromes or, more often, in alcoholism. While folic acid administration is of benefit in nutritional deficiencies associated with alcoholism, this supplementation is often without effect in macrocytosis of alcoholic patients, even in those with low folate levels. A decreased level of vitamin B_{12} is often found in lymphoproliferative disorders, but vitamin B_{12} administration does not improve the anemia or even macrocytosis often present in these disorders.

The knowledge of transport mechanisms of folate and cobalamin derivatives has progressed on the one hand by the utilization of cell lines and, on the other, through studies of some of the disorders in which cobalamin metabolism is involved. Thus, studies of pernicious anemia and achlorhydric gastritis have shown the importance of intrinsic factor and of acid and pepsin secretion in the absorption of cobalamin in food. Pernicious anemia is now well defined and distinctly characterized in the group of achlorhydric gastritis disorders. The frequency of gastric adenocarcinoma in pernicious anemia is, in fact, low; it is less frequent than hyperplasia of endocrine cells of the fundus and probably also less frequent than carcinoid tumors of the fundus.

Congenital defects of cobalamin-binding proteins have led to specification as to which of these are useful for absorption (intrinsic factor) and which for intra-cellular transport (transcobalamin II) of cobalamins. On the other hand, R proteins, despite the high affinity for cobalamins, have a function that is not yet defined. They are degraded by pancreatic enzymes, and insufficient degradation of these R proteins may be the main factor in the malabsorption often observed in exocrine pancreatic insufficiency.

A study of the vitamin status of the French population has shown folic acid deficiency in terms of the recommended dietary allowances and the folate levels usually considered as normal. Minimal needs, however, are very difficult to ascertain. Folate deficiency is, in any case, frequent in certain groups, particularly pregnant women, the elderly, alcoholics, and intensive care patients. Considering the frequency of this deficiency and its possible hematological and extrahematological consequences, folic acid supplements should be considered for these groups.

Whereas the neurological and neuropsychiatric abnormalities associated with cobalamin deficiency are well understood, those resulting from folate deficiency are less well defined and the mechanisms involved still hypothetical.

Congenital abnormalities due to disturbances of folate or cobalamin metabolism are numerous and are related either to an abnormality of absorption or transport, to enzymatic deficiency, or to a defect of synthesis of active forms of these vitamins.

The knowledge of intracellular metabolism and transport of folate has led to a better understanding of the biotransformation pathways and the mechanism of action of methotrexate. This antifolate agent, used in many blood diseases and solid tumors, utilizes the same receptors as folate and is converted into polyglutamate by the same enzymes as folate derivatives. Thus, it has become easier to understand the basis of sequential treatment with high doses of methotrexate combined with folinic acid. The best regimen in clinical use has not yet been well defined in pharmacokinetic and intracellular studies.

The chapters in this book cannot give an exhaustive treatment of the huge scope of folate and cobalamin research, but they provide an overview of the multidisciplinary importance of these two vitamins: they are of significance in hematology, gastroenterology, nutrition, pediatrics, neurology, geriatrics, and oncology. If hematology is the area which is principally involved in confronting the problems in diagnosis of folate or cobalamin deficiencies, this is due to the exteme sensitivity of the hematopoietic system to disturbances induced by vitamin deficiencis, although other systems are, in fact, also affected.

R. and J. ZITTOUN

Table of Contents

Chapter 1
Biochemical Basis of Megaloblastosis 1
A.V. HOFFBRAND and R. G. WICKREMASINGHE

Chapter 2
Diagnosis of Megaloblastic Anemia 21
R. CARMEL

Chapter 3
Cobalamin-Folate Interrelations 41
I. CHANARIN

Chapter 4
The Proteins of Transport of the Cobalamines 53
C. A. HALL

Chapter 5
Cobalamin Absorption and Acquired Forms of Cobalamin Malabsorption 71
J. BELAÏCHE and D. CATTAN

Chapter 6
Current Gastroenterologic Aspects of Pernicious Anemia 85
D. CATTAN, A. M. ROUCAYROL, and J. BELAÏCHE

Chapter 7
Cobalamin Malabsorption in Exocrine Pancreatic Insufficiency in Adults
and Children 105
J. L. GUÉANT, B. MONIN, and J. P. NICOLAS

Chapter 8
Making Sense of Laboratory Tests of Folate Status: Folate Requirements
to Sustain Normality 119
V. HERBERT

Chapter 9
Prevalence of Folic Acid Deficiency in the French Population 129
A. LEMOINE, G. POTIER DE COURCY, S. HERCBERG, and C. LE DEVEHAT

Chapter 10
Effects of some Drugs and Alcohol on Folate and Cobalamin Metabolism 137
J. ZITTOUN

Chapter 11
Neuropsychiatric Illness and Deficiency of Vitamin B_{12} and Folate 145
M. I. BOTEZ

Chapter 12
Folate Deficiency During Pregnancy and Lactation 161
J. METZ

Chapter 13
Folic Acid Deficiency in Developing Nations 171
J. D. RAIN, I. BLOT, and G. TCHERNIA

Chapter 14
Folate Deficiency in Geriatric Patients 179
P. BROCKER and J. C. LODS

Chapter 15
Folate Deficiency in Intensive Care Patients 191
E. DE GIALLULY, B. CAMPILLO, and J. ZITTOUN

Chapter 16
Inherited Defects of Folate Metabolism 199
B. A. COOPER

Chapter 17
The Relationship Between Biopterin and Folate Metabolism 215
J. P. HARPEY

Chapter 18
Inherited Disorders of Colbalamin Metabolism 219
J. ZITTOUN and J. MARQUET

Chapter 19
Transport of Folate Compounds by Hematopoietic Cells 231
G. B. HENDERSON

Chapter 20
Methotrexate and 5-Fluorouracil: Cellular Interactions with Folates 247
J. JOLIVET

Chapter 21
Methotrexate Pharmacokinetic Studies: Their Clinical Use 255
Y. NAJEAN and O. POIRIER

Sachverzeichnis 263

List of Contributors

Jacques BELAÏCHE
Department of Gastroenterology and Hepatology, Faculty of Medicine of
Créteil, Medical Center,
94190 Villeneuve St. Georges, France

Isa BLOT
Laboratoire Central d'Hématologie, Hôpital Antoine Béclère,
Clamart, France

Mihail I. BOTEZ
Neurology Division, University of Montréal, Hôtel Dieu de Montréal,
3840 Saint Urbain, Montréal, Québec H2W 1T, Canada

Patrick BROCKER
Hôpital de Cantaron, CHRU Nice, 251 chemin de la Lauvette,
06730 Saint-André, France

Bernard CAMPILLO
Medical Intensive Care Department, Hôpital Henri Mondor,
94010 Creteil, France

Ralph CARMEL
Division of Haematology, University of souther California, School of
Medicine,
Los Angeles, CA, USA

Daniel CATTAN
Department of Gastroenterology and Hepatology, Faculty of Medicine of
Créteil, Medical Center,
94190 Villeneuve St. Georges, France

Is CHANARIN
Clinical Research Center, Northwick Park Hospital,
Middlesex, London, UK

Bernard A. COOPER
Division d'Hématologie et d'Oncologie médicale, Hôpital Royal Victoria et des Departdments de Medicine interne et de Physiologie, Université McGill, Montréal, Québec, Canada

Eric DE GIALLULY
Medical Intensive Care Department, Hôpital Henri Mondor,
94010 Creteil, France

Jean Louis GUÉANT
Laboratoire de Biochimie médicale et pédiatrique, Faculté de Médecine, Université de Nancy I,
54505 Vandoeuvre-lès-Nancy, France

Charles A. HALL
Nutrition Laboratory for Clinical Assessment and Research, Albany Veterans Administration Medical Center,
Albany, NY 12208, USA

Jean Paul HARPEY
Clinique de Pédiatrie, Génètique Médicale, Unité de Maladies Métaboliques, et de Pathologie anté-natale, Hôpital de la Salpêtrière,
75651 Paris Cedex 13, France

Gary B. HENDERSON
Division of Biochemistry, Scripps Clinic and Research Foundation,
La Jolla, CA 92037, USA

Victor HERBERT
Hematology and Nutrition Research Laboratory, Veterans Administration Medical Center,
130 West Kingsbridge Road, Bronx, NY 10468, USA

Serge HERCEBERG
CNAM, 292 rue Saint Martin, 75003 Paris, France

A. Victor HOFFBRAND
Department of Haematology, The Royal Free Hospital, Pond Street,
London NW3 2QG, UK

Jacques JOLIVET
Institut du Cancer de Montréal et Hôpital Notre-Dame,
Montréal, Québec H2L 4M1, Canada

Claude LE DEVEHAT
Department of Diabetes and Nutritional Disorders, C.H. de Nevers,
58320 Pougues-les-Eaux, France

Alain LEMOINE
Department of Diabetes and Nutritional Disorders, C.H. de Nevers,
58320 Pougues-les-Eaux, France

Jean Claude LODS
Hôpital de Cantaron, CHRU Nice, 254 chemin de la Lauvette,
06730 Saint-André, France

Jeanine MARQUET
Laboratoire Central d'Hématologie, Hôpital Henri Mondor,
94010 Créteil, France

Jack METZ
Department of Haematology, School of Pathology of the South African
Institute for Medical Research and the University of Witwatersrand,
Johannesburg 2000, South Africa

B. MONIN
Laboratoire de Biochimie médicale et pédiatrique, Faculté de Médecine,
University de Nancy I,
54505 Vandoeuvre-lés-Nancy, France

Yves NAJEAN
Service de Médecine Nucléaire et U 204 INSERM, Hôpital St. Louis,
1 Avenue Claude Vellefaux,
75010 Paris, France

Jean Pierre NICOLAS
Laboratoire de Biochimie médicale et pédiatrique, Faculté de Médecine,
Université de Nancy I,
54505 Vandoeuvre-lès-Nancy, France

Odette POIRIER
Service de Médecine Nucléaire et U 204 INSERM, Hôpital St. Louis,
1 Avenue Claude Vellefaux,
75010 Paris, France

Geneviève POTIER DE COURCY
CNRS Nutrition Laboratory, 9 rue Hetzel, 92190 Meudon Bellevue, France

Jean Didier RAIN
Service de Médecine Nucléaire, Hôpital Saint Louis,
1 Avenue Claude Vellefaux,
75010 Paris, France

Anne Marie ROUCARYOL
Laboratory of Pathology, Faculty of Medicine of Créteil, Medical Center,
94190 Villeneuve St. Georges, France

Gilbert TCHERNIA
Laboratoire Central d'Hématologie, Hôpital Antoine Béclère,
Clamart, France

R. Gitendra WICKREMASINGHE
Department of Haematology, The Royal Free Hospital,
Pond Street, London NW3 2QG, UK

Jacqueline ZITTOUN
Laboratoire Central d'Hématologie, Hôpital Henri Mondor et Faculté de
Médecine,
94010 Créteil, France

Chapter 1

Biochemical Basis of Megaloblastosis

A. V. HOFFBRAND and R. G. WICKREMASINGHE

Introduction

Megaloblastic anaemia is caused by disturbances of normal DNA replication in haemopoietic precursor cells. These disturbances are the result of impaired functioning of the pathways that provide nucleotide precursors for DNA synthesis, which in turn are usually the consequence of inadequate provision of either folate or vitamin B_{12}. In this chapter we use the term 'megaloblastic anaemia' to signify anaemias that are due to deficiency of one or other of these vitamins.

In this review we describe first the normal pathways by which the nucleotide precursors of DNA are synthesised and the mechanisms by which nuclear DNA is replicated. We then discuss the effects of folate and vitamin B_{12} deficiency on precursor synthesis and the consequent effects of DNA replication of impaired nucleotide supply. Finally, we attempt te relate the observed molecular lesions to the morphological and cell-kinetic abnormalities characteristic of megaloblastic anaemia.

Nucleotide Precursors for DNA Synthesis

The chromosomes of a human cell consist of long DNA molecules. These molecules are linear polymeters of units called deoxyribonucleotides and consist of a base, sugar (deoxyribose) and phosphate moiety (Fig. 1a). In the chromosome, two such DNA molecules form a double-stranded structure, with the two 'backbones' of alternating sugar-phosphate moieties bound together in a double helix and held together by hydrogen bonds between pairs of bases which protrude towards the interior of the helix. The immediate precursors for DNA replication in human and other mammalian cells are the deoxyribonucleoside triphosphates (dNTPs), which have the general structure shown in Fig. 1b. There are four types of dNTPs, differing only in the base moiety; the pyrimidine dNTPs contain either thymine (deoxythymidine triphosphate, dTTP) or cytosine (deoxycytidine triphosphate, dCTP), whereas the purine dNTPs contain adenine (deoxyadenosine triphosphate, dATP) or guanine (deoxyguanosine triphosphate, dGTP).

Chromosomal DNA molecules replicate by a semi-conservative mechanism; the two strands of the DNA double helix unwind, while a new DNA strand, formed by the polymerisation of dNTP precursors, is synthesised on each pre-

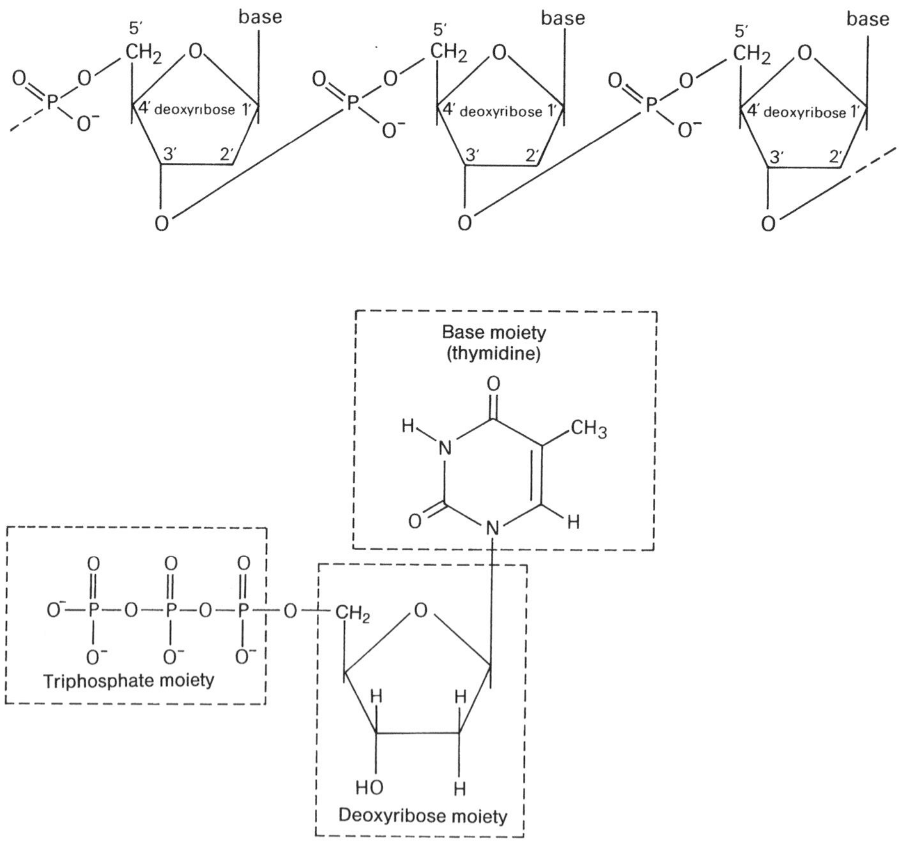

Fig. 1 a, b. a Structure of a DNA molecule. The 'backbone' of the molecule consists of alternating residues of a sugar (deoxyribose) and phosphate groups. An organic base is attached to each deoxyribose. The base can be one of four types: adenine or guanine (purines); cytosine or thymine (pyrimidines). The carbon atoms in each deoxyribose are numbered 1'–5' and a phosphodiester bond links a 3'-hydroxyl group of one sugar to the 5'-hydroxyl group of the next **b** Structure of DNA precursors. Dexyribonucleoside triphosphates (dNTPs) consist of a deoxyribose moiety to which are linked a nitrogenous base and a triphosphate moiety. The example shown is deoxythymidine triphosphate (dTTP), in which the base moiety is thymine. When precursors are incorporated into DNA the two terminal phosphate groups are eliminated, and the remaining phosphate bonds with the free 3'-hydroxyl group of the adjacent deoxyribose

existing strand (Fig. 2). The sequence of nucleotides in a new strand is specified by the sequence of an old strand which thus acts as a template. The base-pairing rules specify that adenine residues pair with thymine and guanine with cytosine. In this way, genetic information encoded in DNA is preserved through successive cycles of replication and cell division. Further details on deoxyribonucleotide structure and DNA replication are given in Kornberg [19].

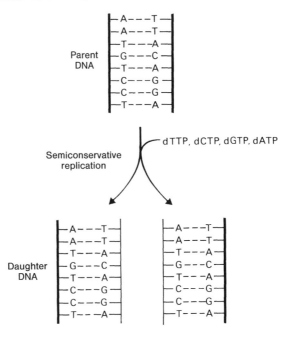

Parent DNA

```
─A───T─
─A───T─
─T───A─
─G───C─
─T───A─
─C───G─
─C───G─
─T───A─
```

Semiconservative replication

dTTP, dCTP, dGTP, dATP

Daughter DNA

```
─A───T─        ─A───T─
─A───T─        ─A───T─
─T───A─        ─T───A─
─G───C─        ─G───C─
─T───A─        ─T───A─
─C───G─        ─C───G─
─C───G─        ─C───G─
─T───A─        ─T───A─
```

Fig. 2. The semi-conservative replication of DNA. In this schematic representation the chain of alternate deoxyribose and phosphate moieties of the DNA backbone are represented by lines: *bold lines* represent parent DNA; *thin lines* newly replicated DNA. The DNA bases are represented by letters: *A*, adenine; *G*, guanine; *C*, cytosine; *T*, thymine. Real DNA molecules are very much larger than those depicted here, consisting of millions of base-pairs. (See text for details of the replication mechanism)

Two Types of Pathway for Synthesis of dNTP

Each dNTP precursor for DNA synthesis is synthesised by two distinct types of pathway. In the salvage pathway, preformed nucleosides (bases attached to deoxyribose) originating from DNA breakdown are phosphorylated in a stepwise fashion by a series of kinase enzymes. Figure 3 identifies the enzyme involved in the salvage of thymidine and the intermediates of the pathway. Salvage synthesis of the remaining dNTPs proceeds by a similar mechanism, the first two enzymes of the sequence being specific for the base. However, the conversion of the diphosphate to the triphosphate is catalysed by a single, non-specific nucleoside diphosphokinase.

The alternative route for dNTP synthesis, the de novo pathway, involves the assembly of the nucleotide bases from small molecules, resulting eventually in the production of dNTPs. An important step in this sequence is the conversion (by reduction) of ribonucleoside diphosphates (containing a ribose sugar), to deoxyribonucleoside diphosphates, containing the deoxyribose sugar necessary for incorporation into DNA. This reaction is catalysed by the enzyme ribonucleotide reductase. The cytotoxic drug hydroxyurea inhibits DNA replication by inactivating ribonucleotide reductase.

In the case of de novo synthesis of dTTP, the first deoxyribonucleoside monophosphate synthesised in the de novo pathway contains the base uracil instead of a thymine residue (Fig. 4). The methylation of the uracil base to yield thymine, in which a methyl group is transferred from 5,10-methylene tetrahydrofolate (THF) polyglutamate to the uracil residue of deoxyuridine monophosphate (dUMP) with the generation of thymidine monophosphate (dTMP), is mediated

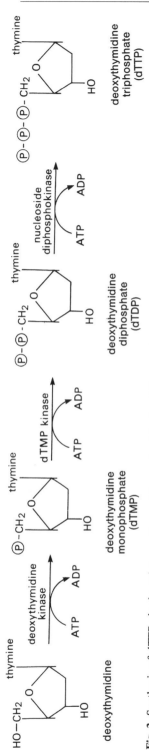

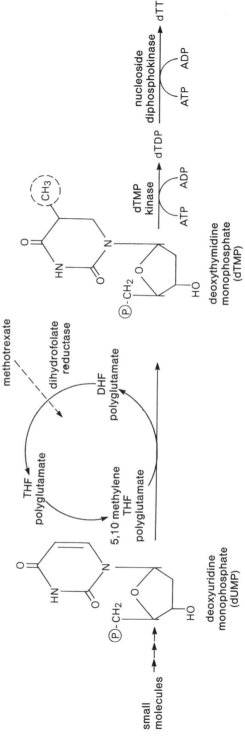

Fig. 3. Synthesis of dTTP via the salvage pathway. The nucleoside thymidine consists of thymine attached to deoxyribose. It is found in extracellular fluids and is formed by DNA breakdown in some tissues. It is converted to dTTP by sequential addition of phosphate groups (*P*) from the donor adenosine triphosphate (ATP). These reactions are catalysed by three separate kinase enzymes

Fig. 4. The final stages of the de novo synthesis of dTTP. (See text for details.) *THF*, tetrahydrofolate; *DHF*, dihydrofolate. Note that these co-factors participate in the reactions in the polyglutamate forms. The methyl group transferred to dTMP to yield dTMP is circled

by the enzyme thymidylate synthase. This reaction is tested in the deoxyuridine suppression test and is of particular importance in understanding the molecular basis of megaloblastic anaemia (Fig. 4). It is the major (but not sole) point at which deficiency of folate or vitamin B_{12} impairs DNA synthesis. The cytotoxic drug methotrexate interferes with DNA replication at the level of thymidylate synthase. This is because the thymidylate synthase catalysed conversion of dUMP to dTMP involves transfer of the methylene group from 5,10-methylene THF polyglutamate as well as two hydrogen atoms from the THF itself, generating dihydrofolate (DHF) polyglutamate (Fig. 4).

The continued operation of the reaction therefore requires continuous cyclical reduction of DHF to THF (catalysed by DHF reductase, DHFR) and the re-addition of the methylene group. By inhibiting the reductase step competitively, methotrexate and its polyglutamate derivatives (see Chap. 20) block the conversion of dUMP to dTMP (Fig. 4). Pyrimethamine also blocks human DHFR, although more weakly, while trimethoprim inhibits the bacterial enzyme but has little action on human DHFR. By causing folate to accumulate in the dihydro form, these DHFR antagonists inhibit not only thymidylate synthesis but all

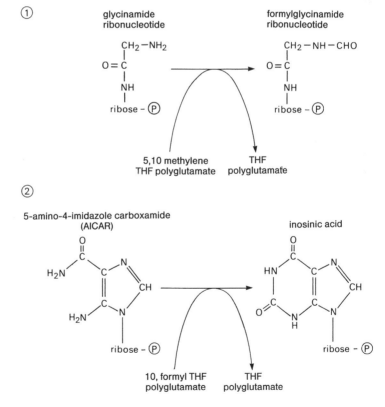

Fig. 5. Two reactions in de novo purine synthesis that require folate co-enzymes to provide single carbon units which provide carbons 8 (reaction 1) and 2 (reaction 2) of the purine ring

other folate-mediated reactions (which require fully reduced, THF co-enzymes), including the two involved in de novo purine synthesis (Fig. 5). Synthesis of dTTP from dTMP is catalysed by the same kinase enzyme utilised in the salvage pathway.

Assembly of dNTP Precursors into DNA Chains

The manner in which the two strands of a DNA molecule unwind to permit synthesis of 'daughter' strands is shown in Fig. 6. The region of unwinding and new DNA synthesis is termed the 'replication fork'. Unwinding is catalysed by an ATP-dependent enzyme, DNA topoisomerase. In eukaryotic cells the polymerisation of dNTPs to form long DNA chains is catalysed by DNA polymerase α. This enzyme assembles a 'daughter' DNA strand by remarkably faithful copying of a 'template' strand, obeying the rules of base pairing. Two further properties of DNA polymerase α are important in understanding the overall mechanism of DNA replication. First, this enzyme cannot initiate synthesis of a DNA strand but can only synthesise DNA by stepwise addition of nucleotide residues to a pre-existing 'primer'. Short primers of RNA are made by the enzyme RNA polymerase (which, unlike DNA polymerase, can initiate chains), and these are extended by DNA polymerase. The primers are then degraded, and the resulting gaps filled by further action of DNA polymerase α (Fig. 6).

Secondly, DNA polymerase is only able to add nucleotides to a growing DNA chain in the 5' to 3' direction (5' and 3' refer to the position of hydroxyl groups on the deoxyribose moieties, see Fig. 1). The two chains of a DNA helix have opposite polarities, with phosphodiester bonds running in the 5' to 3' direction on one chain and in the 3' and 5' direction on the other chain. Consequently, the daughter DNA strand being synthesised in the 5' to 3' direction at a replication fork may be assembled by the continuous action of DNA polymerase. By con-

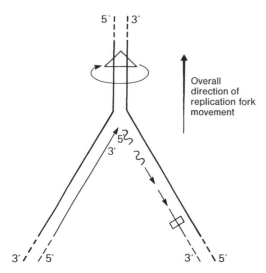

Fig. 6. The replication fork in DNA synthesis. For clarity, individual bases are not shown. *Arrows* at end of DNA strands indicate the direction of movement of DNA polymerase α. (Further details are in the text.) *Triangle,* site of topoisomerase action; *bold line,* template DNA; *fine line,* newly synthesised DNA; *wavy line,* RNA primer; *open box,* site of ligase action

trast, the daughter strand being synthesised in an overall 3' to 5' direction is, of necessity, synthesised in short stretches in the 5' to 3' direction, in a direction opposite to the overall direction of fork movement. These short fragments (consisting of about 200 base pairs) are called Okazaki fragments, and synthesis of each is initiated by a short RNA primer, synthesised by RNA polymerase. These primers are degraded, the gaps between adjacent Okazaki fragments filled in by DNA polymerase and the fragments joined together by the enzyme DNA ligase. In this way, two daughter strands of opposite polarity are synthesised at a single replication fork by a DNA polymerase with a capacity to synthesise DNA in a single direction only. Synthesis of the strand growing in an overall 5' to 3' is 'continuous', whereas synthesis of the strand growing in an overall 3' to 5' direction occurs by a 'discontinuous' mechanism (Fig. 6).

Drugs Directly Affecting DNA Replication

A variety of drugs inhibit DNA replication by interacting with DNA polymerase directly. The nucleoside arabinosylcytosine (AraC) is phosphorylated within cells to give AraCTP, which is similar in structure to the DNA precursor dCTP except that arabinose replaces deoxyribose as the sugar moiety. AraCTP inhibits DNA synthesis by being incorporated into a growing DNA chain which cannot then be extended further by DNA polymerase [3].

The lupenoid aphidicolin is also a potent inhibitor of DNA polymerase α, and serves to distinguish it from DNA polymerase β, which is resistant to inhibition by aphidicolin [15]. Polymerase β is thought to be involved in the repair to DNA and not directly in replication.

Replication of Chromosomal DNA in Replicons

The sites of active DNA synthesis are located on a proteinaceous scaffolding within the nucleus called the nuclear matrix [27]. The single long DNA molecule which comprises each chromosome is not simply replicated from one end to the other. Instead, replication is multifocal, with replication forks originating at multiple points, called replication origins [16], along the DNA molecule. Replication forks proceed bidirectionally from these origins (Fig. 7). The DNA replicated from a single origin is termed a replicon. When the replication forks extending from adjacent origins meet, replication ceases, and the daughter DNA strands produced by individual replicons are joined together by DNA ligase to form the long mature chromosome-sized DNA molecules. In human cells an individual replicon contains 50000–300000 nucleotide pairs. A total of some 10000 replicons are thought to be simultaneously active at any one time during the DNA synthetic (S) phase. This multifocal mode of replication permits the replication of a very large amount of chromosomal DNA within the span of the S phase (about 8 h).

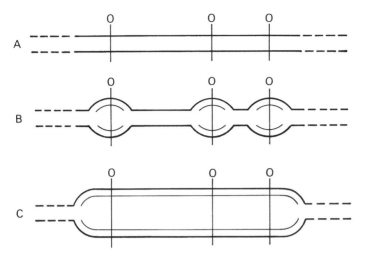

Fig. 7 a–c. DNA replication takes place in units called replicons. **a** Part of a chromosomal DNA molecule, showing the distribution of replication origins. **b** Replication initiates at each origin and replication forks move bidirectionally from these. DNA synthesised by forks starting at a single origin is called a replicon. **c** When forks extending from adjacent origins collide, the DNA newly synthesised within adjacent replicons is joined together by DNA ligase. *O,* Replication origin; *bold line,* template DNA; *fine line,* newly synthesised DNA

Multienzyme Complexes Increase the Efficiency of DNA Replication

The genome of a single diploid human cell consists of some 3×10^9 nucleotide pairs [19]. Replication of this amount of DNA within an S phase of about 8 h requires the provision and incorporation of dNTPs at an average rate of some 10^5 nucleotide molecules per second per cell. Metabolic pathways can be made to proceed with much greater efficiency if the enzymes catalysing the individual reactions are associated in a physical complex rather than existing free in solution [7]. The existence of such complexes permit the transfer of intermediates directly from one enzyme to the next, without the need to build up large free pools of intermediates. The net result is an increased rate of synthesis of the final product.

Much recent evidence shows that the enzymes of DNA precursor synthesis together with DNA polymerase α form such multienzyme complexes in prokaryotes as well as in mammalian cells [22]. Such a multienzyme complex was elegantly demonstrated for bacteriophage T_4 DNA replication by Mathews and his colleagues [23]. They showed that the action of this complex produced high local concentrations of dNTPs at the site of DNA replication, a phenomenon referred to as functional compartmentation. Functional compartmentation of dNTPs in mammalian DNA replication was first demonstrated by Reddy and Pardee [29], who showed that permeabilised Chinese hamster fibroblasts incorporated ribonucleoside diphosphates (rNDPs) into DNA more efficiently than dNTPs, although dNTPs are the immediate precursors for DNA synthesis. Unlabelled dNTPs did not dilute the incorporation of label from rNDPs. However, if hydroxyurea was used to inhibit ribonucleotide reductase, and thus disrupt incorporation of rNDPs, dNTPs were efficiently incorporated into DNA. Reddy and Pardee interpreted these findings in terms of the physical juxtaposition of the enzymes of precursor synthesis and DNA polymerase, resulting in the efficient metabolic channelling of deoxyribonucleotide precursors into DNA. Noguchi et al. [25] subsequently isolated this complex ('replitase') from Chinese hamster embryo cell nuclei. This replitase was a particle of 30-nm diameter and contained many enzymes of DNA synthesis, including ribonucleotide reductase, thymidylate synthetase, DNA polymerase and topoisomerase. However, the precise contribution of this replitase to DNA replication has been questioned by Mathews and Slabaugh [22], since the yield upon partial purification was low. Furthermore, Reddy et al. [30] have shown that a significant proportion (about one-third) of rNDPs were incorporated into RNA and not into DNA.

A multienzyme complex of DNA synthesis has also been isolated in high yield from a lymphoblastoid cell line by Wickremasinghe et al. [38]. This complex was separated from free enzymes of precursor synthesis by gel filtration and was shown to contain DNA polymerase α and several enzymes of precursor synthesis, including thymidylate synthase, thymidine kinase, dTMP kinase and nucleoside diphosphokinase. Kinetic studies showed that this complex chould channel dTMP directly into DNA without the concomitant build-up of a large pool of free dTTP [39]. However, estimation of the effective pool of dTTP at the replication site showed that its concentration was 50 μM, whereas the overall concentration of dTTP pool in the reaction mixture was only 1.7 μM, thus demonstrating that the complex was able to maintain a high degree of functional compartmentation of dNTPs [39]. Furthermore, the flow of nucleotides through the complex was shown to be carefully regulated, since inhibition of DNA polymerase by AraCTP led to a corresponding reduction in synthesis of dTTP, so that the free pool of dTTP did not increase greatly [40, 41]. The co-ordination of the activities of individual enzymes of the complex is also emphasised by the observation that although thymidylate synthase activity was found at all stages of the cell cycle, it was only active in vivo during the S phase, when it was associated with other components of the complex [28]. Thus, the evidence suggests than an exquisitely regulated complex of precursor-synthesising and DNA-replicating enzymes exist in mammalian cells, which facilitates rapid and efficient synthesis of DNA precursors, and which is carefull co-ordinated with the assembly of the precursors into DNA chains.

Evidence for Functional Compartmentation in intact Human Cells

Taheri et al. have demonstrated that DNA precursors are indeed functionally compartmentalised in both human lymphocytes and bone marrow. In these cells only a proportion (25%–60%) of nucleotide precursors of DNA synthesis derived via either the salvage [32] or de novo [33] routes were incorporated into DNA. The remainder was degraded and lost from the cell. Taheri et al. interpreted these data as resulting from the operation of a multienzyme complex of DNA precursor synthesis which channelled precursors into DNA, albeit with less than 100% efficiency. It was suggested that precursors which 'leaked' from the channelling process entered a large diffuse pool of low concentration, and were eventually degraded in order to avoid a toxic build-up of nucleotides (Fig. 8). However, the efficiency of the channelling mechanism was found to be variable, and some cell lines of thymic acute-leukaemia origin were able to channel precursors into DNA with virtually 100% efficiency via both the salvage and de novo pathways [32–34].

Lesions in dTTP Synthesis in Megaloblastic Anaemia

The deoxyuridine suppression test [18] established that an important lesion in megaloblastic anaemia due to vitamin B_{12} or folate deficiency was in provision of dTTP for DNA replication at the level of conversion of dUMP to dTMP in the reaction catalysed by thymidylate synthase. In this test the reduced ability of exogenous unlabelled deoxyuridine to inhibit the incorporation of [^{3}H]thymidine into DNA in cells from patients with megaloblastic anaemia is taken as evidence of impaired functioning of the thymidylate synthase step (Fig. 9). Exoge-

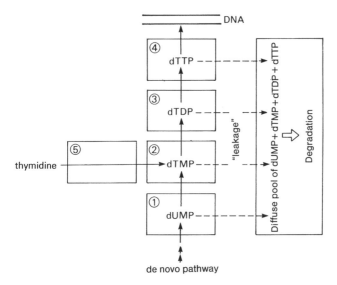

Fig. 8. Functional compartmentation. The figure illustrates how a complex of enzymes of nucleotide synthesis and DNA polymerase catalyses the rapid flow of precursors into DNA. Enzymes are represented as *boxes; 1,* thymidylate synthase; *2,* dTMP kinase; *3,* nucleoside diphosphokinase; *4,* DNA polymerase; *5,* thymidine kinase. The figure also illustrates how 'leakage' from the complex leads to build-up of a large diffuse pool of DNA precursors, which is unavailable for incorporation into DNA and is eventually degraded. Similar complexes may operate for the provision of the other DNA precursors

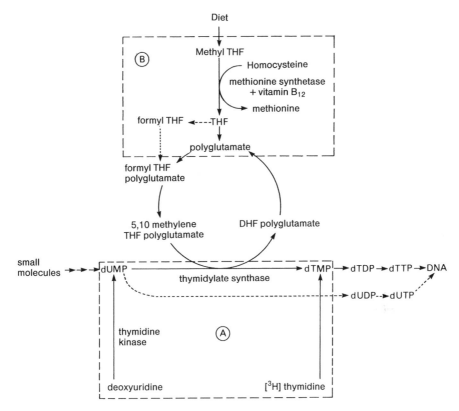

Fig. 9. The de novo pathway for synthesis of dTTP (see Fig. 4 for details). The mechanism of the deoxyuridine suppression test is shown at *A*. The route by which deoxyuridine triphosphate may build up and be misincorporated into DNA is also shown. The interaction of vitamin B_{12} with folate metabolism and its eventual effect on the thymidylate synthese reaction is shown at *B*. (For details, see text)

nously added deoxyuridine is converted to dUMP by thymidine kinase. In bone marrow cells from normal subjects rapid conversion of dUMP to dTMP (via the thymidylate synthase step) expands the dTMP pool and therefore decreases incorporation of exogenous [³H]thymidine into DNA by diluting its specific radioactivity. The rise in dTTP is also thought to feed back inhibit thymidine kinase. The exact biochemical mechanisms underlying the test are probably much more complicated [38]. In bone marrow cells from patients with megaloblastic anaemia, impaired functioning of the thymidylate synthase step reduces conversion of dUMP to dTMP. Therefore, the ability of exogenous deoxyuridine to suppress incorporation of [³H]thymidine is reduced. Depending on the nature of the vitamin deficiency, addition of either methyl-THF or vitamin B_{12} restores a normal deoxyuridine suppression, confirming the diagnosis of megaloblastic anaemia and identifying the vitamin deficiency (see Chapter 2 and 3). The ability of added folate derivates, e.g. formyl-THF, to restore a normal deoxyuridine suppression to megaloblastic cells (whether due to folate or vitamin B_{12} deficiency) provides evidence that reduced folate supply in megaloblastic anaemia impairs thymidylate synthesis.

Direct evidence for impairment of the thymidylate synthase reaction has been obtained by Taheri et al. [24, 35]. When bone marrow cells or lymphocytes from patients with megaloblastic anaemia were labelled with [^{3}H]deoxyuridine (which labels DNA via the de novo pathway), incorporation of label into DNA was reduced and accumulation of labelled dUMP greatly increased compared to cells from healthy individuals. These lesions could be rapidly and dramatically reversed by the addition of folinic acid to the culture medium.

Since neither folate nor vitamin B_{12} plays a role in salvage synthesis of dNTPs, it follows that this pathway should be unaffected in megaloblastic anaemia. The activity of the salvage enzyme, thymidine kinase is indeed increased in megaloblastic cells [14]. Moreover, Taheri et al. [32] have shown that lymphocytes from a patient with megaloblastic anaemia incorporated [^{3}H]thymidine-derived (i.e. salvaged) nucleotides into DNA with enhanced efficiency compared to a control culture to which folinic acid had been added. This suggested that the efficiency of operation of the multienzyme complex could be adjusted in response to reduced availability of the de novo thymine nucleotides, and also that both de novo and salvage pathways are of significance in providing dNTPs for DNA replication. Furthermore, acute inhibition of de novo synthesis of dTTP by the addition of methotrexate to normal phytohaemagglutinin (PHA) stimulated lymphocytes led to 100% incorporation of salvaged [^{3}H]thymidine into DNA within 1 h of drug addition, again emphasising the ability of the precursor synthetic machinery to respond rapidly to the demand for DNA precursors.

Although folate co-enzymes are needed for two steps in purine synthesis (Fig. 5) and these reaction are impaired in patients with megaloblastic anaemia due to folate of B_{12} deficiency, as well as nitrous oxide treated animals [8], there is no evidence that these steps become rate limiting for DNA symthesis in megaloblastic anaemia. Whereas administration of preformed thymidine in vivo leads to haematological remission, no such effect has been shown for purines. Similarly, thymidine in vitro will correct the defects in DNA replication; purines do not. RNA synthesis, which is also dependent on these purine reactions, is not impaired. It is likely that in the bone marrow in vivo these reactions can be bypassed by the uptake of preformed purine nucleosides into the cells.

Interrelations of Vitamin B_{12} and DNA Precursor Biosynthesis

Vitamin B_{12} does not participate directly as a co-factor in any of the reactions of dNTP biosynthesis. In explaining the interactions of vitamin B_{12} with the pathways of dNTP synthesis, the authors favour the hypothesis known as the 'methyltetrahydrofolate trap hypothesis' [4] as modified to include the folate polyglutamate step [2, 20]. The main form of folate available to tissues in the body from plasma is 5-methyl-THF (Fig. 9). This co-factor participates in a reaction catalysed by methionine synthetase, in which the methyl group of 5-methyl-THF is transferred to homocysteine, generating THF and methionine. Vitamin B_{12} is a required co-factor in this reaction. The original trap hypothesis postulated that

5-methyl-THF itself serve as a precursor for synthesis of other folate co-enzymes, but that THF is required. In vitamin B_{12} deficiency, it was proposed, 5-methyl-THF accumulated, and other folate co-enzymes became depleted.

It was subsequently postulated that THF is needed as substrate to synthesise the polyglutamate co-enzyme forms of folate required in cellular metabolism (including the 5,10-methylene THF polyglutamate, the cofactor required in the thymidylate synthase step (reviewed in [38]). Chanarin et al. [1, 2] have modified this to suggest that it is formyl-THF, not THF, which is the correct substrate, and that it is in the generation of the formyl group rather than THF that vitamin B_{12} has its main role.

It may well be that vitamin B_{12} deficiency, by impairing methionine synthesis, affects both the provision of THF from methyl-THF and that of formyl groups from methionine and so leads, by both actions, to failure of folate polyglutamate synthesis. We favour the former action as the dominant one since other sources of formate than methionine from methionine synthesis are available to the cell. The cell level of S-adenosylmethionine (SAM), which regulates the balance between methyl- and methenyl-THF, may be a third level of action, reduced methionine synthesis leading to lowered levels of SAM and so increasing methyl-THF formation at the expense of methenyl-THF (see Chap. 3, this volume).

Molecular Defects in DNA Synthesis in Megaloblastic Anaemia

Since vitamin B_{12} or folate deficiency has been shown to disrupt the provision of dNTP precursors for DNA synthesis, it appears likely that the assembly of these precursors into DNA is also affected in megaloblastic anaemia, and that these lesions may explain the characteristic abnormal haemopoiesis. We therefore performed a series of investigations into the mechanics of DNA replication in megaloblastic anaemia, using as a model PHA-stimulated lymphocytes from untreated patients. These cells display morphological and biochemical abnormalities very similar to those decribed in megaloblastic bone narrow [5, 6]. Furthermore, addition of folinic acid to these cultures leads to the correction of these defects, so each experiment can be provided with 'normal control' lymphocytes derived from the same donor as the 'megaloblastic' lymphocytes.

By using a caesium chloride density gradient technique devised by Painter and Schaeffer [26], we demonstrated that the rate of DNA replication fork movement was significantly reduced in 'megaloblastic' lymphocytes [38]. Whereas replication forks in folinic acid-treated control lymphocytes moved at a rate of 1.0–1.7 μm per minute, the rate was reduced by 8%–50% in the untreated 'megaloblastic' lymphocytes from the same patient. Treatment of PHA-treated lymphocytes from normals donors with either 0.5-mM hydroxyurea or 1-μM methotrexate for 1 h also caused a marked reduction in the rate of fork movement compared to controls. The effect of either of these drugs on the rate of fork movement was far more profound than the effect of vitamin deficiency in megaloblastic anaemia.

The rate of Okazaki fragment joining in megaloblastic lymphocytes was also measured using an indirect chromatographic technique [37]. This method depended on the fact that DNA which is partially single-stranded binds more tightly to a column of benzoylated-naphtoylated diethylaminoethyl cellulose (BND-cellulose) than does normal chromosomal DNA, which is fully double-stranded. We pulse-labelled PHA-stimulated lymphocytes with [^{3}H]thymidine in order to label only the replicating DNA. When the DNA was purified and chromatographed on BND-cellulose, 70%–80% of the label eluted as partially single-stranded DNA, due to the prescence of unfilled gaps between adjacent Okazaki fragments. However, if the [^{3}H]thymidine pulse-labelled lymphocytes were 'chased' by further incubation in unlabelled medium, the labelled DNA became gradually double-stranded, because of the filling of gaps between Okazaki pieces (see Fig. 6). The rate of gap filling and joining of adjacent Okazaki pieces was markedly reduced in megaloblastic lymphocytes in comparison to the rate detected in vitro folinic acid treated controls. Similar studies were also carried out using lymphocytes treated with 1-μM methotrexate or 1-mM hydroxyurea; the drug-treated lymphocytes also showed a clear reduction in the rate of Okazaki fragment joining when compared to controls [37].

In addition to synthesis and joining of Okazaki fragments, DNA replication is also discontinuous at another level, since the DNA newly synthesised in adjacent replicons must also be joined together (see Fig. 7). In another series of experiments we pulse-labelled lymphocyte DNA for 5 min with [^{3}H]thymidine, followed by a 'chase' in unlabelled medium [36]. At intervals, the size of the pulse-labelled DNA was estimated by sedimentation in alkaline sucrose gradients. DNA helices are completely dissociated into single strands under these conditions, and therefore the size of pulse-labelled DNA measured immediately following the pulse corresponded to the size of growing DNA stands within individual replicons, with a broad distribution centered on 15 μm (the average size of a completed replicon in human lymphocytes is 45 μm). During the chase, the size of the pulse-labelled DNA increased, due to the joining up of newly replicated DNA of adjacent replicons. Eventually, all of the label was found in DNA co-sedimenting with mature cellular DNA.

In folinic acid treated 'control' lymphocytes, the conversion of pulse-labelled DNA from replicon-sized pieces to full-sized mature cellular DNA was essentially complete in 6 h. By contrast, this conversion was markedly slower (by about 50%) in untreated 'megaloblastic', lymphocytes. The block was not, however, a permanent one, since by 24 h of chase, the sedimentation patterns of labelled DNA from 'control' and 'megaloblastic' lymphocytes were identical. This finding underlies our belief that the lesions in megaloblastic anaemia are not caused by permanent changes in the structure of DNA but by the slower kinetics of its replication. Earlier studies using isolated nuclei had also suggested that megaloblastic cells had excessive numbers of opened replication sites compared to normal cells [13]. This concept is explored more fully in the following section. We have also found that normal lymphocytes treated with hydroxyurea also showed a markedly reduced rate of joining of replication-sized DNA [36]. This was previously shown in the case of methotrexate-treated cells by Fridland and Brent [9].

Relationship of Defective Precursor Synthesis to Lesions in DNA Synthesis

The simples explanation for our observations on impaired DNA synthesis in megaloblastic anaemia is that a reduced supply of dTTP at the replication fork, resulting from lesions in its biosynthesis, reduce the rate at which DNA polymerase α synthesises new DNA chains. Given the evidence for functional compartmentation of dNTPs in living cells, it would be impossible to measure the effective concentration of dTTP at the site of its utilisation in intact cells. However, we have estimated that an isolated multienzyme complex from HPB-ALL leukaemia cell line was able to maintain an effective dTTP concentration fo 50 µM at the DNA synthesis site [39]. The K_m for dTTP displayed by DNA polymerase is 10 µM. Therefore, if an effective dTTP concentration of about 50 µM were maintained at the replication fork in animal cells, a reduction in this level (due to vitamin deficiency) would be expected to lead to a reduction in the rate of DNA polymerase action. In consequence, this would lead to a reduced rate of synthesis of the continuously synthesised DNA strand as well as a reduced rate of gap filling between and joining of Okazaki pieces. Reduced rate of fork movement and impaired joining up of DNA synthesised within adjacent replications would therefore ensue.

Paradoxically, direct measurement of dTTP levels in bone marrow cells from patients with megaloblastic anaemia have shown increased dTTP pools, compared to normal bone marrow cells [12]. We feel that the explanation of this paradox lies partly in the cell composition of the two since megaloblastic marrows have an increased proportion of primitive cells which are known to contain higher dNTP than mature cells. It also lies in the complexity of the mechanism responsible for functional compartmentation of dTTP, i.e. the multienzyme complex responsible for its synthesis. As described in an earlier section, normal haemopoietic cells appear to channel only a proportion of dTTP synthesised within them into DNA. The remainder enters a degradative pool, whence it was broken down and lost from the cell, presumably to avoid the build-up of toxic levels of nucleotides [32, 33]. We have shown that the system responsible for dTTP synthesis and degradation is able to respond to metabolic stress in megaloblastic anaemia by incorporating a greater proportion of dTTP into DNA with a concomitant reduction in the amount of dTTP that is degraded [32]. This effect is even more marked in lymphocytes acutely deprived of de novo synthesised thymine nucleotides by treatment with methotrexate [32]; in this case, all intracellular dTTP is incorporated into DNA, with virtually no degradation.

We therefore propose that normal cells depend usually on a supply of dTTP from both the de novo and the salvage pathways. In megaloblastic anaemia the supply of dTTP from de novo synthesis is reduced, leading to an overall reduction in dTTP concentration at the sites of DNA replication, which in turn results in the observed lesions in DNA synthesis. In response to this metabolic stress the rate of degradation of dTTP will be reduced, so that the *overall* cellular levels of dTTP may be increased compared to that in normal cells. However, the bulk of this dTTP would be in the low-concentration pools remote from the replication forks and therefore unavailable for DNA synthesis. It is our opinion that this mechanism may also account in part for increased levels of cellular dTTP in cells from patients with megaloblastic anaemia, despite lesions in DNA replication consistent with starvation of this dNTP. It is of interest that we found a slight reduction in the dTTP/dGTP ratio in megaloblastic cells [12] and a more recent study has suggested a much more severe fall in the dTTP/dCTP ratio [17]. This implies a relative overall starvation of dTTP but the exact explanation for this observation remains unclear.

Relationship of Impaired DNA Replication to Morphological and Cell-Kinetic Abnormalities in Megaloblastic Anaemia

The data summarized in the previous section indicates that DNA replication in megaloblastic anaemia proceeds via a normal mechanism, but that the rates at which the various steps of DNA replication occur are reduced compared to normal. In particular, although single-stranded regions in DNA may persist longer during DNA replication in megaloblastic anaemia than in normal cells, they may eventually become fully double-stranded [37]. Likewise, the rate of joining up of replicon-sized newly synthesised DNA is reduced in megaloblastic anaemia, but there appears to be no permanent block in this process [36]. How then can the open chromatin structure of megaloblastic erythroid precursors be explained? In non-replicating human cells the DNA of each nucleus is tightly packed by virtue of its close association with the nucleosomes which are octameric units of histone molecules [19]. The tightness of this packaging is readily appreciated if one considers that the two metres of DNA helix which constitute the human genome is packed into a cell nucleus some 10 μm in diameter. In order to allow access of the DNA replication machinery to DNA in order to replicate it, regions of chromatin, consisting of tandem arrays of replicons, decondense at certain points during S phase and then recondense after replication. This process must occur in an orderly sequence, although the mechanisms which regulate it have remained elusive.

It is our opinion that the reduced rate of DNA replication in megaloblastic anaemia results in an abnormal accumulation of decondensed replicons which cannot recondense because DNA synthesis is incomplete. An abnormal accumulation of decondensed chromatin would be expected to result in the abnormal nuclear morphology associated with megaloblastic anaemia. An increased cytoplasmic to nuclear volume ratio is also characteristic of megaloblastic anaemia. Mitchison [24] has suggested that a reduced rate of DNA synthesis may lead to an uncoupling of the nuclear cycle of DNA replication and the cytoplasmic growth cycle of accumulation of protein. Both these cycles are a prerequisite for normal mitotic division and are closely geared in normal cells. The extension of the nuclear S phase in megaloblastic cells would then be expected to result in an abnormal rate of growth of the cytosol, since the synthesis of RNA and protein appear not to be affected in megaloblastic anaemia, whereas the rate of DNA synthesis is considerably reduced. Whether a greater rate of transcription of messenger RNA than normal also occurs during the S phase because of the more open uncondensed chromatin is unclear.

At the level of analysis employed by us, it appeared that the steps of DNA replication may proceed to completion in megaloblastic cells, albeit as a slower rate than in normal cells. However, it seems likely that in a proportion of megaloblastic cells, small regions of DNA remain in a partially single-stranded form or remain unreplicated. Partially single-stranded DNA would present points of weakness which may be prone to physical damage or to attack by nucleases specific for single-stranded DNA. These events would lead to chromosome fragmentation. The persistance of even short unreplicated regions of chromosomes, which is most likely to occur when vitamin B_{12} or folate deficiency is severe, and supplies of dTTP are most affected, would inevitably lead to chromosome breakage at anaphase, as the mitotic spindle attempts to separate two daughter chromosomes which are still joined together at the unreplicated regions. Both these possibilities would be expected to result in the death of some of the affected cells and may offer an explanation for the ineffective haemopoiesis in megaloblastic anaemia, as well as the random chromosome breaks found in untreated megaloblastic cells.

Uracil Misincorporation Hypothesis

An alternative view of the origins of DNA lesions in megaloblastic anaemia due to folate or vitamin B_{12} deficiency, states that these arise from the misincorporation into DNA of nucleotides containing uracil instead of the usual thymine base, rather than purely from dTTP starvation [10, 21]. Living cells have evolved elaborate safeguards to avoid the incorporation into DNA of dUMP from the dNTP precursor deoxyuridine triphosphate (dUTP) in place of dTMP [19]. Structurally, dUTP and dTTP are very similar, differing only in the presence of a methyl group on the base moiety in dTTP. DNA polymerase α can insert either dUMP or dTMP into DNA, although approximately 100 times less efficiently. Two mechanisms operate to exclude dUMP from DNA. First, the enzyme dUTPase is able to degrade any dUTP that is produced in the cell. Secondly, the enzyme uracil N-glycosidase is able to recognise and excise uracil residues if they become incorporated into DNA. This results in the generation of a sugar residue in the backbone of DNA which is not attached to a base moiety. Such residues are targets for specific endonucleases which nick DNA at these sites. A repair system then uses the intact strand as a template for repair of the resulting gap, with insertion of dTMP in place of the erroneously incorporated dUMP. Inhibition of the thymidylate synthase step in megaloblastic anaemia results in an abnormal accumulation of dUMP, which in turn, leads to high levels of dUTP, synthesised by further phosphorylation of dUMP [33]. This may be expected to result in more extensive misincorporation of dUMP into DNA. This could have two possible consequences. First, persistence of dUMP in DNA may pertub its structure. Second, the extensive nucleolytic cleavage of DNA (which would occur as a result of activation of the uracil excision mechanism by excessive incorporation of dUMP) could have deleterious effects on the cell, particularly if insufficient dTTP is available for repair of the lesions leading to chromosome fragmentation and, possibly, cell death.

Indeed, we have found a significant increase in labelled dUTP when MOLT4 leukaemia cells were laballed with [^{3}H]dU in the presence of 1-μM methotrexate, when compared with cells labelled in the absence of the drug [33]. This increase was attributed to the block in conversion of [^{3}H]dUMP to [^{3}H]dTMP due to methotrexate inhibition of the thymidylate synthase step. The increased pool of [^{3}H]dUMP would in turn result in increased pools of [^{3}H]dUTP. When the DNA of MOLT4 cells labelled with [^{3}H]dU in the presence or absence of methotrexate was purified, hydrolysed to its consistent nucleotides and analysed by paper chromatography, a significant proportion of label (19%) was in fact found in dUMP residues in the DNA prepared from the drug-treated cells, whereas essentially no label was detected in dUMP residues of DNA from the control cells [33]. Similar results have been reported by others who also tested cultured human lymphoid cells [11, 31]. Therefore, it has been established experimentally that at least under conditions of acute inhibition of thymidylate synthase by methotrexate, a sufficiently high level of dUTP can be built up within cells so that detectable and persistent incorporation of dUMP into DNA can occur, by overwhelming the mechanisms which would normally exclude these residues from DNA.

However, we do not feel it justifiable to extrapolate from observations made in methotrexate-treated cells to megaloblastic cells, where the block in the thymidylate synthase step is far less profound. In fact, although we have observed increased levels of [^{3}H]dUMP in megaloblastic lymphocytes labelled with [^{3}H]dU, no significant increases in [^{3}H]dUTP were detected compared to control [33]. This is in contrast to the situation in methotrexate-treated cells. Nucleotide analysis of DNA from [^{3}H]dU labelled megaloblastic bone marrow and lymphocytes also did not show an increased presence of incorporated [^{3}H]dUMP compared to control [33]. Moreover, megaloblastic as well as methotrexate-treated cells have ample dTTP for DNA repair, e.g. of ultraviolet or X-ray damage [37], this process requiring substantially less dTTP than replicative DNA synthesis. Therefore, we consider it unlikely that misincorporation of dUMP into DNA contributes significantly to the cellular abnormalities associated with megaloblastic anaemia, although it is difficult to exclude a minor role. Using a long incubation period Luzzatto et al. [21] reported increased uracil incorporation into megaloblastic cells compared to normal cells, but full details of these experiments have not been published.

A further piece of evidence supports our view that the major cause of cellular abnormalities in megaloblastic anaemia due to folate or vitamin B$_{12}$ deficiency is the reduced rate of DNA replication rather than the misincorporation of dUMP residues. When administered in vivo both methotrexate and hydroxyurea can cause a megaloblastic anaemia morphologically indistinguishable from that produced by vitamin deficiency. Furthermore, both these drugs cause lesions in DNA synthesis as detected in our in vitro lymphocyte model, which are closely similar to those seen in 'megaloblastic' lymphocytes [36, 37]. Whereas methotrexate treatment results in an accumulation of dUTP, treatment with hydroxyurea (which inhibits ribonucleotide reductase) does not. Moreover, many other lesions result in reduced supply of DNA precursors, e.g. inherited deficiency of enzymes of pyrimidine synthesis, with 7-orotic aciduria or administration of the drugs azacytidine or 6-mercaptopurine, or direct inhibition of DNA polymerisation, e.g. by AraC; all cause morphologically identical megaloblastic anaemia. Although it is possible to hypothesise that misincorporation of uracil (and failure of its replacement by thymine) causes identical lesions to those due to precursor starvation, the overwhelming direct evidence leads us to conclude that misincorporation of dUMP into DNA is not a major determinant of the observed abnormalities in megaloblastic anaemia due to vitamin B_{12} or folate deficiency, although it is difficult completely to exclude it as a very minor additional component.

References

1. Chanarin I, Deacon R, Lumb M, Perry J (1980) Vitamin B12 regulates folate metabolism by the supply of formate. Lancet II:505–508
2. Chanarin I, Deacon R, Lumb M, Muir M, Perry J (1985) Cobalamin-folate interaction: A critical review. Blood 66:479–489
3. Cozzarelli NR (1977) The mechanism of action of inhibitors of DNA synthesis. Ann Rev Biochem 46:641–668
4. Das KC, Herbert V (1976) Vitamin B12-folate interrelations. In: Hoffbrand AV (ed) Megaloblastic anaemia. Clin Haematol 5:697–725
5. Das KC, Hoffbrand AV (1970) Studies of folate uptake by phytohaemagglutinin-stimulated lymphocytes. Br J Haematol 19:203–221
6. Das KC, Hoffbrand AV (1970) Lymphocyte transformation in megaloblastic anaemia: Morphology and DNA symnthesis. Br J Haematol 19:459–468
7. Davis RH (1972) Metabolite distribution in cells. Science 178:835–840
8. Deacon R, Chanarin I, Lumb M, Perry J (1985) Role of folate dependent transformylase in synthesis of purine in bone marrow of man and in bone marrow and liver of rats. J Clin Pathol 38:1349–1352
9. Fridland A, Brent TP (1975) DNA replication in methotrexate-treated human lymphoblasts. Eur J Biochem 57:379–385
10. Goulian M, Bleile B, Tseng BY (1980) The effect of methotrexate on levels of dUTP in animal cells. J Biol Chem 255:10630–10637
11. Goulian M, Bleile B, Tseng BY (1980) Methotrexate-induced misincorporation of uracil into DNA. Proc Natl Acad Sci USA 77:1956–1960
12. Hoffbrand AV, Ganeshaguru K, Lavoie A, Tattersall MHN, Tripp E (1974) Thymidylate concentrations in megaloblastic anaemia. Nature 248:602–604
13. Hooton JWL, Hoffbrand AV (1977) DNA synthesis in isolated lymphocyte nuclei: Effects of megaloblastic anaemia due to folate or vitamin B12 deficiency or antimetabolite drugs. Biochem Biophys Acta 477:250–258
14. Hooton JWL, Hoffbrand AV (1976) Thymidine kinase in megaloblastic anaemia. Br J Haematol 33:527–537
15. Huberman JA (1981) New views of the biochemistry of eucaryotic DNA replication revealed by aphidicolin, an unusual inhibitor of DNA polymerase. Cell 23:647–648
16. Huberman JA, Riggs AD (1968) On the mechanism of DNA replication in mammalian chromosomes. J Mol Biol 32:327–341
17. Iwata N, Omine M, Yamauchi H, Maekawa T (1982) Characteristic abnormality of deoxyribonucleoside triphosphate metabolism in megaloblastic anaemia. Blood 60:918–923
18. Killman SA (1964) Effect of deoxyuridine on incorporation of tritiated thymidine: difference between normoblasts and megaloblasts. Acta Med Scand 175:483–488
19. Kornberg A (1980) DNA replication. Freeman, San Francisco

20. Lavoie A, Tripp E, Hoffbrand AV (1974) The effect of vitamin B12 deficiency on methylfo-late metabolism and pteroylpolyglutamate synthesis in human cells. Clin Sci Mol Med 47:6117–630

21. Luzzatto L, Faluis AO, Jory EA (1981) Uracil in DNA in megaloblastic anaemia. N Eng J Med 305:1156–1157

22. Mathews CK, Slabaugh MB (1986) Eukaryotic DNA metabolism. Are deoxyribonucleo-tides channelled to replication sites? Exp Cell Res 162:285–295

23. Mathews CK, North TW, Reddy GPV (1978) Multienzyme complexes in DNA precursor biosynthesis. Adv Enzyme Regul 17:133–156

24. Mitchison J (1983) The biology of the cell cycle. Cambridge University Press, New York

25. Noguchi H, Reddy GPV, Pardee AB (1983) Rapid incorporation of label from ribonucleo-side diphospates into DNA by a cell-free high molecular weight fraction from animal cell nuclei. Cell 32:443–451

26. Painter RB, Schaefer AW (1969) Rate of synthesis along replicons of different kinds of mammalian cells. J Mol Biol 45:467–479

27. Pardoll DM, Vogelstein B, Coffey DS (1980) A fixed site of DNA replication in eucaryotic cells. Cell 19:527–536

28. Reddy GPV (1982) Catalytic function of thymidylate synthase is confined to S phase due to its association with replitase. Biochem Biophys Res Commun 109:908–915

29. Reddy GPV, Pardee AB (1986) Multienzyme complex for metabolic channelling in mam-malian DNA replication. Proc Natl Acad Sci USA 77:3312–3316

30. Reddy GPV, Klinge EM, Pardee AB (1986) Ribonucleotides are channelled into a mixed DNA-RNA polymer by permeabilized hamster cells. Biochem Biophys Res Commun 135:340–346

31. Sedwick WD, Kutler M, Brown OE (1981) Antifolate-induced misincorporation of deoxy-uridine monophosphate into DNA: Inhibition of high molecular weight DNA synthesis in human lymphoblastoid cells. Proc Natl Acad Sci USA 78:917

32. Taheri MR, Wickremasinghe RG, Hoffbrand AV (1981) Alternative metabolic fates of thy-mine nucleotides in human cells. Biochem J 194:451–461

33. Taheri MR, Wickremasinghe RG, Hoffbrand AV (1981) Metabolism of thymine nucleotides synthesised via the "de novo" mechanism in normal, megaloblastic and methotrexate treated human cells and in a lymphoblastoid cell line. Biochem J 196:225–235

34. Taheri MR, Wickremasinghe RG, Hoffbrand AV (1982) Functional compartmentation of DNA precursors in human leukaemoblastoid cell lines. Br J Haematol 52:401–409

35. Taheri MR, Wickremasinghe RG, Jackson BFA, Hoffbrand AV (1982) The effect of folate analogues and vitamin B12 on provision of thymine nucleotides for DNA synthesis in me-galoblastic anemia. Blood 59:634–640

36. Wickremasinghe RG, Hoffbrand AV (1979) Defective DNA synthesis in megaloblastic anaemia: Studies employing velocity sedimentation in alkaline sucrose gradients. Biochim Biophys Acta 563:46–48

37. Wickremasinghe RG, Hoffbrand AV (1980) Reduced rate of DNA replication fork move-ment in megaloblastic anemia. J Clin Invest 65:26–36

38. Wickremasinghe RG, Hoffbrand AV (1982) Megaloblastic anemia. In: Hoffbrand AV (ed) Recent advances in haematology. Churchill Livingstone, Edinburgh, pp 25–44

39. Wickremasinghe RG, Hoffbrand AV (1983) Inhibition by aphidicolin and dideoxythymid-ine triphospate of a multienzyme complex of DNA synthesis from human cells. FEBS Lett 159:175–179

40. Wickremasinghe RG, Yaxley JC, Hoffbrand AV (1982) Solubilization and partial character-ization of a multienzyme complex of DNA synthesis from human lymphoblastoid cells. Eur J Biochem 126:589–596

41. Wickremasinghe RG, Yaxley JC, Hoffbrand AV (1983) Gel filtration of a complex of DNA polymerase and DNA precursor synthesizing enzymes from a human lymphoblastic cell line. Biochim Biophys Acta 740:243–248

Chapter 2

Diagnosis of Megaloblastic Anemia

R. CARMEL

The recognition of megaloblastic anemia rests on clinical and hematologic features that have been known for many years. The diagnostic tools, however, have become more refined and reliable, thus allowing specific diagnosis to be made more accurately and in more subtle kinds of disorders than before.

In order to take full advantage of these tools, two diagnostic principles must be followed. First, it is necessary in all cases to establish as firmly as possible the specific vitamin abnormality responsible for the megaloblastic anemia. In the vast majority of cases, this is deficiency of either folic acid or cobalamin (vitamin B_{12}), or sometimes of both vitamins. The second principle is that it is as important to diagnose the underlying disorder that produced the deficiency as it is to identify the vitamin deficiency.

This chapter reviews the diagnostic aspects of megaloblastic anemia itself, and then the diagnostic details of the two principles mentioned.

Megaloblastic Anemia

Whether the characteristic megaloblastic anemia is directly responsible for the patient's seeking medical attention or not, it is usually the feature that prompts consideration of the diagnosis. Very mild or partially masked abnormalities, however, often make the diagnostic task difficult.

Although most patients are anemic, and many have pancytopenia, sometimes anemia is absent or very mild. Indeed, 19–25% of patients with megaloblastosis have normal hemoglobin values [37]. The mechanism of anemia is one of ineffective erythropoiesis, with a hemolytic component in many cases. Therefore, evidence of increased erythroid destruction, such as indirect hyperbilirubinemia and elevated lactate dehydrogenase levels, is common; however, it is not invariable [27].

Macrocytosis is a major clue to the existence of megaloblastic anemia. Due to the widespread use of electronic cell counters and the fact that development of macrocytosis precedes the development of anemia [56], attention to the mean corpuscular volume (MCV) is very helpful diagnostically [16]. Elevated MCV is the most accessible marker for megaloblastic anemia. However, it has limitations. Table 1 lists the conditions in which macrocytosis may occur. Most surveys suggest that the more striking the macrocytosis, the greater is the likelihood that megaloblastic anemia is responsible for it [75]. However, numerous excep-

Table 1. Causes of macrocytosis

1. Megaloblastic anemia
2. Drugs
   - Alcohol
   - Chemotherapeutic and immunosuppressive agents (e.g., azathioprine)
3. Hematological disorders
   - Aplastic anemia
   - Pure red cell aplasia
   - Myeloproliferative disease
   - Leukemia
   - Multiple myeloma
   - Refractory anemia
   - 5q- syndrome
   - Hemolytic anemia
   - Hereditary hydrocytosis
4. Nonhematological diseases
   - Liver disease (usually, but not invariably, alcoholic)
   - Hypothyroidism
5. Physiological
   - MCV is normally elevated in the first 4 weeks of life
6. Idiopathic
   - Pregnancy[a]
   - Old age[a]
   - Chronic lung disease, smoking[a]
   - Cancer[a]
7. Artifact
   - Cold agglutinins
   - Severe hyperglycemia
   - Hyponatremia
   - Stored blood

[a] Higher MCV levels have been observed in pregnancy, old age, smoking, and cancer, but the mechanism is unknown, and the possibility of subtle megaloblastosis as the explanation has not been entirely excluded.

tions occur, and reliance on such a rule can never substitute for careful examination of the hematologic and clinical picture. Furthermore, it is not unusual for megaloblastic anemia to coexist with one of the other causes of macrocytosis, particularly with chronic alcoholism.

Despite the great usefulness of the MCV, macrocytosis is not invariably seen in megaloblastic anemia. Normal or even low MCV may result when disorders that produce microcytosis coexist with megaloblastic anemia [53, 97]. Normal MCV has been said to occur in about 10% of patients with megaloblastic anemia [37, 88]. However, recent reports have noted an incidence of 33%–40% in patients with pernicious anemia [18a, 23, 98]. The explanation for this higher incidence may lie in the high proportion of black patients in these series, but the earlier recognition of disease with the widespread use of more sensitive cobalamin assays may also have contributed to this startling incidence. In many cases, the normal MCV remains unexplained.

It is also useful to remember that children have lower MCV than adults. "Microcytosis" is the norm from the age of 4–6 months to the age of 10 years or so

[66]. Thus, for example, macrocytosis in a 3-year-old child may produce an MCV of only 90 fl.

Examination of the blood smear is necessary in all cases. Macroovalocytes are a typical but not specific finding. The important morphological feature is hypersegmentation of the neutrophil nuclei. This finding is a much more sensitive and specific marker of megaloblastic anemia than is macrocytosis [72]. It also precedes the appearance of macrocytosis [56].

Besides its occurrence in nearly all cases of megaloblastic anemia, hypersegmentation has been described in chronic myelogenous leukemia, myelofibrosis, steroid therapy [44], iron deficiency [4], renal failure [55], and as a congenital finding [2]. Some of these entities may conceivably have subtle megaloblastosis confined to single cell lines, as has been found in iron deficiency [39], but it is otherwise unexplained. It is also important not to be misled by the bizarre radial segmentation of neutrophil nuclei that may occur in necrotizing diseases [81] and in heat stroke [49]. Hypersegmentation appears to be absent in megaloblastic anemia mainly when there is also marked neutropenia or a "shift to the left" because of coexisting infection [72].

Examination of the bone marrow aspirate is advisable whenever megaloblastic anemia cannot be diagnosed unequivocally from the peripheral blood smear. Two features are sought: nuclear-cytoplasmic asynchrony and enlarged cells. There is marked hyperplasia in the erythroid series, although sometimes hypoplasia or even aplasia accompanies megaloblastosis [87]. The typical megaloblastic appearance of the erythroid series may sometimes be masked when iron deficiency coexists. However, even in such circumstances the megaloblastic appearance of the myeloid series – with giant metamyelocytes and band forms, often showing large serpentine nuclei with budding projections – persists.

Identifying the Vitamin Deficiency

Serum Vitamin Levels

Some basic theoretical difficulties mark the connection between serum levels and vitamin status in the body. However, practical considerations make serum levels our main diagnostic tool, and indeed this tool usually works surprisingly well. Because of the close biochemical and physiological interaction between folate and cobalamin, and because the two deficiencies sometimes coexist, it is advisable to obtain levels of both in all patients. Many commercially available methods now assay both vitamins simultaneously.

Cobalamin levels can be measured either microbiologically or by radioassay. Microbiological assays depend on the growth requirement of various organisms for cobalamin. Although each microorganism recognizes with different affinity the various forms of cobalamin, none of them utilizes nonfunctional analogues of the vitamin much. Microbiological assay measures biologically active cobalamin fairly reliably and is widely viewed as the "gold standard" for measuring cobalamin levels. The disadvantage lies chiefly in the demanding nature of microbiological techniques and in the delays caused by the long assay incubation periods. A lesser problem is the inhibitory effect of some antibiotics, and occasionally other substances present in serum, on the growth of the microorganisms. Radioassays, on the other hand, produce falsely low results if the blood is radioactive [15].

The radioassays, because of their simplicity and the aggressive marketing of assay kits, have become by far the most common means of measuring vitamin levels, at least in the United States. However, considerable controversy surrounds them, and the mere existence of so many different radioassay techniques attests to the dissatisfaction. A major problem was defined in 1978 when it was noted that normal assay results were produced in up to 20% of cases of pernicious anemia by the radioassays then in use [35, 67]. This was shown to be largely due to the use of a binding protein, called R binder, which bound not only cobalamin but also various inactive analogues of cobalamin. Thus, large amounts of analogue in a serum sample could mask a low cobalamin content [67]. This problem has been corrected, the most popular means of correction being the use of pure intrinsic factor as the binding protein in the assay. Nevertheless, controversy persists [57, 106]. It is not clear that analogues are responsible for all cases of discrepantly high results [6, 40].

Despite these unresolved issues, the revised radioassays have been widely adopted and seem to have largely eliminated the problem of falsely normal results in cobalamin-deficient patients. It is important to remember that an equally serious problem with radioassays, especially commercially produced ones, remains the old problem of quality control by the laboratories using them [40]. Some laboratories have modified the methods, consciously or not, and many have not established the normal range in their own hands, relying instead on the manufacturer's stated normal range. Careful attention by the laboratory remains the main determinant of reliability [32, 90]. Indeed, variations between laboratories using the same technique can be greater than variations between different techniques [80].

Table 2 presents a list of possible causes that can be useful in interpreting low serum cobalamin results. A rather lengthy list of conditions has been associated with low values. It has been suggested that the revised assays now produce many falsely low cobalamin levels [12]. However, the frequency of low results with the revised radioassays seems to equal that with microbiological assay.

Some of the conditions listed in Table 2 appear not to involve actual cobalamin deficiency. Nevertheless, it is always wise to consider the possibility that deficiency may exist. Recent reports suggest that careful assessment often reveals subtle evidence of deficiency and atypical malabsorptive states in cases that appear to represent falsely low levels at first glance [8, 22, 26]. Many of the entities now listed in Table 2 under categories 3 and 4 may, with further study, be moved to category 1 or 2 in the future. Although the lower the serum cobalamin level, the more it is likely to represent overt deficiency states such as pernicious anemia, enough exceptions exist to make it unwise to ignore cobalamin levels that are only minimally subnormal [18a, 23]. It is a good rule to approach each patient with a low serum cobalamin level with the attitude that it represents deficiency until proven otherwise. This is particularly true in the elderly, in whom levels are

Table 2. Causes of low serum cobalamin levels

1. Cobalamin deficiency
2. Compromised cobalamin status without overt evidence of deficiency
   - Latent pernicious anemia or other malabsorptive states
   - Postgastrectomy status without anemia or neurological dysfunction
   - Vegetarianism (overt evidence of deficiency is only rarely displayed)
3. Unexplained but perhaps associated with compromised cobalamin status
   - Old age
   - Gastric achlorhydria
   - Chronic dialysis for renal failure
   - Hydantoin therapy
   - Cancer
4. Presumed not to represent cobalamin deficit
   - Pregnancy
   - Folate deficiency
   - Transcobalamin I deficiency
   - Multiple myeloma and Waldenström's macroglobulinemia
   - Acquired immune deficiency syndrome
   - Aplastic anemia
   - Hairy cell leukemia
   - Oral contraceptive use
   - Severe iron deficiency
5. Artifact
   - Radioactive serum
   - High-dose ascorbic acid ingestion

often decreased but should not be automatically assumed to be falsely low [8, 22, 47, 59].

Misleadingly normal cobalamin levels are less of a problem with the new radioassay methods, as already discussed. Still, deficient patients can fail to show decreased serum levels when their disorder involves cobalamin transport in the blood or impaired metabolic utilization. For example, hereditary transcobalamin II deficiency is usually accompanied by normal serum levels despite severe cellular depletion (although recent cases have been described where the serum cobalamin level is actually low [24, 78]). The same is true of hereditary metabolic disorders, such as congenital methylmalonic aciduria-homocystinuria, and in acquired metabolic blocks, such as that induced by nitrous oxide anesthesia (although decreased serum cobalamin levels may occur [69]). Since cobalamin is largely bound to transcobalamin I in the blood, diseases with elevated transcobalamin I levels, such as chronic myelogenous leukemia, can mask the low cobalamin levels of pernicious anemia when the two disorders coexist [10, 36]. Falsely increased radioassay results also occur in patients given chloral hydrate [79]. And, of course, a falsely normal serum level commonly occurs in the deficient patient recently given a cobalamin injection.

Serum folate, like serum cobalamin, can be measured microbiologically or by radioisotopic dilution. Many of the considerations discussed for cobalamin assay apply to folate also. Comparisons of the radioisotopic and microbiological folate assays have not reached consistent conclusions [40, 68, 74, 94], although in

Table 3. Conditions associated with low serum folate levels

1. Folate deficiency
2. Low serum levels apparently not reflecting overt folate deficiency
   - Poor dietary intake without actual folate deficiency
   - Sickle cell anemia and other chronic hemolytic states
   - Drugs (e.g., oral contraceptives, acetylsalicylic acid, alcohol)
   - Idiopathic
3. Artifact
   - Improper handling of specimen (e.g., long storage)
   - Radioactive blood[a]
   - Folate-binding protein abnormality in renal failure[a]
   - Drugs that inhibit microbial growth (e.g., antibiotics)[b]

[a] Source of artifact only in radioisotopic assay.
[b] Source of artifact only in microbiologic assay.

certain clinical settings the isotopic methods may yield the more representative serum levels [3, 14].

Folate measurement is often less reliable than cobalamin measurement because of the lability of folates. Folates are unstable and subject to errors due to improperly handled specimens [84]. However, most of the fluctuation is physiological; serum folate levels undergo considerable short-term fluctuation and thus may not mirror body status accurately. Folate circulating in the blood is in equilibrium with absorbed folate. Thus, diminished intake of only a few days duration can lead to diminished serum levels, which do not actually reflect the much more stable tissue stores. Also noteworthy is the acute drop in serum folate induced by alcohol ingestion [58]. Table 3 lists the conditions in which low serum folate levels are seen.

Serum folate levels also rise under conditions that do not mirror tissue stores. Sometimes this produces falsely normal results. This may happen when the patient has eaten folate-rich food (but not following a normal meal) or has been given vitamin supplements. Artifact may also occur in a hemolyzed blood sample, since red cell folate content is so much greater than serum content, enough folate can flood the serum to significantly raise the level. Finally, certain deficiency states occur at the cellular level and do not depress the serum folate content. This occurs in some inborn errors of folate metabolism and in acquired metabolic blocks induced by drugs. It also occurs in some critically ill patients who develop acute megaloblastic changes in the bone marrow, apparently due to acute folate deficiency not reflected by serum levels [5].

Tissue Vitamin Levels

Tissue biopsy should reflect body vitamin status more reliably than does the serum level, provided that the biopsy is a representative one. The only practical target currently is the red blood cell. For folate, red blood cell assay has become a valuable tool. Red blood cell folate is less subject than serum folate to the various fluctuations described earlier. It is not affected acutely by diet, nor is it affected by therapy until a new population of cells is generated. It does not become subnormal until deficiency is well established.

The shortcomings of red blood cell folate measurement are few but can be imposing. The most important drawback is that the values are also subnormal in about half of patients with cobalamin deficiency [1, 34, 54, 60]. This can create confusion unless one also measures serum levels. Sometimes, too, unexplained low levels are found in patients known not to be deficient [1]. An obvious problem in interpreting red blood cell results also occurs in the patient who has been transfused. Sometimes levels can also be falsely raised by significant reticulocytosis, because reticulocytes are richer in folate than older cells [60]. Red blood cell folate levels also tend to be raised in iron deficiency [86], but the extent of this as a problem is not clear. The oxygenation status of the red cells also affects their assayable folate content [82].

A practical obstacle to the wider use of the red blood cell folate content is technical. The microbiological assay is not widely available, and radioisotopic assays have technical shortcomings that require careful attention by the laboratories that use them [74].

Red blood cell cobalamin has proven to be subject to too many variables to be a useful measurement of cobalamin status [85]. It is, therefore, of very limited clinical value.

Tests of Metabolic Status

Metabolic intermediates accumulate in cobalamin or folate deficiency and can be measured in the blood or, more commonly, in the urine. Since folates are involved in many reactions, several metabolites have been proposed as indices of deficiency. These include urinary formiminoglutamic acid, 5-amino-4-imidazole carboxamide ribonucleotide, and serum homocysteine. However, since these are sometimes also elevated in cobalamin deficiency and in other conditions [30, 99], they are not widely used clinically.

Cobalamin deficiency causes increased urinary excretion of methylmalonic acid, an intermediate in propionate metabolism. This pathway is not affected in folate deficiency and, hence, the test has good specificity. However, elevated excretion can also occur in congenital defects of the mutase enzyme unrelated to cobalamin. Unexplained methylmalonic aciduria has also been noted in some adults [51]. Nevertheless, sensitivity appears to be a greater problem than specificity, since not all cobalamin-deficient patients have measurable methylmalonic aciduria [31]. The sensitivity can be enhanced by giving the patient a loading dose of valine or isoleucine orally to "stress" the pathway. The biggest obstacle to the wider diagnostic use of methylmalonic acid assay has been technical. Recently, mass spectrometry techniques have been applied and sensitivity has been greatly enhanced [73a, 83, 99a].

A very interesting diagnostic tool, the deoxyuridine suppression test, is currently limited to research centers. Two excellent reviews of this test have appeared recently [76, 105]. The test measures the incorporation of radioactive thymidine into DNA, which is normally suppressed by preincubating the cells with deoxyuridine. Since deoxyuridylate conversion to thymidylate requires methylene tetrahydrofolate, folate-deficient cells produce abnormal results in which deoxyuridine fails to adequately suppress radioactive thymidine incorporation into DNA [64]. Cobalamin deficiency produces a similar abnormality because cobalamin deficiency traps folate as methyl tetrahydrofolate, thus creating secondarily a deficiency of methylene tetrahydrofolate. Cobalamin and folate deficiencies can be differentiated in most cases by adding vitamins in vitro to aliquots of the cells and observing whether they correct the deoxyuridine suppression abnormality or not [77, 108].

The test is usually done on bone marrow cells. It can also be done on peripheral blood lymphocytes [38]. The practical advantages of the latter are obvious. Moreover, testing lymphocytes can be used to explore deficiency limited to a single cell line [39]. However, the long-lived dormant lymphocytes may not accurately reflect the patient's current vitamin status [38] and the test is not as reliable as testing marrow cells [73b].

Although normal deoxyuridine suppression-test results occur periodically despite cobalamin or folate deficiency, such as when iron deficiency coexists [103], the test appears to be very sensitive. Abnormalities have been detected in cobalamin-deficient patients whose peripheral blood and bone marrow cells were normoblastic in appearance [8, 19, 22, 26]. In vitro vitamin additives markedly enhance the sensitivity of the test [19, 22, 26]. A particularly important additive may be methyl tetrahydrofolate, which sometimes unmasks very subtle cobalamin deficiency even when the baseline suppression result is normal [19, 26]. In vitro additives not only enhance the sensitivity of the test but also help identify the specific vitamin abnormality. They can also define acquired or hereditary metabolic blocks where serum vitamin levels are often normal.

Identifying the Specific Disorder Responsible for Vitamin Deficiency

This diagnostic goal is ignored by many physicians, who may simply prescribe the vitamin without investigating the cause of the deficiency [23]. This unfortunate practice leaves unanswered the important questions of how long to treat and of whether the vitamin should be given orally or parenterally. It also may prevent the identification of an underlying disorder that needs independent treatment or has prognostic importance. Therefore, the best approach is to consider the diagnostic work-up of cobalamin or folate deficiency to be incomplete until the underlying cause for the deficiency is also found.

Cobalamin Deficiency

Many different disorders can cause cobalamin deficiency. Thinking about them can be simplified considerably by keeping in mind the physiologic cycle of how

Table 4. Disorders producing cobalamin deficiency

Phases of cobalamin metabolism in man that are subject to disruption	Disorders	Diagnosis established by
I. Dietary intake	1. Veganism	1. Rule out absorptive defects and demonstrate response to small dose of oral cobalamin.
	2. Newborn infant of vegan mother	2. Rule out absorptive and metabolic defects and identify maternal dietary restrictions.
II. Gastrointestinal transport and absorption: A. Gastric phase	1. Pernicious anemia (Addisonian)	1. Abnormal Schilling test corrected with oral intrinsic factor *or* Absent gastric secretion of intrinsic factor *or* Positive serum anti-intrinsic factor antibody
	2. Pernicious anemia (congenital)	2. Schilling test and gastric analysis for intrinsic factor as above, *and* normal gastric acid secretion and gastric biopsy.
	3. Abnormal intrinsic factor	3. As in 2, and demonstration of nonfunctional intrinsic factor.
	4. Postgastrectomy	4. Schilling test as above, and history of gastric surgery.
B. Luminal phase	1. Bacterial overgrowth	1. Abnormal Schilling test corrected only after antibiotic therapy.
	2. Pancreatic insufficiency (rarely associated with actual deficiency of cobalamin)	2. Abnormal Schilling test corrected only with pancreatic enzyme.
	3. Diphyllobothrium latum	3. Abnormal Schilling test with demonstration of the parasite.
C. Ileal absorption	1. Ileal disease or surgery	1. Abnormal Schilling test not corrected by any maneuver.
	2. Drugs affecting ileal absorption	2. As in 1, and history of long-standing ingestion of drug.
III. Serum transport	1. Transcobalamin II deficiency	1. Demonstrate absence of transcobalamin II in serum.
	2. Abnormal transcobalamin II	2. Demonstrate abnormally functioning transcobalamin II.
IV. Cellular uptake and metabolism	1. Nitrous oxide anesthesia	1. History of exposure, and rule out absorptive defects.
	2. Hereditary metabolic defects	2. Cellular studies (specialized laboratory needed).
V. Increased requirement	?	
VI. Increased losses	?	

cobalamin is taken in, absorbed (with the crucial role of intrinsic factor), and utilized. Table 4 provides a basic list following this logical sequence, along with the essential diagnostic steps.

The diagnostic task is made easier by the fact that malabsorption is nearly always responsible for cobalamin deficiency, especially in adults [71]. Thus, a test of cobalamin absorption, such as the urinary excretion test (Schilling test), forms the backbone of the diagnostic approach. The initial approach can be further simplified because most cases are due to pernicious anemia [71]. Pernicious anemia is a gastroenterologically defined entity rather than a hematologic one. It signifies the disorder in which the stomach fails to secrete intrinsic factor. Pernicious anemia may be discovered in some patients before cobalamin deficiency has actually developed.

The diagnosis of pernicious anemia is most often made by the Schilling test. The most common diagnostic problem arises when the test fails to show correction with an oral dose of instrinsic factor, suggesting an intestinal defect instead of pernicious anemia. This can occur in nearly half of cases, especially when the test with intrinsic factor is done before the patient has been adequately treated with cobalamin [20, 73]. Such a finding is transient; most patients revert to the expected pattern when retested a few days or weeks (but sometimes a few months) later. Nevertheless, it is worthwhile to remember that pernicious anemia sometimes coexists with intestinal disease [11], in particular with bacterial overgrowth of the intestine [73].

Alternative ways of diagnosing pernicious anemia are often very useful. The most accurate way is to demonstrate that properly collected gastric juice (i.e., after stimulation of gastric secretion) does not contain intrinsic factor [52, 63]. This directly meets the absolute criterion by which the disease is defined, but few laboratories assay intrinsic factor. Gastric analysis for acid secretion, on the other hand, may be helpful but is never diagnostic.

Anti-intrinsic factor antibody in the serum is the most useful marker for pernicious anemia [62, 107]. The blocking, or type I, antibody is present in 60%–75% of patients with the disease. In some types of patients, such as individuals with the juvenile form of the disease [29] or black women [21, 98], the incidence may exceed 90%. False positive results are extremely uncommon. Most often, when the antibody is discovered in a patient without pernicious anemia, the disease develops a few months to a few years later [104]. Nevertheless, a few patients with the antibody never display pernicious anemia [91]. This phenomenon is seen most often in women with insulin-dependent diabetes, hyperthyroidism, or myasthenia gravis [91, 109]. However, the most common cause of false-positive results is artifactually induced, by using blood drawn within 24 hours of a cobalamin injection. Despite these problems, the usefulness of the test is such that a *cobalamin-deficient* patient who has anti-intrinsic factor antibody can be assumed with confidence to have pernicious anemia.

A less conclusive but sensitive test that can often suggest the diagnosis of pernicious anemia is the serum gastrin assay [18b, 50]. Gastrin levels are elevated in 65%–95% of patients with this disease. The levels are often very high, especially in women [25]. However, increased gastrin levels may occur in achlorhydria and/or gastritis without pernicious anemia [50], as well as in many other conditions (including the Zollinger-Ellison syndrome, which itself causes cobalamin malabsorption). Pepsinogen I and pepsinogen II levels, and their ratios, may be the most sensitive serum markers for pernicious anemia [18b] but, like gastrin, have limited specificity.

The antiparietal cell microsomal antibody is frequently positive in pernicious anemia. However, it is a poor diagnostic tool because it lacks specificity. It is a marker for gastritis rather than for pernicious anemia per se [62]. Moreover, the incidence of this antibody in pernicious anemia may no longer exceed that of the much more specific anti-intrinsic factor antibody, at least in certain populations [18b, 25].

As valuable as tests like the anti-intrinsic factor antibody are, the absorption test remains central to the work-up of cobalamin deficiency. It should be done in all cases of cobalamin deficiency (unless the cause has been established by other methods) and is also indicated when deficiency is suspected but not proven. The most common way to assess the amount absorbed is to test urinary excretion: the Schilling test. Absorption can also be determined from hepatic uptake of the dose of radioactive cobalamin, whole-body retention, fecal excretion, or by measuring the absorbed amount circulating in plasma [61]. Unlike the Schilling test, these do not require the injection of a large "flushing" dose of cobalamin (thus leaving the patient in an untreated state), and they avoid the problem of relying on a complete collection of urine. However, they have various drawbacks of their own, such as requiring specialized facilities and equipment, and so are not widely used.

The Schilling test, as mentioned, is subject to problems, some common to all the absorption tests and some unique to it. The problems unique to it revolve around the need for a complete 24-h or 48-h urine collection, free of contamination by stool, in a patient whose renal function must be normal. A failure in any of these respects can produce a false result.

When absorption is abnormal, the next step is to determine what maneuver will correct the malabsorption. The first attempt, of course, is with intrinsic factor. The disorders defined by the results are listed in Table 5. Attempts have been made to simplify the Schilling test by combining in one procedure the baseline testing and the testing with intrinsic factor. Not only does this save time and effort, but incomplete urine collection need not invalidate the results because the ratios of the two results can sometimes be diagnostic. However, the test has a distressing tendency to produce normal results in as many as 40% of patients with pernicious anemia, which appears to be due to an isotope exchange in vivo between the components of the test [9, 42, 48]. It has been suggested that this problem can be ameliorated by giving the two components 2 hours apart [110].

Some uncertainty exists about patients with suspected cobalamin deficiency who display normal Schilling test results. Such results are regarded as indicating one of several possibilities: (a) the patient's cobalamin status is normal; (b) the patient has cobalamin deficiency due to a nonintestinal cause, such as dietary insufficiency; (c) the normal absorption result is an error. Considerable evidence now suggests a fourth possibility: these patients may malabsorb only protein-bound cobalamin, perhaps because their ability to split cobalamin from its natu-

Table 5. Cobalamin disorders classified according to results in the cobalamin absorption test

Abnormal Schilling test corrected with intrinsic factor
- Addisonian pernicious anemia
- Congenital pernicious anemia
- Abnormal intrinsic factor
- Postgastrectomy state
- Diphyllobothrium latum infestation (some cases)

Abnormal Schilling test uncorrected with intrinsic factor
- Ileal disease
- Ileal resection
- Cobalamin deficiency (transient abnormality corrected after cobalamin therapy)
- Bacterial overgrowth (test corrected after antibiotics)
- Diphyllobothrium latum infestation (some cases)
- Chronic pancreatic insufficiency (test corrected with oral pancreatic enzymes)[a]
- Folate deficiency and/or alcohol ingestion (some cases)[a]
- Drugs (para-amino salicylic acid, colchicine, neomycin, anticonvulsants, potassium salts, phenformin, metformin)[a]
- Zollinger-Ellison syndrome (test result may be corrected with bicarbonate)
- Transcobalamin II deficiency
- Artifacts (e.g., inactive intrinsic factor preparation)

[a] Not usually associated with cobalamin deficiency despite the malabsorption.

ral binding to proteins in the food is impaired [43]. This step is not measured by the Schilling test, which provides the test dose of cobalamin in a free form. The tests measuring protein-bound cobalamin absorption usually are done exactly like the Schilling test, except that the small oral dose of radioactive cobalamin is first combined with egg yolk or with a source of cobalamin-binding proteins such as chicken serum.

Table 6 lists the conditions in which this subtle form of malabsorption has been noted in numerous conditions despite the intact ability to absorb free cobalamin [7, 13, 17, 26, 26a, 28, 41, 43, 65, 93, 96, 101, 102].

Such patients usually have normal gastric intrinsic factor secretion, but this newly recognized form of cobalamin malabsorption sometimes marks a "prepernicious anemia" state, with pernicious anemia developing 1–2 years later [17, 26].

Table 6. Conditions in which malabsorption of protein-bound cobalamin has been found despite normal absorption of free cobalamin

- Gastric achlorhydria or hypochlorhydria
- "Prepernicious anemia"
- Postgastrectomy state
- Following vagotomy for gastric ulcer
- H$_2$-receptor blocking drugs
- Iron deficiency anemia (?)
- Chronic Alcoholism (?)
- Folate deficiency (?)
- Idiopathic

Clinically significant cobalamin deficiency can also result from malabsorption confined to protein-bound cobalamin alone [13, 26, 41, 65]. However, the natural history of this disorder still has to be defined. It is not clear how often protein-bound cobalamin deficiency actually leads to cobalamin deficiency, and therefore whether it needs to be treated immediately or observed. Moreover, it is not known whether such malabsorption always arises from disorders of gastric secretion, or whether oral, salivary, intestinal, or other abnormalities may contribute to it.

Folate Deficiency

Determination of the cause of folate deficiency involves much less testing than is the case with cobalamin deficiency and in fact is often not even attempted. This is because most cases of folate deficiency appear to arise from dietary insufficiency [71], and because no widely available test of folate absorptive capacity exists. Several absorption tests have been devised [45, 46, 92]. These measure either serum or urinary folate levels following a standard oral dose of folic acid, or they involve administration of radioactive folates much as is done with cobalamin in the Schilling test. These tests are limited currently to a few research centers. Better understanding and use of folate absorption testing may identify a wider range of disorders in the future than we are currently aware of.

The clinical diagnosis of folate malabsorption is usually established indirectly, by demonstrating that intestinal disease exists. The search for malabsorption needs to be made in all patients who do not have an obvious dietary cause. A drug history is also very important [100]. In addition, critically ill patients may develop acute folate deficiency that can be diagnosed only from the megaloblastic marrow appearance (and, when tested, abnormal deoxyuridine suppression) [5]. As for attributing folate deficiency to a state of increased utilization [70], such as in chronic hemolytic anemia, this should be done with caution. Although such patients often have low serum folate levels, evidence of deficiency is absent in most cases and folate supplementation seems to make no difference [89]. In those patients who actually develop folate deficiency, increased utilization can usually be shown not to be the only factor. Coexisting dietary inadequacy, alcoholism, or other contributory factors must be carefully sought. Increased requirement, however, seems to be a significant contributor in pregnant women with folate deficiency [33].

Megaloblastic Anemia in Children

Several factors dictate a different approach to the child with megaloblastic anemia than is used in adults. The diagnostic probabilities form a different spectrum in children than in adults. There is a greater likelihood of hereditary disorders. This is especially true in the first year of life, although the presentation of hereditary disorders may be delayed until the 2nd decade of life or even later [18,

19, 26b, 95]. The consequences of vitamin deficiency can be far more devastating in the growing and developing child, and the need to treat is thus usually more urgent. Therefore, the approach often requires obtaining the appropriate test samples and beginning therapy before all the results are in, especially in very young children. As a minimum in such cases, it is advisable to obtain serum folate and cobalamin levels, a red blood cell folate level, serum transcobalamin II determination, and urine for methylmalonic acid, homocystine, and orotic acid. It is also often desirable to obtain serum vitamin levels in the mother, even if she is asymptomatic.

One must also bear in mind that in many of the hereditary metabolic and transport disorders serum vitamin levels remain normal. This paradox, combined with the fact that cobalamin deficiency can respond at least partially to large therapeutic doses of folate, has led to occasional misdiagnosis. Finally, megaloblastic anemia unrelated to abnormalities of either folate or cobalamin, such as orotic aciduria or thiamine-responsive megaloblastic anemia, must also be considered in children.

Concluding Comments

The diagnostic approach to the megaloblastic anemias sometimes appears to be a victim of the ease and success with which these disorders can be treated. Too often there is a tendency to treat the deficiency without precise diagnosis of either the deficiency itself or the disorder that caused it. At the same time, considerable expansion has occurred in the diagnostic tools available, although some of these have not achieved universal availability. It is now possible to define precisely the cause of megaloblastic anemias of many diverse causes. As important, increasingly subtle disorders can now be identified at very early stages, and new entities are being discovered continually. This diagnostic sophistication shows promise of future growth, as clinicians become increasingly aware that subtle or early deficiency may cause very subtle changes rather than the florid abnormalities commonly described in textbooks. Proper clinical assessment depends on sound clinical judgment, enhanced by an informed use of both the basic diagnostic tools and the appropriate resort to highly specialized techniques.

References

1. Bain BJ, Wickramasinghe SN, Broom GN, Litwinczuk RA, Sims J (1984) Assessment of the value of a competitive protein binding radioassay of folic acid in the detection of folic acid deficiency. J Clin Pathol 37:888–894
2. Barbier PF (1958) Un cas particulier d'hypersegmentation constitutionnelle des noyaux des neutrophiles chez l'homme. Acta Haematol (Basel) 19:121–128
3. Baril L, Carmel R (1978) Comparison of radioassay and microbiological assay for serum folate, with clinical assessment of discrepant results. Clin Chem 24:2192–2196
4. Beard MEJ, Weintraub LR (1969) Hypersegmented granulocytes in iron deficiency anaemia. Br J Haematol 16:161–163

5. Beard MEJ, Hatipov CS, Hamer JW (1980) Acute onset of folate deficiency in patients under intensive care. Crit Care Med 8:500–503

6. Begley JA, Hall CA (1981) Forms of vitamin B_{12} in radioisotope dilution assays. J Clin Pathol 34:630–636

7. Belaiche J, Zittoun J, Marquet J, Nurit Y, Yvart J (1983) Effect de la ranitidine sur la secretion de facteur intrinseque gastrique et sur l'absorption de la vitamine B_{12}. Gastroenterol Clin Biol 7:381–384

8. Blundell EL, Matthews JH, Allen SM, Middleton AM, Morris JE, Wickramasinghe SN (1985) Importance of low serum vitamin B_{12} and red cell folate concentrations in elderly hospital inpatients. J Clin Pathol 38:1179–1184

9. Briedis D, McIntyre PA, Judisch J, Wagner HN (1973) An evaluation of a dual-isotope method for the measurement of vitamin B_{12} absorption. J Nucl Med 14:135–141

10. Britt RP, Rose DP (1966) Pernicious anemia with a normal serum vitamin B_{12} level in a case of chronic granulocytic leukemia. Arch Intern Med 117:32–33

11. Brody EA, Estren S, Herbert V (1966) Coexistent pernicious anemia and malabsorption in four patients. Ann Intern Med 64:1246–1251

12. Brynskov J, Andersen K, Gimsing P, Hippe E (1983) False low serum vitamin B_{12} values with radiodilution assays using blocked R binders. Lancet 1:1104–1105

13. Carmel R (1978) Nutritional vitamin B_{12} deficiency: possible contributory role of subtle vitamin B_{12} malabsorption. Ann Intern Med 88:647–649

14. Carmel R (1978) Effects of antineoplastic drugs on *Lactobacillus casei* and radioisotopic assays for serum folate. Am J Clin Pathol 69:137–139

15. Carmel R (1978) Artifactual radioassay results due to serum contamination by intravenous radioisotope administration. Falsely low serum vitamin B_{12} and folic acid results. Am J Clin Pathol 70:364–367

16. Carmel R (1979) Macrocytosis, mild anemia and delay in the diagnosis of pernicious anemia. Arch Intern Med 139:47–50

17. Carmel R (1982) Subtle cobalamin malabsorption in a vegan patient. Evolution into classic pernicious anemia with anti-intrinsic factor antibody. Arch Intern Med 142:2206–2207

18. Carmel R (1983) Gastric juice in congenital pernicious anemia contains no immunoreactive intrinsic factor molecule: study of three kindreds with variable ages at presentation, including a patient first diagnosed in adulthood. Am J Hum Genet 35:67–77

18a. Carmel R (1988) Pernicious anemia. The expected findings of very low serum cobalamin levels, anemia and macrocytosis are often lacking. Arch Intern Med 148:1712–1714

18b. Carmel R (1988) Pepsinogens and other serum markers in pernicious anemia. Am J Clin Pathol 90:442–445

19. Carmel R, Goodman SI (1982) Abnormal deoxyuridine suppression test in congenital methylmalonic aciduria-homocystinuria without megaloblastic anemia: divergent biochemical and morphological bone marrow manifestations of disordered cobalamin metabolism in man. Blood 59:306–311

20. Carmel R, Herbert V (1967) Correctable intestinal defect of vitamin B_{12} absorption in pernicious anemia. Ann Intern Med 67:1201–1207

21. Carmel R, Johnson CS (1978) Racial patterns in pernicious anemia. Early age at onset and increased frequency of intrinsic factor antibody in black women. N Engl J Med 298:647–650

22. Carmel R, Karnaze DS (1985) The deoxyuridine suppression test identifies subtle cobalamin deficiency in patients without typical megaloblastic anemia. JAMA 253:1284–1287

23. Carmel R, Karnaze DS (1986) Physician response to low serum cobalamin levels. Arch Intern Med 146:1161–1165

24. Carmel R, Ravindranath Y (1984) Congenital transcobalamin II deficiency presenting atypically with a low serum cobalamin level: studies demonstrating the coexistence of a circulating transcobalamin I (R binder) complex. Blood 63:598–605

25. Carmel R, Ozturk G, Johnson CS, Tung KSK, Terasaki PI (1981) Profiles of black and Latin-American patients having pernicious anemia. HLA antigens, lymphocytotoxic antibody, anti-parietal cell antibody, serum gastrin levels, and ABO blood groups. Am J Clin Pathol 75:291–296

26. Carmel R, Sinow R, Karnaze DS (1987) Atypical cobalamin deficiency. J Lab Clin Med 109:454–463

26a. Carmel R. Sinow RM, Siegel ME, Samloff IM (1988) Food cobalamin malabsorption occurs frequently in patients with unexplained low serum cobalamin levels. Arch Intern Med 148:1715–1719

26b. Carmel R, Watkins D, Goodman SI, Rosenblatt DS (1988) Hereditary defect of cobalamin metabolism (cbl G mutation) presenting as a neurologic disorder in adulthood. N. Engl J Med 318:1738–1741

27. Carmel R, Wong ET, Weiner JM, Johnson CS (1985) Racial differences in serum total bilirubin levels in health and in disease (pernicious anemia). JAMA 253:3416–3418

28. Cattan D, Belaiche J, Zittoun J, Yvart J, Chagnon JP, Nurit Y (1982) Role de la carence en facteur intrinseque dans la malabsorption de la vitamine B$_{12}$ liee aux proteines, dans les achlorhydries. Gastroenterol Clin Biol 6:570–575

29. Chanarin I (1969) The megaloblastic anaemias, 1st edn. Blackwell, Oxford, pp 732–733

30. Chanarin I, Bennett MC, Berry V (1962) Urinary excretion of histidine derivatives in megaloblastic anaemia and other conditions and a comparison with folic acid clearance test. J Clin Pathol 15:269–273

31. Chanarin I, England JM, Mollin C, Perry J (1973) Methylmalonic acid excretion studies. Br J Haematol 25:45–53

32. Christensen RH, Dent GA, Tuszynski A (1985) Two radioassays for serum vitamin B$_{12}$ and folate determination compared in a reference interval study. Clin Chem 31:1358–1360

33. Cooper BA (1973) Folate and vitamin B$_{12}$ in pregnancy. Clin Haematol 2:461–476

34. Cooper BA, Lowenstein L (1964) Relative folate deficiency of erythrocytes in pernicious anemia and its correction with cyanocobalamin. Blood 24:502–521

35. Cooper BA, Whitehead VM (1978) Evidence that some patients with pernicious anemia are not recognized by radiodilution assay for cobalamin in serum. N Engl J Med 299:816–818

36. Corcino JJ, Zalusky R, Greenberg M, Herbert V (1971) Coexistence of pernicious anaemia and chronic myeloid leukaemia: an experiment of nature involving vitamin B$_{12}$ metabolism. Br J Haematol 20:511–520

37. Croft RF, Streeter AM, O'Neill BJ (1974) Red cell indices in megaloblastosis and iron deficiency. Pathology 6:107–117

38. Das KC, Herbert V (1978) The lymphocyte as a marker of past nutritional status: persistence of abnormal lymphocyte deoxyuridine suppression test and chromosomes in patients with past deficiency of folate and vitamin B$_{12}$. Br J Haematol 38:219–233

39. Das KC, Herbert V, Colman N, Longo DL (1978) Unmasking covert folate deficiency in iron-deficient subjects with neutrophil hypersegmentation: dU suppression tests on lymphocytes and bone marrow. Br J Haematol 39:357–375

40. Dawson DW, Delamore IW, Fish DI, Flaherty TA, Gowenlock AH, Hunt LP, Hyde K, MacIver JE, Thornton JA, Waters HM (1980) An evaluation of commercial radioisotope methods for the determination of folate and vitamin B$_{12}$. J Clin Pathol 33:234–242

41. Dawson DW, Sawers AH, Sharma RK (1984) Malabsorption of protein bound vitamin B$_{12}$. Br Med J 288:675–678

42. Domstad PA, Choy YC, Kim EE, DeLand FH (1981) Reliability of dual-isotope Schilling test for the diagnosis of pernicious anemia or malabsorption syndrome. Am J Clin Pathol 75:723–726

43. Doscherholmen A, Swaim WR (1973) Impaired assimilation of egg ^{57}Co vitamin B$_{12}$ in patients with hypochlorhydria and achlorhydria and after gastric resection. Gastroenterology 64:913–919

44. Eichacker P, Lawrence C (1985) Steroid-induced hypersegmentation in neutrophils. Am J Hematol 18:41–45.

45. Elsborg L (1981) The folic acid absorption test compared with other laboratory tests for malabsorption. Acta Med Scand 209:323–325

46. Elsborg L, Bastrup-Madsen P (1976) Folic acid absorption in various gastrointestinal disorders. Scand J Gastroenterol 11:333–335

47. Elsborg L, Lund V, Bastrup-Madsen P (1976) Serum vitamin B$_{12}$ levels in the aged. Acta Med Scand 200:309–314

48. Fairbanks VF, Wahner HW, Phyliky RL (1983) Tests for pernicious anemia: the "Schilling Test". Mayo Clin Proc 58:541–544
49. Friedman EW, Williams JC, Prendergast E (1982) Polymorphonuclear leucocyte hypersegmentation in heat stroke. Br J Haematol 50:169–170
50. Ganguli PC, Cullen DR, Irvine WJ (1971) Radioimmunoassay of plasma gastrin in pernicious anaemia, achlorhydria without pernicious anaemia, hypochlorhydria, and in controls. Lancet 1:155–158
51. Giorgio AJ, Trowbridge M, Boone AW, Patten RS (1976) Methylmalonic aciduria without vitamin B_{12} deficiency in an adult sibship. N Engl J Med 295:310–313
52. Gottlieb C, Lau KS, Wasserman LR, Herbert V (1965) Rapid charcoal assay for intrinsic factor (IF), gastric juice unsaturated B_{12} binding capacity, antibody to IF, and serum unsaturated B_{12} binding capacity. Blood 25:875–884
53. Green R, Kuhl W, Jacobson R, Johnson C, Carmel R, Beutler E (1982) Masking of macrocytosis by alpha thalassemia in blacks with pernicious anemia. N Engl J Med 307:1322–1325
54. Hansen HA, Weinfeld A (1962) Metabolic effects and diagnostic value of small doses of folic acid and B_{12} in megaloblastic anaemias. Acta Med Scand 172:427–443
55. Hattersley PG, Engels JL (1974) Neutrophilic hypersegmentation without macrocytic anemia. West J Med 121:179–184
56. Herbert V (1962) Experimental nutritional folate deficiency in man. Trans Assoc Am Physicians 75:307–320
57. Herbert V, Colman N, Palat D, Manusselis C, Drivas G, Block E, Akerkar A, Weaver D, Frenkel E (1984) Is there a "gold standard" for human serum vitamin B_{12} assay? J Lab Clin Med 104:829–841
58. Hillman RS, McGuffin R, Campbell C (1977) Alcohol interference with the folate enterohepatic cycle. Trans Assoc Am Physicians 90:145–156
59. Hitzhusen JC, Taplin ME, Stephenson WP, Ansell JE (1986) Vitamin B_{12} levels and age. Am J Clin Pathol 85:32–36
60. Hoffbrand AV, Newcombe BFA, Mollin DL (1966) Method of assay of red cell folate activity and the value of the assay as a test for folate deficiency. J Clin Pathol 19:17–28
61. International Committee for Standardization in Hematology (1981) Recommended methods for the measurement of vitamin B_{12} absorption. J Nucl Med 22:1091–1093
62. Irvine WJ (1965) Immunologic aspects of pernicious anemia. N Engl J Med 273:432–438
63. Irvine WJ, Davies SH, Hayes RC, Scarth L (1965) Secretion of intrinsic factor in response to histamine and gastrin in the diagnosis of Addisonian pernicious anaemia. Lancet 2:397–401
64. Killmann SA (1964) Effect of deoxyuridine on incorporation of tritiated thymidine: difference between normoblasts and megaloblasts. Acta Med Scand 175:483–488
65. King CE, Leibach J, Toskes PP (1979) Clinically significant vitamin B_{12} deficiency secondary to malabsorption of protein-bound vitamin B_{12}. Dig Dis Sci 24:397–402
66. Koerper MA, Mentzer WC, Brecher G, Dallman PR (1976) Developmental change in red blood cell volume: implication in screening infants and children for iron deficiency and thalassemia trait. J Pediatr 89:580–583
67. Kolhouse JF, Kondo H, Allen NC, Podell E, Allen RH (1978) Cobalamin analogues are present in human plasma and can mask cobalamin deficiency because current radioisotope dilution assays are not specific for true cobalamin. N Engl J Med 299:785–792
68. Kubasik NP, Volosin MT, Sine HE (1975) Comparison of commercial kits for radioimmunoassay. Clin Chem 21:1922–1926
69. Layzer RB (1978) Myeloneuropathy after prolonged exposure to nitrous oxide. Lancet 2:1227–1230
70. Lindenbaum J (1977) Folic acid requirement in situations of increased need. In: Folic acid: biochemistry and physiology in relation to the human nutrition requirement. National Academy of Sciences, Washington DC, pp 256–276
71. Lindenbaum J (1979) Aspects of vitamin B_{12} and folate metabolism in malabsorption syndromes. Am J Med 67:1037–1048
72. Lindenbaum J, Nath BJ (1980) Megaloblastic anaemia and neutrophil hypersegmentation. Br J Haematol 44:511–513
72a. Lindenbaum J, Healton EB, Savage DG, Brust JCM, Garrett TJ, Podell ER, Marcell PD,

Stabler SP, Allen RH (1988) Neuropsychiatric disorders caused by cobalamin deficiency in the absence of anemia or macrocytosis. N Engl J Med 318:1720–1728

73. Lindenbaum J, Pezzimenti JF, Shea N (1974) Small intestinal function in vitamin B_{12} deficiency. Ann Intern Med 80:326–331

73a. Matchar DB, Feussner JR, Millington DS, Wilkinson RH, Watson DJ, Gale D (1987) Isotope-dilution assay for urinary methylmalonic acid in the diagnosis of vitamin B_{12} deficiency. Ann Intern Med 106:707–710

73b. Matthews JH, Wickramasinghe SN (1988) The deoxyuridine suppression test performed on phytohemagglutinin-stimulated peripheral blood cells fails to reflect in vivo vitamin B_{12} or folate deficiency. Eur J Haematol 40:174–180

74. McGown EL, Lewis CM, Dong MH, Sauberlich HE (1978) Results with commercial radioassay kits compared with microbiological assay of folate in serum and whole blood. Clin Chem 24:2186–2191

75. McPhedran P, Barnes MG, Weinstein JS, Robertson JS (1973) Interpretation of electronically determined macrocytosis. Ann Intern Med 78:677–683

76. Metz J (1984) The deoxyuridine suppression test. CRC Crit Rev Clin Lab Sci 20:205–241

77. Metz J, Kelly A, Swett VC, Waxman S, Herbert V (1968) Deranged DNA synthesis by bone marrow from vitamin B_{12} deficient humans. Br J Haematol 14:575–592

78. Meyers PA, Carmel R (1984) Hereditary transcobalamin II deficiency with subnormal serum cobalamin levels. Pediatrics 74:866–871

79. Mitchell CA, Tesar PJ, Maynard JH, Choong SH (1981) Chloral hydrate interferes with radioassay of vitamin B_{12}. Clin Chem 27:1480–1481

80. Mollin DL, Hoffbrand AV, Ward PG, Lewis SM (1980) Interlaboratory comparison of serum vitamin B_{12} assay. J Clin Pathol 33:243–248

81. Neftel KA, Muller OM (1981) Heat-induced radial segmentation of leucocyte nuclei: a nonspecific phenomenon accompanying inflammatory and necrotizing diseases. Br J Haematol 48:377–382

82. Nelson NL, Klausner JS, Branda RF (1983) Oxygenation alters red cell folate levels. Br J Haematol 55:235–242

83. Norman EJ, Martelo OJ, Denton MD (1982) Cobalamin (vitamin B_{12}) deficiency detection by urinary methylmalonic acid quantitation. Blood 59:1128–1131

84. O'Broin JD, Temperley IJ, Scott JM (1980) Erythrocyte, plasma and serum folate: specimen stability before microbiological assay. Clin Chem 26:522–524

85. Omer A, Finlayson NDC, Shearman DJC, Samson RR, Girdwood RH (1970) Erythrocyte vitamin B_{12} activity in health, polycythemia and in deficiency of vitamin B_{12} and folate. Blood 35:73–82

86. Omer A, Finlayson NDC, Shearman DJC, Samson RR, Girdwood RH (1970) Plasma and erythrocyte folate in iron deficiency and folate deficiency. Blood 35:821–828

87. Pezzimenti JF, Lindenbaum J (1972) Megaloblastic anemia associated with erythroid hypoplasia. Am J Med 53:748–754

88. Prentice AG, Evans IL (1979) Megaloblastic anaemia with normal mean cell volume. Lancet 1:606–607

89. Rabb LM, Grandison Y, Mason K, Hayes RJ, Serjeant B, Serjeant GR (1983) A trial of folate supplementation in children with homozygous sickle cell disease. Br J Haematol 54:589–594

90. Raniolo E, Phillipou G, Paltridge G, Sage RE (1984) Evaluation of a commercial radioassay for the simultaneous estimation of vitamin B_{12} and folate, with subsequent derivation of the normal reference range. J Clin Pathol 37:1327–1335

91. Rose MS, Chanarin I, Doniach D, Brostoff J, Ardeman S (1970) Intrinsic factor antibodies in absence of pernicious anaemia. 3–7 year follow-up. Lancet 2:9–12

92. Rosenberg IH (1976) Absorption and malabsorption of folates. Clin Haematol 5:589–618

93. Salom IL, Silvis SE, Doscherholmen A (1982) Effect of cimetidine on the absorption of vitamin B_{12}. Scand J Gastroenterol 17:129–131

94. Shane B, Tamura T, Stokstad ELR (1980) Folate assay: a comparison of radioassay and microbiological results. Clin Chim Acta 100:13–19

95. Shinnar S, Singer HS (1984) Cobalamin C mutation (methylmalonic aciduria and homocystinuria) in adolescence. N Engl J Med 311:451–454

96. Sinow RM, Carmel R, Siegel ME (1985) The egg ^{57}Co B$_{12}$ absorption test in the evaluation of patients with low serum B$_{12}$. J Nucl Med 26:P96
97. Spivak JL (1982) Masked megaloblastic anemia. Arch Intern Med 142:2111–2114
98. Solanki D, Jacobson RJ, Green R, McKibbon J, Berdoff R (1981) Pernicious anemia in blacks. Am J Clin Pathol 75:96–99
99. Stabler SP, Marcell PD, Podell ER, Allen RH, Savage DG, Lindenbaum J (1988) Elevation of total homocysteine in the serum of patients with cobalamin of folate deficiency detected by capillary gas chromatography mass spectrometry. J Clin Invest 81:466–474
99a. Stabler SP, Marcell PD, Podell ER, Allen RH, Lindenbaum J (1986) Assay of methylmalonic acid in the serum of patients with cobalamin deficiency using capillary gas chromatography-mass spectrometry. J Clin Invest 77:1606–1612
100. Stebbins R, Scott J, Herbert V (1973) Drug-induced megaloblastic anemias. Semin Hematol 10:235–251
101. Steinberg WM, King CE, Toskes PP (1980) Malabsorption of protein-bound cobalamin but not unbound cobalamin during cimetidine administration. Dig Dis Sci 25:188–192
102. Streeter AM, Duraiappah B, Boyle R, O'Neill BJ, Pheils MT (1974) Malabsorption of vitamin B$_{12}$ after vagotomy. Am J Surg 128:340–343
103. Van der Weyden MB, Rother M, Firkin B (1972) Megaloblastic maturation masked by iron deficiency: a biochemical basis. Br J Haematol 22:299–307
104. Wangel AG, Schiller KFR (1966) Diagnostic significance of antibody to intrinsic factor. Br Med J 1:1274–1276
105. Wickramasinghe SN (1981) The deoxyuridine suppression test: a review of its clinical and research applications. Clin Lab Haematol 3:1–18
106. Zacharakis REA, Muir M, Chanarin I (1981) Comparison of serum vitamin B$_{12}$ estimation by saturation analysis with intrinsic factor and with R-protein as binding agents. J Clin Pathol 34:357–360
107. Zittoun J, Debril J, Jarret J, Sultan C, Zittoun R (1975) La recherche des anticorps antifacteur intrinseque dans le diagnostic de l'anemie de Biermer. Sem Hop Paris 51:227–232
108. Zittoun J, Marquet J, Zittoun R (1978) Effect of folate and cobalamin compounds on the deoxyuridine suppression test in vitamin B$_{12}$ and folate deficiency. Blood 51:119–128
109. Zittoun J, Tulliez M, Estournet B, Goulon M (1979) Humoral and cellular immunity to intrinsic factor in myasthenia gravis. Scand J Haematol 23:442–448
110. Zuckier LS, Chervu LR (1984) Schilling evaluation of pernicious anemia: current status. J Nucl Med 25:1032–1039

Chapter 3

Cobalamin–Folate Interrelations

I. CHANARIN

While the biochemical roles of cobalamin and folate have been well defined for many years, the complex manner in which they interrelate in vivo and the mechanism by which cobalamin deficiency produces megaloblastosis in man remains unclear. Cobalamin is a coenzyme in two well-studied reactions in mammals: the synthesis of methionine by methylation of homocysteine and the conversion of methylmalonyl-CoA to succinyl-CoA by an intra-molecular rearrangement. Folate is a coenzyme in a variety of reactions by which single C units (C-1) are transferred from carbon donors such as serine, formate, methionine and histidine and are donated in the synthesis of purines, pyrimidines and methionine.

Evidence for Cobalamin–Folate Interrelations

The common ground where folate and cobalamin interact is, *first,* in the methionine synthetase reaction. Here C-1 as the methyl of methyltetrahydrofolate is transferred to cob (I) alamin to form methylcobalamin, and in turn the methyl is transferred to homocysteine to yield methionine:

$$HS-CH_2-CH_2-CH(NH_2)-COOH + 5\text{-}CH_3-H_4PteGlu \quad \rightarrow$$
$$\text{homocysteine} \hspace{6cm} \text{cobalamin}$$
$$CH_3-S-CH_2-CH_2-CH(NH_2)-COOH + H_4PteGlu$$
$$\text{methionine}$$

(1)

Secondly, at a clinical level cobalamin-deficient patients, such as those with untreated pernicious anaemia, regularly respond haematologically to treatment with folic acid, despite the fact that the cobalamin deficiency persists unchanged. However, patients with megaloblastic anaemia due to folate deficiency do not respond in any significant way to cobalamin therapy.

Thirdly, serum levels of both cobalamin and folate are affected by deficiency of only one of these nutrients: in cobalamin deficiency the serum folate level rises [1, 2], while in folate deficiency serum cobalamin falls [3]. The rise in serum folate in pernicious anaemia has been attributed to impaired cellular uptake of methylfolate from serum [4, 5]; this has been demonstrated in experimental animals [6, 7]. The reason for the low serum cobalamin level in one-third of folate-deficient megaloblastic anaemias is not known.

Fourthly, the administration of substances that yield C-1 units can produce haematological responses in pernicious anaemia; these substances include histidine, choline and serine [8]. Responses also occur by administration of substances that are thought to bypass metabolic blocks resulting from cobalamin or folate deficiency. These substances include pyrimidine precursors such as carbamoylaspartic acid, orotic acid, uridylic acid, cytidylic acid and uracil and a purine, inosinic acid.

Finally, cobalamin deficiency commonly produces neuropathy, which may be aggravated or precipitated by long-term folate therapy [9]. This effect of folate has been confirmed in the experimental animal [10].

The Methylfolate Trap

The hypothesis termed the methylfolate trap was put forward 25 years ago to account for the observed relationship between folate and cobalamin [11, 12]. It depended, first, on the observation in *in vitro* studies that the reaction by which methylfolate was formed was largely irreversible [13].

$$5,10\text{-CH}_2\text{—H}_4\text{PteGlu} \rightarrow 5\text{-CH}_3\text{—H}_4\text{PteGlu} \tag{2}$$

Secondly, the serum folate concentration (largely 5-methyl-H_4PteGlu) was elevated in some patients with pernicious anaemia. And, thirdly, both cobalamin and folate were involved in methionine synthesis. Normally the methyl group on methylfolate is passed on to cobalamin, and in this way frees H_4folate to take up other C-1 units (reaction 1 and Fig. 1). In cobalamin deficiency it is postulated that the methyl group cannot be shifted, and hence H_4folate is trapped as methylfolate. H_4folate is thus unavailable to take up other C-1 units needed for purine and pyrimidine (thymidine) synthesis. The raised serum folate level in pernicious anaemia was advanced as evidence for the accumulation of methylfolate. It has proved very difficult to test this hypothesis in any meaningful way, and observations that have been reported as evidence in its favour have merely been compatible with such a hypothesis rather than providing positive proof.

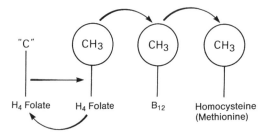

Fig. 1. The synthesis of methyl groups occurs by reduction to methyl of a single carbon taken up by tetrahydrofolate. The methyl group is transferred to cobalamin and then to homocysteine to form methionine. The enzyme mediating this methyl transfer is methionine synthetase.

Observations Incompatible with the Methylfolate Trap Hypothesis

Figure 2 shows the results of the deoxyuridine suppression test in patients with megaloblastic anaemia due to cobalamin deficiency and folate deficiency. The abnormal result in cobalamin deficiency is fully corrected with folinic acid (5-CHO—H$_4$PteGlu) but not with methylfolate. Nor is methylfolate used very well by folate-deficient marrow cells [14]. The most striking finding, however, is the failure of cobalamin-deficient marrow cells to use untrapped tetrahydrofolate itself [15, 16]. This is not explained by the methylfolate trap. The failure to use H$_4$PteGlu has been confirmed in the cobalamin-inactivated experimental animal [14, 17].

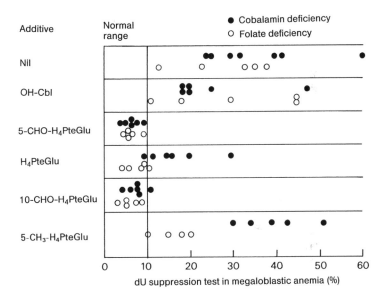

Fig. 2. In the deoxyuridine (*dU*) suppression test, marrow cells are incubated with dU. The cells take up dU, convert it to thymidine monophosphate by the addition of a methylene group (synthetic path), and thymidine is incorporated into DNA. After incubation of cells with dU has proceeded, [³H]thymidine is added, and any requirement for thymidine not already met by synthesis is met by using the preformed [³H]thymidine (salvage path). DNA is extracted and radioactivity counted. The control is an aliquot of marrow cells incubated only with [³H]thymidine. The counts in DNA of the control cells constitute a 100% value, and the results with cells with dU are expressed as a percentage of the control. Normoblastic marrow cells give results of less than 10%, while cells from patients with megaloblastic anaemia give results greater than 10%. The results in patients with megaloblastic anaemia due to cobalamin deficiency (*filled circles*) and folate deficiency (*open circles*) are shown. In addition, compounds can be added to the incubation mixtures to see whether they are able to correct the defect. The addition of cobalamin partially corrected the defect in cobalamin-deficient cells. Formyltetrahydrofolate (5-CHO-H$_4$PteGlu and 10-CHO-H$_4$PteGlu) fully corrected the defect in both cobalamin- and folate-deficient cells. Tetrahydrofolate (H$_4$PteGlu) corrected the defect in folate deficiency but not in cobalamin deficiency. Methyltetrahydrofolate (5-CH$_3$-H$_4$PteGlu) did not correct the defect in cobalamin deficiency and was not very effective in folate deficiency. (From [48])

Secondly, in vivo observations indicate that reaction 2 is readily reversed, that is, the methyl group can be oxidized back to methylene, formate and, if necessary, disposed of as CO_2. Thus, methyl group transfer is not the only way to utilize the methyl group of methylfolate [51].

Cobalamin can be completely oxidized and inactivated by exposure to the anaesthetic gas, nitrous oxide (N_2O) [18]. In man, indeed, N_2O exposure, if prolonged, leads to fatal megaloblastic anaemia [19, 20], and chronic exposure leads to cobalamin neuropathy [21]. In both man and animal, methionine synthetase activity virtually ceases following N_2O inhalation [22]. Animals exposed to N_2O have provided an outstanding experimental animal for studies on cobalamin–folate interrelations.

When rats are given methylfolate with a ^{14}C label in the methyl group and a ^{3}H label in the pteridine portion, it becomes possible to follow the turnover of the methyl group relative to the rest of the molecule [23]. In the liver of healthy rats half the methyl group is transferred from methylfolate in 2 h. Here the methyl is passed on to homocysteine (reations 1). In the cobalamin-inactivated rat there was no methyl group transfer for 72 h after N_2O exposure, and thereafter there was a slower methylfolate turnover. The mechanism was different from that in controls. The cobalamin-inactivated rat used methylfolate itself as the substrate for forming folate polyglutamate, which is the active coenzyme. When six or more glutamic acid residues had been added, there was rapid oxidation of the methyl group to formate and CO_2 [23, 24].

The rate of methylfolate oxidation in both control and cobalamin-deficient rats depends on the level of methionine. A large dose of methionine leads to the oxidation of all methylfolate in the liver within 30 min [24], and even a small physiological dose will remove 80% of the methyl group within 30 min (unpublished observations). The delay of 72 h before methylfolate is utilized in the cobalamin-inactivated rat is due to the time for partial recovery of methionine levels following induction of the betaine methyltransferase enzyme. Reaction 2, by which methylfolate is formed, is thus freely reversible in vivo, and methylfolate trapping is thus not possible.

Other observations can also be explained only on the basis of methyl group oxidation. In 1972 it was found that methylfolate could serve as the methyl donor for the methylation of biogenic amines [25]. Clarification of the reaction showed that the methyl group was oxidized to formate, formate transferred to the biogenic amine, and formate thereafter reduced back to methyl. The enzyme involved in methyl group oxidation was shown by several different groups to be methylenetetrahydrofolate reductase (reaction 2) [26, 27, 28].

Thorndike and Beck [29] studied the production of CO_2 from methylfolate by lymphocytes and could find no difference in CO_2 production by control cells and by cells from one patient with untreated pernicious anaemia.

Experimental Cobalamin "Deficiency" Using N_2O

Oxidation of cobalamin to an inactive form by inhalation of N_2O can be used to produce experimental cobalamin "deficiency". Such studies were done largely in Sprague-Dawley rats.

Methionine Synthetase. The enzyme methione synthetase, of which cobalamin is a coenzyme, is inactivated within 30 min of N_2O exposure [22] and remains inactive as long as N_2O inhalation continues. Full recovery of enzyme activity occurs only 4 days after N_2O withdrawal, suggesting a need to resynthesize apoenzyme as well as replacing irreversibly oxidized cobalamin.

Methylmalonyl-CoA-Mutase. N_2O has no effect on urinary methylmalonic acid excretion, even after stressing the pathway by an injection of propionic acid [22]. After several weeks of N_2O exposure there is a decline in mutase activity in rat liver [30]. It is probable that reduced cobalamin is not required for the mutase reaction, hence its resistance to oxidation. The long-term fall in activity is probably due to overall loss of tissue cobalamin.

Plasma and Tissue Folate. Inactivation of cobalamin is accompanied by a prompt rise in plasma methylfolate [6]. This in turn leads to a marked increase in urinary folate excretion [31]. The result of the considerable and sustained folate diuresis is a rapid loss of tissue folates, so that liver folates fall to 20% of starting values after 10 days exposure to N_2O [32]. Withdrawal of N_2O results in a restoration of tissue folates over several days, but only if folates are available from the diet [33]. The loss of folate affects all folate analogues, including methylfolate. However, in the first 12–24 h after N_2O exposure there is a rise in liver methylfolate polyglutamate [32], presumably because the methyl group can no longer be passed on to homocysteine, but this rise is replaced by a loss of methylfolate thereafter.

Folate Polyglutamate Synthesis. The active folate coenzymes in tissues have not one, but up to seven glutamic acid residues. Not only do these folate polyglutamates mediate C-1 transfers, but differences in glutamic acid chain length also regulate rates of enzyme activity. The cobalamin-inactivated rat cannot make any folate polyglutamate from PteGlu, $H_4PteGlu$ or $5\text{-}CH_3\text{—}H_4PteGlu$ [34]. There is induction of the enzyme folate polyglutamate synthetase, which forms polyglutamates [35]. However, if the cobalamin-inactivated rat is given folate containing a formyl group such as 5- or $10\text{-}CHO\text{—}H_4PteGlu$ or $5,10\text{-}CH\text{=}H_4PteGlu$ as substrate, then normal polyglutamate synthesis ensues. Thus two conclusions can be drawn: (a) formyltetrahydrofolates are the required substrates for making polyglutamate and (b) the effect of loss of cobalamin function is to prevent formylation of folate.

Source of Formate. The block to folate polyglutamate synthesis in the cobalamin-inactivated rat is overcome by giving methionine [7]. Injection of 16 μM me-

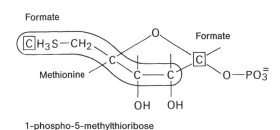

Formate

1-phospho-5-methylthioribose

Fig. 3. 5-Methylthioribose is principally reconverted to methionine. Each mole of methionine formed also yields one mole of formate. In addition, methionine can yield formate by oxidation of the methyl group. With $[^{14}C]H_3$-labelled methionine the oxidation of the $[^{14}C]$ to $[^{14}C]O_2$ is is almost completely suppressed by the addition of formate [50]

thionine (about 2 mg) fully restores folate polyglutamate synthesis in a 100-g Sprague-Dawley rat. The normal daily dietary intake of methionine in the rat exceeds 40 mg. Methionine can yield formate in two ways (Fig. 3): (a) by oxidation of the methyl group and (b) by conversion to S-adenosylmethionine, 5-methylthioadenosine and 5-methylthioribose. The latter is rapidly reconverted to methionine, and for each mol of methionine formed 1 mol of formate is released [36]. Methylthioadenosine is the most potent compound tested in reversing the effect of cobalamin inactivation; it is significantly more active than methionine itself and is more active than S-adenosylmethionine [7]. Current dogma states that serine is the major source of C-1 units and hence of formate. The cobalamin-inactivated animal cannot use serine for reasons that are unclear. Thus the onus for providing formate falls on methionine, the supply of which is limited in cobalamin deficiency, as shown by a steady fall in S-adenosylmethionine concentration in the liver [37]. Some of these pathways are illustrated in Fig. 4.

Thymidine Synthesis. The enzyme thymidylate synthetase, converts deoxyuridine to deoxythymidine by transfer of a methylene ($—CH_2$) group from 5,10-methylenetetrahydrofolate. The methylene group ($—CH_2—$) is further reduced to methyl ($—CH_3$) during the reaction. Cobalamin does not have a direct part in the reaction. Nevertheless, the reaction, as assessed in the deoxyuridine suppression test using marrow cells, is impaired within 60 min of N_2O exposure [15]. Thymidylate synthetase is induced, and enzyme activity in both marrow in pernicious anaemia [38] and in marrow from N_2O-exposed rats is several times greater than in controls [39]. The impaired thymidine synthesis must thus derive from a failure of supply of C-1 units. This would normally be supplied as a methylene group from serine. Although the enzyme serine transhydroxymethylase is unaffected during N_2O exposure [40], serine is unable to donate the appropriate C-1 unit. Thus, this must come from another source of C-1 units in cobalamin deficiency, and present evidence suggests it comes from methionine. Methionine in appropriate dose corrects the abnormal deoxyuridine suppression test [41], although in larger dose methionine is toxic and has the opposite effect [42].

Enzymes Involved in Interconversion of Folate Analogues. C-1 units may be fully oxidized (formate, $—CHO$) or fully reduced (methyl, $—CH_3$), or they may be at an intermediate state of reduction (methylene, $—CH_2—$). For purine synthesis C-1

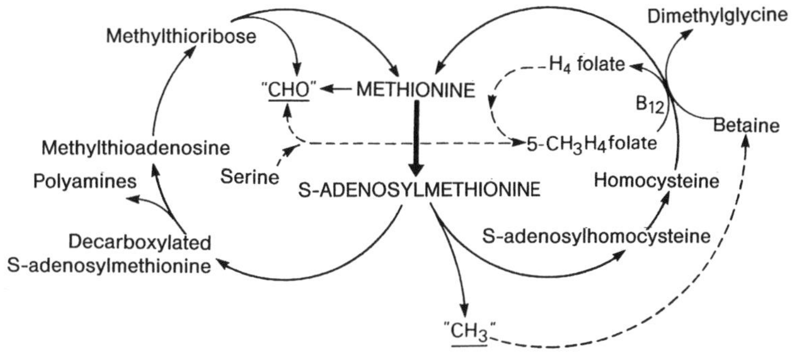

Fig. 4. Some interrelationships between methionine, its derivatives and cobalamin-folate metabolism.

units must be at the formate level of oxidation, that is, 10-formyl-H_4PteGlu$_n$; for thymidine synthesis they must be at the methylene level of oxidation, that is, 5,10-methylene-H_4PteGlu$_n$; and for methionine synthesis they must be at the methyl level of oxidation, that is, 5-methyl-H_4PteGlu$_n$. C-1 units that are surplus to requirements are oxidized further to CO_2. There is a further form of C-1 units, namely, methenyl, —CH$=$, which is an intermediary between formyl and methylene but is not itself used in synthetic pathways.

Cobalamin inactivation has important effects on the enzymes involved in changing the state of oxidation of C-1 units, and these affect the formyl end of the sequence (Table 1). Following exposure to N_2O there is a rise in formyltetrahydrofolate synthetase activity and a fall in activity of the enzyme which reduces formyl to methenyl, namely, 5,10-methenyltetrahydrofolate cyclohydrolase [43]. The synthetase adds formate to H_4PteGlu. The data have been interpreted as indicating a response to a shortage of formyltetrahydrofolate. Formyl-H_4PteGlu appears to be the preferred substrate for the addition of glutamic acid residues in the synthesis of the folate coenzyme. These data, too, point to an impaired supply of formate resulting from cobalamin deficiency. The other enzymes concerned in C-1 interconversion remain unaffected by cobalamin inactivation.

Table 1. Effect of inactivation of cobalamin by N_2O on the activity of enzymes in liver, regulating the state of reduction of single carbon units transferred by the folate coenzymes

	CHO	
Formyl-H_4PteGlu synthetase	$\downarrow$	Increased
	10-CHO—H_4PteGlu	
5,10 methenyl-H_4PteGlu cyclohydrolase	$\updownarrow$	Decreased
	5,10-CH$=$$H_4$PteGlu	
5,10 methylene-H_4PteGlu dehydrogenase	$\updownarrow$	No change
	5,10-CH$_2$—H_4PteGlu	
5,10 methylene-H_4PteGlu reductase	$\updownarrow$	No change
	5-CH$_3$—H_4PteGlu	

Purine Synthesis. Carbons 2 and 8 of the purine nucleus are supplied as 10-formyl-H_4PteGlu. It was not unexpected to find that the folate-dependent transformylase enzymes mediating purine synthesis were affected by cobalamin inactivation. GAR transformylase in rat liver, adding C-8 to glycinamide ribonucleotide, declines in activity, and AICAR transformylase, adding C-2 to compete the purine ring, increases in activity on N_2O exposure [44]. Thus, all the major folate-dependent synthetic pathways are affected by cobalamin inactivation, namely, thymidine synthesis, purine synthesis and methionine synthesis.

Adaptation to Cobalamin Inactivation

Methyl-H_4PteGlu is no longer metabolized along the normal route to methionine. Instead, it is converted to methylfolatehexaglutamate, and the methyl group is oxidized to methylene, methenyl, formyl and, if necessary, to CO_2. There is some recovery of methionine levels in plasma and liver due to induction of an alternative pathway for methionine synthesis, namely, betaine homocysteine methyltransferase [37], where betaine provides the methyl group to convert homocysteine to methionine instead of methylfolate. Betaine methyltransferase activity remains elevated as long as cobalamin remains inactivated. However, the supply of methionine remains less than required. The evidence for this is that (a) the level of S-adenosylmethionine in liver continues to fall [37, 45] and (b) the rate of methylfolate oxidation, which is regulated by the level of methionine, remains low.

Other enzyme inductions include folate polyglutamate synthetase, thymidylate synthetase, AICAR transformylase and formyltetrahydrofolate synthetase.

The Formate Starvation Hypothesis

The data that have been reviewed about the effects of cobalamin inactivation by N_2O suggest that cobalamin is required for a normal supply of C-1 units at the formate levels of oxidation [46]. In the healthy animal, serine may be the major source of such units. However, serine seems to be unavailable as a source of C-1 units in cobalamin deficiency. The alternative source, perhaps even the main source, is methionine. Methionine ameliorates or reverses many of the effects of dietary-induced cobalamin deficiency [47] as well as N_2O-induced cobalamin inactivation, including neuropathy [10, 48]. Methionine yields formate by oxidation of the methyl group and via methylthioribose, a product of S-adenosylmethionine, along the polyamine pathway.

Cobalamin Deficiency and Megaloblastosis

Only man develops a megaloblastic anaemia as a result of cobalamin deficiency. A neuropathy due to cobalamin deficiency has been produced in the monkey and in the fruit bat as well as in man. Other species do not show any of these manifestations, although all species show essentially the same biochemical changes as a result of cobalamin deficiency or inactivation.

It is probable that megaloblastosis arises as a result of impairment of thymidylate synthesis. The result of the deoxyuridine suppression test is a measure of the loss of the thymidylate synthetic pathway in cobalamin deficiency. In the fruit bat this is less than 3% [49], in the rat it may reach 23% [14], but in man may rise to over 60% [14]. Man thus appears to be more dependent on the synthetic pathway for supply of thymidylate and less dependent on the salvage pathway than are other species. This may explain man's unique development of megaloblastosis when the supply of cobalamin fails.

References

1. Waters AH, Mollin DL (1961) Studies on the folic acid activity of human serum. J Clin Pathol 14:335–344
2. Herbert V (1961) The assay and nature of folic acid activity in human serum. J Clin Invest 40:81–91
3. Chanarin I (1979) The megaloblastic anaemias, 2nd edn. Blackwell, Oxford, pp 138–140
4. Das KC, Hoffbrand AV (1970) Studies of folate uptake by phytohaemagglutinin-stimulated lymphocytes. Br J Haematol 19:203–221
5. Tisman G, Herbert V (1973) B_{12}-dependence of cell uptake of serum folate: an explanation for the high serum folate and cell folate depletion in B_{12} deficiency. Blood 41:465–469
6. Lumb M, Perry J, Deacon R, Chanarin I (1981) Change in plasma folate levels in rats inhaling nitrous oxide. Scand J Haematol 26:61–64
7. Perry J, Chanarin I, Deacon R, Lumb M (1983) Chronic cobalamin inactivation impairs folate polyglutamate synthesis in the rat. J Clin Invest 71:1183–1190
8. Chanarin I (1979) The megaloblastic anaemias, 2nd edn. Blackwell, Oxford pp297–298
9. Chanarin I (1979) The megaloblastic anaemias, 2nd edn Blackwell, Oxford pp293–295
10. Van der Westhuyzen J, Fernandes-Costa F, Metz J (1982) Cobalamin inactivation by nitrous oxide produces severe neurological impairment in fruit bats: protection by methionine and aggravation by folates. Life Sci 31:2001–2010
11. Herbert V, Zalusky R (1962) Interrelationship of vitamin B_{12} and folic acid metabolism: folic acid clearance studies. J Clin Invest 41:1263–1267
12. Noronha JM, Silverman M (1962) On folic acid, vitamin B_{12}, methionine and formiminoglutamic acid. In: Heinrich HC (ed) Vitamin B_{12} und intrinsic factor 2. Enke, Stuttgart, pp728–736
13. Kutzbach C, Stokstad ELR (1971) Mammalian methylenetetrahydrofolate reductase, partial purification, properties and inhibition by S-adenosylmethionine. Biochim Biophys Acta 250:459–477
14. Deacon R, Chanarin I, Perry J, Lumb M (1981) The effect of folate analogues on thymidine utilization by human and rat marrow cells and the effect on the deoxyuridine suppression test. Postgrad Med J 57:611–616
15. Deacon R, Chanarin I, Perry J, Lumb M (1982) A comparison of tetrahydrofolate and 5-formyltetrahydrofolate in correcting the impairment of thymidine synthesis in pernicious anaemia. Br J Haematol 28:289–292
16. Taheri MR, Wickremasinghe RG, Jackson BFA, Hoffbrand AV (1982) The effect of folate analogues and vitamin B_{12} on provision of thymidine nucleotides for DNA synthesis in megaloblastic anaemia. Blood 59:634–640
17. Deacon R, Chanarin I, Perry J, Lumb M (1980) Impaired deoxyuridine utilization in the B_{12}-inactivated rat and its correction by folate analogues. Biochem Biophys Res Commun 93:516–520
18. Banks RGS, Henderson RJ, Pratt JM (1968) Reactions of gases in solution. Part III. Some reactions of nitrous oxide with transition-metal complexes. J Chem Soc (A) 2886–2889

19. Lassen HCA, Henriksen E, Neukirch F, Kristensen AS (1956) Treatment of tetanus. Severe bone marrow depression after prolonged nitrous-oxide anaesthesia. Lancet 1:527–530

20. Amess JAL, Burman JF, Rees GM, Nancekievill DG, Mollin DL (1978) Megaloblastic haemopoiesis in patients receiving nitrous oxide. Lancet ii:339–342

21. Layzer RB (1978) Myeloneuropathy after prolonged expsure to nitrous oxide. Lancet ii:1227–1230

22. Deacon R, Lumb M, Perry J, Chanarin I, Minty B, Halsey MJ, Nunn JF (1978) Selective inactivation of vitamin B_{12} in rats by nitrous oxide. Lancet ii:1023–1024

23. Lumb M, Chanarin I, Perry J, Deacon R (1985) Turnover of the methyl moiety of 5-methyltetrahydropteroylglutamic acid in the cobalamin-inactivated rat. Blood 66 (5):1171–1175

24. Brody T, Watson JE, Stokstad ELR (1982) Folate pentaglutamate and folate hexaglutamate mediated one-carbon metabolism. Biochemistry 21:276–282

25. Laduron P (1972) N-methylation od dopamine to epinine in brain tissue using N-methyltetrahydrofolic acid as the methyl donor. Nature 238:212–213

26. Leysen J, Laduron P (1974) Characterization of an enzyme yielding formaldehyde from 5-methyltetrahydrofolic acid. FEBS Lett 47:299–303

27. Taylor RT, Hanna ML (1975) 5-Methyltetrahydrofolate aromatic alkylamine N-methyltransferase: an artefact of 5,10-methylenetetrahydrofolate reductase activity. Life Sci 17:111–120

28. Pearson AGM, Turner AJ (1975) Folate-dependent 1-carbon transfer to biogenic amines mediated by methylenetetrahydrofolate reductase. Nature 258:173–174

29. Thorndike J, Beck WS (1977) Production of formaldehyde from $N5$-methyltetrahydrofolate by normal and leukemic leukocytes. Cancer Res 37:1125–1132

30. Kondo H, Osborne ML, Kolhouse JF, Binder MJ, Podell ER, Utley CS, Abrams RS, Allen RH (1981) Nitrous oxide has multiple deleterious effects on cobalamin metabolism and cause decreases in activities of both mammalian cobalamin-dependent enzymes in rats. J Clin Invest 67:1270–1283

31. Lumb M, Perry J, Deacon R, Chanarin I (1982) Urinary folate loss following inactivation of vitamin B_{12} by nitrous oxide in rats. Br J Haematol 51:235–242

32. Lumb M, Deacon R, Perry J, Chanarin I, Minty B, Halsey MJ, Nunn JF (1980) The effect of nitrous oxide inactivation of vitamin B_{12} on rat hepatic folate. Implications for the methylfolate trap hypothesis. Biochem J 186:933–966

33. Lumb M, Perry J, Deacon R, Chanarin I (1981) Recovery of tissue folate after inactivation of cobalamin by nitrous oxide. The significance of dietary folate. Am J Clin Nutr 34:2418–2422

34. Perry J, Chanarin I, Deacon R, Lumb M (1979) The substrate for folate polyglutamate biosynthesis in the vitamin B_{12}-inactivated rat. Biochem Biophys Res Commun 91:678–684

35. Perry J, Chanarin I, Deacon R, Lumb M (1985) Folate polyglutamate synthetase activity in the cobalamin-inactivated rat. Biochem J 227:73–77

36. Trackman PC, Abeles RH (1981) The metabolism of 1-phospho-5-methyl-thioribose. Bio chem Biophys Res Commun 103:1238–1244

37. Lumb M, Sharer N, Deacon R, Jennings P, Purkiss P, Perry J, Chanarin I (1983) Effects o nitrous oxide-induced inactivation of cobalamin on methionine and S-adenosylmethionin metabolism in the rat. Biochim Biophys Acta 756:354–359

38. Sakamoto S, Niina M, Takaku F (1975) Thymidylate synthetase activity in bone marrow cells in pernicious anemia. Blood 46:699–704

39. Deacon R, Perry J, Lumb M, Chanarin I (1981) The effect of nitrous oxide-induced inacti vation of vitamin B_{12} on thymidylate synthetase activity of rat bone marrow cells. Biochem Biophys Acta 102:215–218

40. Deacon R, Perry J, Lumb M, Chanarin I (1980) The effect of nitrous oxide-induced inacti vation of vitamin B_{12} on serine transhydroxymethylase. Biochem Biophys Res Commo 97:1324–1328

41. Sourial NA, Brown I (1983) Regulation of cobalamin and folate metabolism by methionin in human bone marrow cultures. Scand J Haematol 31:413–423

42. Cheng FW, Shane B, Stakstad ELR (1975) The antifolate effect of methionine on bone marrow of normal and vitamin B_{12} deficient rats. Br J Haematol 31:323–336

43. Perry J, Deacon R, Lumb M, Chanarin I (1980) The effect of nitrous-oxide induced inactivation of vitamin B_{12} on the activity of formyl-methenyl-methylenetetrahydrofolate synthetase, methylenetetrahydrofolate reductase and formiminotetrahydrofolate transferase. Biochem Biophys Res Commun 97:1329–1333

44. Deacon R, Perry J, Lumb M, Chanarin I (1983) Effect of nitrous oxide-induced inactivation of vitamin B_{12} on glycinamide ribonucleotide transformylase and 5-amino-4-imidazole carboxamide transformylase. Biochem Biophys Res Commun 112:327–331

45. Gawthorne JM, Smith RM (1974) Folic acid metabolism in vitamin B_{12}-deficient sheep. Effects of injected methionine on methotrexate transport and the activity of enzymes associated with folate metabolism in liver. Biochem J 142:119–126

46. Chanarin I, Deacon R, Lumb M, Perry J (1980) Vitamin B_{12} regulates folate metabolism by the supply of formate. Lancet ii:505–508

47. Chanarin I, Deacon R, Lumb M, Muir M, Perry J (1985) Cobalamin-folate interrelations – a critical review. Blood 66 (3):479–489

48. Scott JM, Dinn JJ, Wilson P, Weir DG (1981) Pathogenesis of subacute combined degeneration: a result of methyl group deficiency. Lancet ii:334–337

49. Miller H, Fernandes-Costa F, Metz J (1980) Synthesis of DNA as shown by the deoxyuridine suppression test is normal in the vitamin B_{12}-deficient fruit bat *(Rosettus aegyptiacus)*. Br J Nutr 44:229–435

50. Benevenga J, Radcliffe BC, Egan AR (1983) Tissue metabolism of methionine in sheep. Aust J Biol Sci 36:475–485

51. Lumb M, Chanarin I, Deacon R, Perry J (1988) In vitro oxidation of the methyl group of hepatic 5-methyltetrahydrofolate. J Clin Path 41:1158–1162

Chapter 4

The Proteins of Transport of the Cobalamins

C. A. HALL

Overview of the Complete Transport System

The body's system for transport of vitamin B_{12} (cobalamin, Cbl) is so vital to man that all physicians must understand its fundamentals. The functioning parts of the system are a series of carrier proteins and receptors (Table 1) that move the Cbl from the food in the stomach to the cells of the body. The molecule of each soluble protein has a single binding site for Cbl and a separate binding site for its receptor. The system also recycles some of the hepatic Cbl via the bile and recycles the body Cbl among tissues via the blood. Even *Escherichia coli* has a complex transport system for Cbl, and why a highly specialized system for transporting this one vitamin has persisted throughout the evolution to man, although much modified, is an intriguing mystery.

The haptocorrin (HC) of salviv is the first binder of Cbl to come in contact with Cbl, entering through the physiologic route. Our understanding of HC, known initially as R binder, comes principally from a long series of investigations in the laboratory of Gräsbeck [61]. This glycoprotein is found in human plasma, gastric juice, bile, cerebrospinal fluid, milk, tears, amniotic fluid, and saliva, as well as in leukocytes and platelets [61]. The amino acid composition of HC is similar regardless of source, all HC reacts with the same antibody, and when congenitally absent, it is absent from all of the usual sources [11]. The amount of sialic acid [61] and other carbohydrates [8] of HC varies among sources. The HC from any one source can be fractionated into two or several components by techniques that separate according to differences in molecular charge, largely determined by the sialic acid. Regardless of this variability in the carbohydrate, HC is essentially a single protein. Intrinsic factor (IF), also a glycoprotein, follows in sequence. The first of the specific membrane binders is the receptor for IF-Cbl on the enterocytes of the ileum, which transfers the Cbl to the third soluble carrier, transcobalamin II (TC II), not a glycoprotein. The fifth participating substance is the specific receptor for TC II-Cbl. The remainder of this chapter is devoted to showing how the pieces of the transport puzzle fit together, and how Cbl transport is disturbed in disease. The physicochemical properties of these macromolecules of Cbl transport are not given in depth here, but can be found in recent reviews [18, 47, 51, 59].

The binding affinity for Cbl is, however, one physicochemical property of the carrier proteins that affects function. Binding is optimal at a range of pH spanning neutrality, and TC II always functions well within this range. With the

Table 1. Macromolecules that participate in human Cbl Transport

Substance	Principal synonyms[a]	Where observed	Cell of origin	Transport function
Haptocorrin (HC)	R binder, cobalophilin; in serum, TC I and TC III	Saliva, gastric juice, bile, granulocytes, serum, and other fluids	Several types	Unknown
Intrinsic factor (IF)	–	Gastric juice, lumen of small intestine	Gastric parietal cell	Transport of Cbl from proximal small intestine to ileum
Receptor for IF-Cbl	–	Ileal mucosa	Ileal enterocyte	Binding and internalization of IF-Cbl
Transcobalamin II (TC II)		Serum, trace amounts in several fluids	Potentially several	Transport of Cbl from within enterocyte to plasma and in circulation
Receptor for TC II-Cbl	–	Several types of body cell	Presumably each body cell where observed	Binding and internalization of TC II-Cbl

HC, IF, and, TC II are soluble proteins; the two receptors are membrane bound, but can be solubilized in vitro.
[a] For a more complete list of synonyms, especially those no longer in use, see [18, 51].

sharp change in pH between the stomach and the duodenum, the effect of pH becomes more important for the functions of HC and IF [1]. At pH 2 the affinity between CN-Cbl and HC is much higher than between CN-Cbl and IF (Table 2a). HC loses much of its advantage at pH 8. The affinity of each of the three binders is about the same for the forms of Cbl known to be present in the blood and in the cells. Man does, however, encounter corrinoids related to Cbl that may be inert or even toxic [43]. The affinities for three typical analogs are given in Table 2b. HC may have nearly as high an affinity for these forms as for CN-Cbl, while that of TC II often is considerably lower. IF appears to be the most fastidious and selective of the binders.

Intestinal Transport

Intestinal transport is only summarized here because the essentials are well established and have been documented in previous reviews [8, 47, 59]. Cbl of food is bound and exists in several biologically active forms [16]. Inactive or even harmful corrinoids are also possible. Although HC is abundant in saliva, it cannot bind food Cbl until the Cbl is freed by the gastric phase of digestion. The role of HC was not known until the in vitro studies of Allen et al. [1] established the following: (a) initial binding of Cbl to HC at the acid pH of the stomach, (b) digestion of the HC of HC-Cbl by pancreatic proteases in the proximal small intestine and, (c) rebinding of the freed Cbl to IF in the more nearly neutral environment of the small intestine. Marcoullis et al. [48] then demonstrated by the sampling of jejunal contents that this sequence does in fact take place in

Table 2. Affinity between corrinoids and binding proteins at 37°C

a. Effect of pH on binding of CN-Cbl [1]

| | Association Constant (K_a) as pM^{-1} | |
	Saliva HC	IF
pH 2	1.0	0.02
pH 8	1.6	0.6

b. Binding of different corrinoids at pH 7.5 [43]

| | Binding by: | | |
	Granulocyte HC	IF	TC II
K_a for CN-Cbl	7.5	1.9	0.5
K_a CN-Cbl [e-OH]/K_a CN-Cbl	0.5	4×10^{-3}	2×10^{-2}
K_a [Ade] CN-Cba/K_a CN-Cbl	0.1	7×10^{-5}	0.1
K_a [CN, OH] Cbi/K_a CN-Cbl	0.1	$<1 \times 10^{-6}$	7×10^{-4}

Cbl, cobalamin; HC, haptocorrin; IF, intrinsic factor; TC II, transcobalamin II
[Ade] CN-Cba: 1 adenine replaces 5-6 dimethylbenzimidazole
[CN, OH] Cbi: no nucleotide moiety.

man. They observed recently ingested Cbl to be bound to IF in healthy persons, but when pancreatic proteases were much reduced or absent, the Cbl was bound to HC. Gastric HC-Cbl has never been demonstrated directly, but it can be assumed from these experiments.

The HC to which Cbl binds in the stomach can come from either the salivary or gastric gland. It resists gastric digestion. This apo-HC, HC with open binding sites for Cbl, is even more susceptible to pancreatic digestion than is holo-HC. Thus, no HC remains to rebind Cbl in the intestine. Whereas the gastric transport of Cbl by HC does occur, the binding is not essential for the absorption of Cbl. Five persons known to have a complete deficiency of "functioning" HC all absorbed Cbl normally. IF can survive both gastric and pancreatic digestion, a critical feature of the transport system. Substances, including food, that stimulate the release of apo-IF by the human parietal cells are known, but the in vivo control of the IF phase of transport has not been well studied. Enough apo-IF is released daily to bind up to ten times the daily intake of Cbl [51].

Reabsorption of the Cbl in bile may also be an important part of intestinal transport. Human bile contains Cbl and also contains HC. This Cbl would be lost to the individual if it could not be released from any binding to HC and rebound to IF prior to reaching the ileum. Fragmentary observations over the years from several laboratories have been made of the relationships between biliary function and Cbl absorption, but no comprehensive studies have defined the physiology and impact of the enterohepatic circulation of Cbl in vivo.

IF-Cbl passes intact to the ileum, where it encounters the specific receptors for IF-Cbl. These receptors of man and other mammals have been solubilized and studied. They are distributed within the ileum, generally increasing in numbers from proximal to distal. Neither TC II-Cbl, HC-Cbl, nor free Cbl compete with IF for the receptor. The internalization of Cbl, which is probably by endocytosis, requires energy. Whether the complex of IF-Cbl enters the enterocyte intact, is still disputed, but a recent analysis of receptor activity in the dog suggests that the receptor IF-Cbl complex does penetrate, but that it does not remain intact for long [45]. There is, however, a 3-h lag before the Cbl enters the portal blood. The function of IF appears to end once the Cbl is released from it in the enterocyte. This sequence is known to occur in human Cbl transport and has a counterpart in other species. The cell of origin and some of the properties of IF may differ, but other species do absorb Cbl via IF and specific receptors. The phase of binding to HC in the stomach, with rebinding to IF in the intestine, has so far been examined only in man.

Plasma Transport

Plasma transport is presented in more detail here, not because it is more important, but because there are more unanswered questions and areas of controversies. TC II is essentially a plasma protein, although minute amounts are found in saliva, cerebrospinal fluid, milk, and semen. Representative levels of both TC II and HC of serum are given in Table 3. The total mass of the protein can be subdivided into the holo form, that actively carrying Cbl, and the apo form, that

Table 3. Compartments of serum binding of Cbl as ng Cbl/l

	Total Serum Cbl (n = 29)	Holo-TC II (n = 12)	Holo-HC (n = 10)	Total TC II (n = 20)	Total HC (n = 10)	Apo-TC II (n = 9)	Apo-HC (n = 9)
Mean	480	90	325	971	603	908	67
SD	142.6	31.2	87.6	227.3	115.7	123.0	22.0
Range	190–777	47–173	256–432	586–1413	416–800	739–1084	39–118

Serum: Separated from cells immediately [2]. With the exception of the apo binding in paired samples, the measurements were made in different groups of sera from healthy persons.
Assays: All described in [2]. Total serum Cbl, bioassay, *Euglena gracilis*; holo binders, removal of binder Cbl by antibody coupled to Sepharose, elution, and bioassay; total binders, by radioimmunoassay; apo binders, by saturation with CN-[^{57}Co] Cbl and separation by gel filtration.

with open binding sites for Cbl. Unfortunately the distinction is often ignored in the literature, and levels represented as total are really those of the apo form only, which is much easier to measure. Little error is induced when measuring TC II since about 90% of the total is in the apo form (Table 3), but when this method of measurement is applied to the HC of serum, more than 80% of that present may be missed because it is in the holo form. The amounts of Cbl binding proteins are customarily expressed as the weight of Cbl that can be bound per unit volume of serum, although molecular units are preferable for scientific purposes. Based on the data of Table 3, the mean level of serum TC II, as a protein, would be 720×10^{-12} mol/l, or 27µg/l. Comparable levels for HC would be 440×10^{-12} mol/l, or 17 µg/l. These are truly trace proteins. The sum of the holo binders of serum Cbl make up the total serum Cbl, an important measurement of clinical chemistry. When the holo forms are measured directly by reliable methods, a mean of 25%–30% of serum Cbl is bound to TC II [33], although values may range from 10% to 45% in individual sera. The remainder is largely holo-HC, but minor amounts of Cbl may be bound to unidentified proteins in some situations [50], a fraction that becomes more pronounced after injection of therapeutic hydroxocobalamin [34]. TC II has by far the greater number of open Cbl binding sites of the circulation, but it is the lesser carrier of circulating Cbl. This approximately 100 µg TC II-Cbl is, however, essential for life.

It is probably premature to discuss the differences in TC II and HC levels among different populations and through the life cycle because in none of the studies have more than apo levels been measured. Nevertheless, apo-TC II may be higher in women, with a further net rise of apo-TC II and apo-HC during pregnancy [18]. Blacks appear to have higher levels than whites [18]. There is further evidence of a genetic determination of levels of apo-TC II [46, 52].

Nomenclature of the plasma carriers remains a problem. The term transcobalamin was introduced for substances in plasma that actively transported Cbl in vivo [30]. TC II is properly named. Originally TC I, the first Cbl binding protein identified, appeared to be an essential carrier because it carries about three-fourths of the circulating Cbl. A person can, however, be perfectly healthy without any demonstrable TC I [11]. As discussed above, all HC, including that of the serum, can be fractionated into two to a half dozen or more compenents of varying molecular charge. One-step fractionation of serum HC gives two fractions, which have been designated TC I and TC III. TC I is that fraction of HC with the lower pI, the more acidic, whereas TC III includes mainly those components with the higher pI, the more basic. TC I and TC III are definitely not distinct proteins, nor are they subunits of a macromolecule in the sense in which factor VIII is made up of subunits. TC 0 is another class of Cbl carriers. When Cbl binding proteins are separated by gel filtration, large binding substances elute in the void volume (Vo) – thus the term TC 0. This component may consist of aggregates of TC II and HC, complexes, or unidentified binding substances. This, usually minor, heterogenous binding component is not known to be physiologically important. The functions of TC II begin in the enterocyte where the absorbed Cbl is released from IF, binds to TC II, and enters the portal circulation as TC II-Cbl [12, 42, 57]. The origin of the TC II is unclear, but in a guinea pig model the enterocyte did not appear to be the cell of synthesis [57]. TC II appears to be in some way essential to the processes of absorption because absorption is impaired when TC II is congenitally deficient [25]. Absorption of Cbl was, however, normal in a person with a TC II that could bind Cbl but was inert as a promoter of cell entry of Cbl [37]. The plasma transport by TC II continues beyond the postabsorptive phase [7].

The cell or cells from which the circulating TC II originates have not been identified. The only pertinent in vivo study suggests that in man at least part of the circulating TC II comes from cells originating in the bone marrow [21]. In order to interpret this study it must be appreciated that normal TC II exists as genetically determined variants [20]. Determination of the type of TC II was applied to the serum TC II of a group of patients who had received bone marrow transplants after suppression of their own bone marrow. In half of those in whom the type of TC II differed between host and donor there was a partial or complete shift to donor type TC II. Subsequently human bone marrow in culture was shown to synthesize and release apo-TC II [22]. TC II has been demonstrated in, or to be released by, several tissues, perfused organs, or isolated cells of different species, and some of these experiments have led to important advances [54]. However, release does not prove synthesis, and even the demonstration of synthesis of TC II by a cell does not prove that the cell is a source of TC II in the intact man.

More is known about the regeneration of apo-TC II than about its origins. Injection of saturating amounts of CN-Cbl immediately converts all of the apo-TC II to holo-TC II, induces a movement of holo-TC II into tissues, and reduces the total TC II by about one-half. Over the next 6–8 h much of the apo-TC II is restored. The technique of these experiments can be most useful in the study of the function of TC II and its receptors as well as of the body's ability to replenish plasma TC II [3, 15, 44].

The one known function of TC II is to enhance the entry of Cbl into cells. This principle was observed [14] before it could be associated with TC II [19], and time has not changed the significance of these early observations. Human

TC II promotes the attachment of Cbl to human cells, followed by entry into the cell, a property not shared by HC from any source [19]. TC II-Cbl binds rapidly to specific, high-affinity receptors on the cell [62]. Holo-TC II binds with a higher affinity than apo-TC II [31, 62], free Cbl does not compete [35, 52], and binding is partially species specific [38, 62]. As an example of binding dynamics, the human hepatocyte in culture has about 10^4 receptors per cell which bind TC II-Cbl with a K_a of 6–8 mM^{-1} [35]. Receptor number varies, however, with the phase of the cell cycle [36].

The TC II-Cbl enters human cells by endocytosis, and the complex is split within the lysosomes [62]. The Cbl is converted to the coenzymes adenosyl-Cbl (AdoCbl) and methyl-Cbl (MeCbl) [35] which bind to the respective apoenzymes [49]. The distribution of TC II-Cbl to many tissues was observed after the injection of either human or rabbit TC II-Cbl into the rabbit [58].

The release of Cbl from cells is one of the least understood aspects of the chain of Cbl transport. Formal studies of human Cbl dynamics [24] and observations of the slow decline of body Cbl in deficiencies leave no doubt that Cbl leaves cells, is retained, and recycles, but the mechanism of exit remains unknown.

The studies on Cbl entry and exit in human cells in culture must be interpreted cautiously. They can describe processes and show which cells can perform each process, but they may not reflect what takes place in the intact person. Although cultured human hepatocytes can express receptors for TC II-Cbl [35], others could not demonstrate receptors on fresh isolated rat hepatocytes [60]. Receptor activity declines when cell division is suppressed [26, 34], and probably other factors also induce changes in receptors. The pattern of Cbl uptake among cells in vivo may differ from that suggested by isolated data collected in vitro.

Caution is also necessary in the application of animal models to the human system. Systematic analyses of the complete transport process in species other than man have been rare. The serum Cbl of common laboratory mammals ranges widely, from 300 ng/l in the dog to 16 000 ng/l in the mouse, and the number of open binding sites for Cbl varies as much [39]. These same animals seem to function with little if any plasma HC [39], cell entry being facilitated by a protein resembling TC II [38]. However, the fact that a protein resembles human TC II in some properties does not always mean that it performs the same functions in vivo. These differences suggest major variations in the overall plasma transport system among mammals although there may be principles in common.

Man himself is the most reliable model for the study of human Cbl transport. The earliest experiments showed that Cbl bound to TC II was cleared in vivo much more rapidly than Cbl bound to HC [29], an observation correlating with the active binding of TC II-Cbl to specific receptors [19]. Studies quantitating the dynamics of Cbl turnover followed [40], but the calculations were based on the assumption that the initial rapid loss of injected TC II-Cbl from the circulation represented distribution in the extracellular fluids. This may not be the complete explanation. All the injected Cbl accumulating in the liver of intact man in 15 min was taken up in the first 2 min, and much of it on one pass through the liver [19]. The uptake of canine TC II-Cbl by isolated, perfused canine organ was equally rapid [55]. It may be that receptors on cell surfaces bind and remove much TC II-Cbl on first contact. After Cbl is taken into the body, and at a time when the newly absorbed Cbl bound to TC II is declining in the circulation, the amount bound to HC increases [24, 30, 40]. The ratio of new holo-HC to new holo-TC II reaches equilibrium within a few days [24]. No active transfer of Cbl between TC II and HC has been identified, and for the present it must be assumed that after Cbl separates from the TC II in the cell and is subsequently metabolized, free Cbl enters the circulation, where some of it is again bound to apo-TC II and to apo-HC. The holo-TC II recycles through the receptor system on cells, while the holo-HC clears more slowly by an unknown mechanism. The sources of apo-HC and apo-TC are unknown. Much apo-HC can be released by granulocytes, and these cells are a logical source. However, the HC of plasma and the HC granulocytes differ in carbohydrate content, and if the theory is valid, there must be additional processing of the HC. The new holo-HC appears to assume the sialic acid composition of the already circulating Cbl within hours [24].

Whereas experiments in the laboratory or clinic have told us what TC II can do, only experiments in nature, the congenital defects, reveal how vital TC II really is. The present discussion relates principally to the physiology of transport, and the reader should consult recent reviews [20, 25, 50] for detailed descriptions of the syndromes. A total of 22 persons from 19 families have been diagnosed as having a congenital deficiency of TC II, and 5 of their siblings died of what almost certainly was the unrecognized defect. The presentations have been remarkably similar. The child is born healthy and without any evidence of Cbl deficiency, but soon, usually within the first 2 months of life, it becomes seriously ill. The most prominent symptoms are failure to thrive, vomiting and/or diarrhea, small ulcers of the oral mucus membrane, and recurrent infections. Megaloblastosis of the bone marrow is always present, with peripheral cytopenias, usually pancytopenia. The entire sequence suggests that at birth the tissues contain adequate amounts of Cbl, but that the bone marrow rapidly becomes deficient. The oral ulcers are evidence that the mucosa of the gut is affected in the same way, and the vomiting and diarrhea probably originate in part from the same abnormality. The repeated infections suggest that cells of the immune system also fail to obtain Cbl after birth. Defects in the function of both lymphocytes and granulocytes were demonstrable in the one affected infant that has been studied in depth. Low immunoglobulin levels were observed in all four of the other untreated infants studied adequately at the time of symptoms.

The prompt and complete response to the administration of large amounts of Cbl leave little doubt that the symptoms of TC II deficiency are produced by a cellular deficiency of the coenzymes of Cbl. The defect is apparently superceded by creation of enough free Cbl in the circulation to enter cells, although inefficiently [32]. The extent of the Cbl deficiency is not yet clear. The fact that serum Cbl is usually normal, and that development in utero is normal suggest that much total body Cbl is still present in these desperately ill infants. There is no apo-TC II that can bind Cbl, and, what is even more important, there is no holo-TC II [25, 50]. Thus, the child probably cannot move Cbl from one tissue to another as Cbl is released from cells into the circulation. Although TC II deficiency produces a form of Cbl deficiency, there are some puzzling, atypical aspects.

Affected infants, fully symptomatic, do not excrete excessive homocystine and methylmalonic acid (MMA) as do infants with an acquired deficiency of Cbl or those congenitally incapable of formation of the coenzyme forms of Cbl [25]. (One infant, not fully reported, did excrete large amounts of MMA). Malfunctions of the nervous system, typical of acquired Cbl deficiency, are conspiciously absent. Only three TC II deficient persons have clearly expressed an effect on this system, and each is in a special category. They were diagnosed late, having been kept alive for years by nonspecific treatment which relied heavily on folic acid [25, 50]. One child transiently lost the deep tendon reflexes of her legs during a period of inadequate maintenance treatment [63]. Apparently then, TC II deficiency can affect the nervous system, but the effects are not evident at the time of initial presentation.

The expression of TC II deficiency is selective, and is it possible that some cells can, under physiologic circumstances, obtain Cbl in the absence of TC II. There has been neither a systematic study of the distribution of receptors for TC II-Cbl among human cells nor a determination of which cells actually depend upon the receptors. Another pertinent question is whether some protein other than TC II can mediate the entry of Cbl into cells. As discussed above, neither HC-Cbl nor IF-Cbl have reacted with receptors for TC II-Cbl in any model tested to date. When Cbl, especially OH-Cbl, is added to plasma in amounts much in excess of the binding capacities of TC II and HC, there is some binding to other serum proteins, but the Cbl so bound does not appear to enter cells [34]. Cbl bound to HC may, however, be able to enter hepatocytes.

In a series of comprehensive and provocative studies, Burger et al. [8, 9] showed that the hepatocytes of the intact rabbit can take up Cbl bound to human HC. Uptake was clearly the receptors, confined to hepatocytes, and responsive to glycoproteins with a terminal galactose (asialoglycoproteins). The ultimate distribution of the Cbl was through the rabbit TC II, but it seems possible that the liver could obtain cellular Cbl by this mechanism. To answer this question we examined the binding and internalization of human TC II-Cbl and HC-Cbl by cultured human hepatocytes derived from a hepatoma [35]. The receptor system for TC II-Cbl functioned as expected. The cells also contained receptors for asialoglycoproteins in general and did bind HC-Cbl. Even with the standard manipulations to expose galactose residues, the binding was of low affinity and inconsistent when compared to the binding of TC II-Cbl to the specific receptors for TC II-Cbl. Isolated HC-Cbl was internalized and converted to coenzyme forms. The process was, however, inefficient. For reasons in addition to the low affinity, it does not seem likely that the liver obtains much of its cellular Cbl from HC-Cbl. The HC-Cbl observed in the human circulation is more "acidic" than the glycoproteins most readily taken up by hepatocytes [24]. Desialation, which enhances the process, may not occur in vivo. In contrast to the highly specific receptors for TC II-Cbl, many other glycoproteins compete with HC for their set of receptors. Moreover, the person completely deficient in HC can maintain a normal liver content of Cbl [27, 35].

Another experiment of nature, the congenital absence of HC, suggests that

HC performs no essential function in vivo. It also should be recalled that several mammals other than man have no measurable circulating HC. Five persons, one unreported, are known to have the now classical form of complete deficiency of HC [11, 25, 27]. They reached the ages of 46–72 before diagnosis. All were afflicted with acquired illnesses which brought them to the hospital where the diagnosis was made as part of a general evaluation. Their illnesses did not appear to be related to Cbl metabolism and failed to respond to treatment with Cbl. They absorbed Cbl normally. Cytopenias of the peripheral blood were present in some, but not megaloblastosis of the bone marrow, and the cytopenias did not respond to Cbl. Affected persons did not excrete increased amounts of MMA nor in the one patient studied, of homocystine. AdoCbl and MeCbl were present in normal amounts in the liver of one patient [35].

Merely because no in vivo function of HC has been observed, it cannot be assumed that there is none. As shown in Table 2b, corrinoids other than Cbl bind preferentially to HC. In the gut harmful corrinoids could be sequestered by HC and denied to IF. The HC-corrionoids would be excreted [43]. In the circulation apo-HC could keep similar substances from binding to TC II and being distributed. The HC-corrinoids could be taken up by the liver and excreted in the bile [43].

IF and TC II clearly interface in the enterocyte, and gastric HC and IF almost certainly do in the upper intestine, but interactions between TC II and HC in the circulation are obscure. The apo-HC of serum may be low in the congenital deficiency of TC II [32, 50]. Presumably, affected persons have abundant apo-HC in their granulocytes, and if granulocytes are the usual source of serum apo-HC, HC levels should be normal. TC II is not, however, depressed in the patient with the classic congenital deficiency of HC. We have recently reported a patient with mild tissue deficiency of Cbl, with low TC II-Cbl and low HC-Cbl but normal amounts of the apo forms [6]. This pattern is seen in acquired Cbl deficiency [33], but there was no evidence of an acquired deficiency. There may be interrelationships between plasma TC II and HC yet to be discovered, and patients with congenital disorders of Cbl may be the best model for study.

The transport of Cbl in the fetus is another still unclear component of the system. The fetus can make IF but obviously does not need an intestinal phase of transport. Fetal tissues normally contain Cbl [56], and presumably the tissues of the TC II deficient fetus do as well. If the fetus needs TC II in utero, and it certainly does postpartum, one could assume that maternal TC II enters the fetal circulation and distributes the Cbl. Direct evidence, however, suggests that fetal TC II is of fetal origin [5, 53].

All of the above presentation is put in schematic form in Fig. 1.

Measurement of the Transport Proteins

IF is no longer measured directly except in the evaluation of rare and complex disorders of Cbl absorption. Measurement is simple, but collection of the gastric juice is not appealing either to the physician or to the patient. The principle of the assay of IF is to first measure the capacity of the gastric juice to bind radioactive Cbl. The binding capacity is measured again after blocking the binding

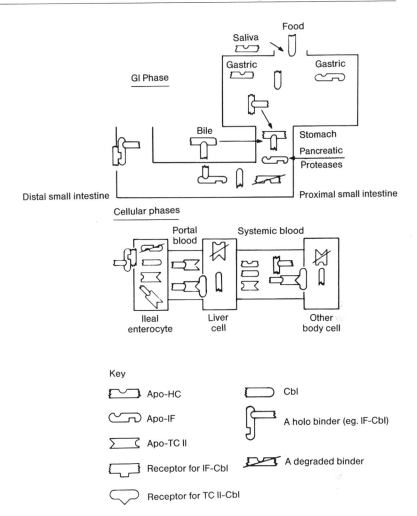

Fig. 1. Summary of cobalamin (*Cbl*) transport. After release of the ingested Cbl from food by gastric digestion, the Cbl is bound to apo-haptocorrin (*HC*) derived from both salivary and gastric secretions. Binding to HC is not essential, but absorption cannot proceed until the holo-HC, and any apo-HC, are degraded by pancreatic digestion. Pancreatic digestion is also essential to free the Cbl bound to HC in the bile. The Cbl then binds to intrinsic factor (*IF*) in the proximal intestine and is thus carried to the receptors for IF-Cbl on the enterocyte of the ileum. The Cbl is released early in the penetration and transferred to apo-transcobalamin (*TC II*) within the enterocyte to be released as TC II-Cbl into the portal circulation. The TC II-Cbl is then taken up by specific receptors for TC II-Cbl in many tissues. To this point, the key steps in the process are reasonably well identified. TC II is also an essential carrier of Cbl among tissues. Three-fourths of the Cbl is bound to HC, which is nonessential. Probably many or most cells are dependent on TC II for their Cbl, but the most rapidly dividing and turning-over cells are the most dependent on an immediate supply of TC II-Cbl

to IF by specific antibodies or after blocking the HC by cobinamide. The difference is the amount of the unblocked substance in units of Cbl bound [23]. The presence of IF is more commonly detected by measuring the absorption of radioactive Cbl. IF, release of Cbl from HC, and receptors must all be present, while inhibiting substances such as bacteria in blind loops or the fish tapeworm must be absent for absorption to occur normally. If malabsorption of Cbl is corrected by providing IF with the test dose of radioactive Cbl, failure to release adequate IF to combine with the Cbl is the basis of the malabsorption. These principles can be applied to any of several measures of absorption, but the semi-quantitative Schilling test is the most widely used. Others are the spot feces test, whole body counting, and total fecal collection. The specific details of the performance of these procedures can be found in specialized references [23].

The measurement of HC in gastric juice needs to be considered only in the attempt to assay IF. Measurement of the HC in saliva can be most valuable in the diagnosis of the congenital absence of HC. Since virtually all of the Cbl binding capacity of saliva comes from the HC, and there is little holo-HC, simple measurement of the Cbl binding capacity is a practical way of measuring salivary HC. HC in any source can also be measured by radioimmunoassay.

For the measurements of the TC II and HC of serum a step must be applied to distinguish between the two. The apo forms are first labeled with radioactive Cbl. Some physicochemical property then forms the basis of separation. TC II filters on gel filtration as if a M_r of 38000 while HC with substances of M_r 120000. (The true relative molecular weight of HC is half of this, but gel filtration is influenced by the high carbohydrate content). TC II does not bind Cbl at pH 1.5–2.0, while binding to HC persists. TC II is insoluble in $2.0M$ $(NH_4)_2SO_4$, while HC is soluble. Differential adsorption to zirconium phosphate gel, to silica powder, or to charged cellulose have been incorporated into separation techniques. The latter two techniques can be influenced by hydrophobicity and may not be applicable to the TC II of some species other than man that differ in hydrophobic interactions [4]. Electrophoresis and ion exchange chromatography can be applied, but separation of a portion of the HC from TC II may be difficult. Critiques and detailed descriptions of the techniques are available [2, 18, 23, 41]. All the above techniques can be used to advantage, but the user must take care to become familiar with the method selected and to establish reference values. The method of collection of the blood is a factor, and the user must know whether any anticoagulant is incompatible with the separation techniques to be applied. Serum is preferred although contact between the clotted cells and the serum should be kept at a minimum and be standardized. During contact, HC escapes from the granulocytes. This does not influence the results when measuring TC II, but it increases the amount of apo-HC and alters the composition by increasing the more basic components.

The total amount of either HC or TC II can be measured by specific radioimmunoassays [2, 20, 22]. All molecules of the respective substance will react, whether carrying Cbl or not. Specific antibodies can be applied in other ways, such as in removal of TC II or HC from serum, the blocking of binding, or in final confirmation of the nature of a binder.

To date, measurements of holo-TC II and HC have been made only in research laboratories. Since TC II-Cbl is the crucial component of serum Cbl, there are valid reasons for measuring it. The technique requires the combination of one of the above methods for the separation of TC II and HC, followed by an assay of the Cbl. One approach is to assay the total serum Cbl, remove the TC II-Cbl or HC-Cbl by adsorption or precipitation, assay again the remainder, and obtain the binder Cbl not measured by differences. Any errors introduced by these two assays for Cbl are compounded in assigning a value to the component not measured. Another method is to isolate directly the desired holo form, TC II-Cbl or HC-Cbl, dissolve or concentrate it to the desired volume, and assay it for Cbl directly. Having tried several methods [33] we prefer to isolate TC II-Cbl, or HC-Cbl, by immunoadsorption and measure directly the desired substance by bioassay [2]. The techniques used respectively for the separation of TC II from HC and the assay for Cbl must be compatible. That is, the former cannot introduce substances that interfere with the latter.

Diagnosis of Congenital Defects

Congenital disorders of transport are rare but constitute the main reason for measuring TC II or HC. The known syndromes affecting HC and TC II and the respective abnormalities of Cbl levels are illustrated in Table 4. The congenital abnormalities of IF and those of the receptors for IF-Cbl which present in childhood as acquired deficiencies of body Cbl are discussed in Chap. 5 of this volume. HC deficiency is suspected when a distinctly low serum Cbl is observed, often incidentally, in the absence of any confirming evidence of Cbl deficiency. To date, all affected persons have been either Black, heterozygous for hemoglo-

Table 4. Compartments of serum cobalamin (Cbl) in the congenital disorders of plasma Cbl transport

Genotype	Phenotype	Serum Cbl	TC II		HC		Immunoreactive	
			Holo	Apo	Holo	Apo	TC II	HC
Normal								
TC2/TC2	TC2	N	N	N	N	N	N	N
HC/HC	HC	N	N	N	N	N	N	N
HC Deficiency								
HC*NY/HC*NY	HC NY	Low	N	N	Nil	Nil	N	Nil
TC II Deficiencies								
TC2*SEA/TC2*SEA	TC2 SEA	N	Nil	Nil	N	Present	Nil	Present
TC2*DEN/TC2*DEN	TC2 DEN[a]	N	Nil	Nil	N	Present	N	Present
TC2*CZA/TC2*CZA	TC2 CZA[b]	High	High	High	N	N	High	N

HC, haptocorrin; TC II, transcobalamin II; N, normal; NY, New York; SEA, Seattle; DEN, Denver; CZA, Cardeza.

[a] Not observed in the homozygous form. The data are derived from a single case of the genotype TC2*DEN/TC2*SEA and heterozygotes.

[b] Data from a single case where the TC II failed to react with normal receptors for TC II-Cbl.

Further details and the observations in those heterozygous for a binder defect are given in [25].

bin S, or both [11] (and personal observation). All have been male, but the limited observation in the only family available for study do not suggest X-linked transmission [27]. Total, holo-, and apo-HC are all absent from the serum. Caution is necessary when interpreting apparently absent apo-HC because levels may be quite low in healthy persons. It is essential to establish the absence of HC from some fluid or cell in addition to serum. Absence of apo-HC from saliva confirms the diagnosis, and, as noted, the simple measure of Cbl binding capacity suffices [2].

The congenital abnormalities of TC II present early in infancy with the fulminating syndrome discussed above. Serum Cbl has usually been normal but can be low-normal or low [50]. Those with the genotype TC 2*SEA/TC 2*SEA show no demonstrable total, holo-, or apo-TC II. although in reported cases the diagnosis has usually been confirmed by more than one measurement, the absence of apo-TC II by a reliable test may establish the diagnosis [63]. The child with no TC II may have been treated with Cbl before the diagnosis is known, and only posttreatment serum is available. If the serum was collected a few hours after an injection of Cbl, all binding sites may be blocked and the failure to detect apo-TC II erroneous. By 48 h after even large injection, sites are open in persons who have TC II [3, 6]. Either independently or perhaps associated with therapy, complexes may form that obscure the apo binding pattern [28, 32, 50]. Both apo and holo binding may be atypical and not recognized when evaluated by the usual methods. Gel filtration is helpful here because the complexes can be isolated by their molecular size. Some may incorporate HC or, when present, TC II, which can be identified by reaction with specific antibodies [2]. Persons heterozygous for TC II deficiency have partial deficiencies of all components of TC II [28, 83] and are asymptomatic.

Radioimmunoassay can detect TC II whether carrying Cbl or not, but there is some dispute over the results of radioimmunoassay in TC II deficiency. Most observers have not been able to detect any immunoreactive TC II in the classic type, TC 2 SEA [25], but one investigator has routinely found an immunoreactive substance, although much reduced [20]. It would not be surprising if a disturbed molecule of nonfunctioning TC II did exist in all TC II defects, and in two cases there was direct evidence. TC 2 DEN has not been observed in the homozygous states but paired with TC 2 SEA as a double heterozygote [25]. TC 2 DEN neither binds Cbl nor functions but is recognized immunologically. TC 2 CZA, also reported only once, does bind Cbl and is immunoreactive but does not transport Cbl into cells [37]. This defect is recognized by testing the TC II as a promoter of the entry of Cbl into some standard cells such as fibroblasts or hepatocytes [6, 13, 35, 62]. Finally, one can postulate a congenital defect of the tissue receptor for TC II-Cbl, although none has been identified. The defect could be detected by reacting skin fibroblasts cultured from the affected person with normal TC II-Cbl.

Congenital defects of the TC II system must be recognized because they are lethal if unrecognized but easily treated once they are known. The known defects can all be readily detected by examination of the serum only (Table 4), although fibroblasts can be applied where a culture has been established as part of a genetic evaluation. Fibroblasts normally synthesize and release apo-TC II into the culture medium, but cells cultured from persons deficient in TC II do not release recognizable TC II [22].

Abnormalities in Acquired Disorders

The plasma carriers of Cbl are increased in several disorders not clearly related to Cbl metabolism. For the most part, these increases are not helpful to the clinician and may sometimes be confusing. Secondary disease is not likely to mask a frank deficiency of TC II or HC in a person congenitally unable to synthesize it, but could conceivably mask heterozygosity for a defect. The levels of HC and TC II, principally in the apo forms, in various diseases have been well published; for complete lists and specific references the reader should consult recent reviews [10, 18, 41, 64]. Certain groups of diseases in which abnormalities are the greatest or form a pattern are however, discussed briefly here.

The increase of HC of the myeloproliferative states [64] can be striking, may be useful diagnostically, and may reflect the origins of HC. The fact that granulocytes and their precursors contain and release apo-HC may well be the basis for the sometimes great increase seen in neoplasia of the granulocytes. It is not clear, however, whether the HC levels reflect an exaggeration of the normal source of plasma HC or the superimposition of an abnormal source upon the normal. The composition of the HC varies with the type of myeloproliferative states, showing a shift to the more acidic components in myeloid leukemia and to the more basic in polycythemia vera [64]. This observation could be helpful in differential diagnosis if the patterns were more specific and other better markers of the respective disease not available. The amount and composition of HC could be used to monitor the course of a changing myeloproliferative state, but longitudinal studies are needed to evaluate this potential application. HC is often increased in leukemoid reactions, especially those in persons with neoplasia, and increased levels therefore do not distinguish leukemia from the leukemoid. The level and composition of HC probably are, however, valuable in separating primary polycythemia from secondary [64] and certainly from relative polycythemia. This diagnostic application should be studied further and include longitudinal evaluation in an effort to predict changes in the basic disease by changes in HC. Any new studies of HC in the myeloproliferative states must rigorously control the in vitro release of HC from the blood as it is prepared. The importance of this factor was not recognized in the studies of more than a decade ago, and some of the observations reported then are invalid.

The principal increases of TC II concern another group of disorders, those of the macrophage/histiocyte [17]. Levels are high in both benign and malignant proliferation of these cells and in Gaucher's disease. Perhaps related are the usually more modest increases in autoimmune disease, inflammation, lymphoproliferative states, and multiple myeloma. As discussed above, part of the normal circulating human TC II comes from cells of bone marrow origin [21] and can come from related cells of other species [22]. The increased TC II of macrophage/histiocyte disorders may represent an increased activity of an expected source of TC II or may be the superimposition of an unusual source upon the usual. Increased levels of a binder could, of course, reflect impaired removal rather than increased production. Although this explanation might pertain for some increase in HC, it does not seem likely in cases in which TC II is increased. In fact, increased turnover could not be demonstrated in patients with systemic lupus and increased TC II [44].

References

1. Allen RH, Seetharam B, Podell E, Alpers DH (1978) Effect of proteolytic enzymes on the binding of cobalamin to R protein and intrinsic factor. J Clin Invest 61:47–54
2. Begley JA (1983) The materials and processes of plasma transport. In: Hall CA (ed) Methods in hematology. The cobalamins. Churchill Livingstone, Edingburgh, pp109–133
3. Begley JA, Morelli TA, Hall CA (1977) B$_{12}$ binding proteins after injection of cyanocobalamin. N Engl J Med 297:614–615
4. Begley JA, Heckman SM, Hall CA (1983) Hydrophobic interactions of transcobalamin II (TC II) from mammalian sera. Proc Soc Exp Biol Med 172:370–374
5. Begley JA, Hall CA, Scott CR (1984) Absence of transcobalamin II from cord blood. Blood 63:490
6. Begley JA, Burkart P, Hall CA (1986) A patient with the inability to maintain in vivo levels of bound cobalamin (Cbl) and manifestations of tissue deficiency in Cbl. Am J Hematol 22(1):69–78
7. Benson RE, Rappazzo ME, Hall CA (1972) Late transport of vitamin B$_{12}$ by transcobalamin II. J Lab Clin Med 80:488–495
8. Burger RL, Mehlman CS, Allen RH (1975) Human plasma R-type vitamin B$_{12}$ binding proteins. I. Isolation and characterization of transcobalamin I, transcobalamin III and the normal granulocyte vitamin B$_{12}$ binding protein. J Biol Chem 250:7700–7706
9. Burger RL, Schneider RJ, Mehlman CS, Allen RH (1975) Human plasma R-type vitamin B$_{12}$ binding proteins. II. The role of transcobalamin I, transcobalamin II and the normal granulocyte vitamin B$_{12}$-binding protein in the plasma transport of vitamin B$_{12}$. J Biol Chem 250:7707–7713
10. Carmel R (1981) Cobalamin binding proteins of man. In: Silber R (ed) Contemporary hematology oncology, vol 2. Plenum, New York, pp 79–129
11. Carmel R (1983) R-binder deficiency. A clinically benign cause of cobalamin pseudodeficiency. JAMA 250:1886–1890
12. Chanarin I, Muir M, Hughes A, Hoffbrand AV (1978) Evidence for an intestinal origin of transcobalamin II during vitamin B$_{12}$ absorption. Br Med J 1:1453–1455
13. Cooper BA (1983) The entry of cobalamin into tissues. In: Hall CA (ed) Methods in hematology. The cobalamins. Churchill Livingstone, Edinburgh, pp 148–160
14. Cooper BA, Paranchych W (1961) Selective uptake of specifically bound cobalt-58 vitamin B$_{12}$ by human and mouse tumor cells. Nature 191:393–395
15. Donaldson RM, Brand M, Serfilippi D 61977) Changes in circulating transcobalamin II after injection of cyanocobalamin. N Engl J Med 196:1427–1430
16. Farquharson J, Adams JF (1976) The forms of vitamin B$_{12}$ in foods. Br J Nutr 36:127–135
17. Fehr J, DeVecchi P (1985) Transcobalamin II. A marker for macrophage/histiocyte proliferation. Am J Clin Nutr 84:291–296
18. Fernandes-Costa F, Metz J (1982) Vitamin B$_{12}$ binders (transcobalamins) in serum. Crit Rev Clin Sci 18 (1):1–30
19. Finkler AE, Hall CA (1967) Nature of the relationship between vitamin B$_{12}$ binding and cell uptake. Arch Biochem Biophys 120:79–85
20. Fräter-Schröder M (1983) Genetic patterns of transcobalamin II and the relationships with congenital defects. Mol Cell Biochem 56:5–31
21. Fräter-Schröder M, Nissen C, Gmur J, Hitzig WH (1980) Bone marrow participates in the biosynthesis of human transcobalamin II. Blood 56:560–563

22. Fráter-Schröder M, Porck HF, Erten J et al. (1985) Synthesis and secretion of the human vitamin B_{12}-binding protein, transcobalamin II by cultured skin fibroblasts and by bone marrow cells. Biochim Biophys Acta 845:421–427

23. Gräsbeck R, Kouvonen I (1983) The materials and processes of intestinal transport. In: Hall CA (ed) Methods in hematology. The cobalamins. Churchill Livingstone, Edinburgh, pp 79–108

24. Hall CA (1975) Transcobalamin I and II as natural transport proteins of vitamin B_{12}. J Clin Invest 56:1125–1131

25. Hall CA (1981) Congenital disorders of vitamin B_{12} transport and their contributions to concepts. II. Yale J Biol Med 54:485–495

26. Hall CA (1984) The uptake of vitamin B_{12} by human lymphocytes and the relationships to the cell cycle. J Lab Clin Med 103:70–81

27. Hall CA, Begley JA (1977) Congenital deficiency of human R-type binding proteins of cobalamin. Am J Hum Genet 1977; 29:619–626

28. Hall CA, Begley JA (1982) Atypical cobalamin binding in the serum of congenital deficiency of transcobalamin II. Br J Haematol 51:65–71

29. Hall CA, Finkler AE (1962) In vivo plasma vitamin B_{12} binding in B_{12} deficient and nondeficient dubjects. J Lab Clin Med 60:765–776

30. Hall CA, Finkler AE (1965) The dynamics of transcobalamin II. A Vitamin B_{12} binding substance in plasma. J Clin Med 65:459–468

31. Hall CA, Green PD (1978) Competition between apo and holo transcobalamin II (TC II) for the TC II-mediated uptake process. Proc Soc Exp Biol Med 158:206–209

32. Hall CA, Hitzig WH, Green PD, Begley JA (1979) Transport of therapeutic cyanocobalamin in the congenital deficiency of transcobalamin II (TC II). Blood 53:251–263

33. Hall CA, Begley JA, Carmel R (1980) Serum transcobalamin II-cobalamin in pernicious anemia (Abstract 417). Abstracts of the 18th international society of hematology meeting in Montreal, Canada, 16–22 Aug 1980, p 98

34. Hall CA, Begley JA, Green-Colligan PD (1984) The availability of therapeutic hydroxocobalamin to cells. Blood 63:335–341

35. Hall CA, Green-Colligan PD, Begley JA (1985) The metabolism of cobalamin bound to transcobalamin II and to glycoproteins that bind Cbl in HepG2 cells (Human Hepatoma). J Cell Physiol 124:507–515

36. Hall CA, Green-Colligan PD, Begley JA (1987) Cyclic activity of the receptors of cobalamin bound to transcobalamin II J Cell Physiol 133:187–191

37. Haurani FI, Hall CA, Rubin R (1979) Megaloblastic anemia as a result of an abnormal transcobalamin II (Cardeza). J Clin Invest 64:1253–1259

38. Haus M, Green PD, Hall CA (1979) Species specificity in the immunologic reactions and biological functions of transcobalamin II. Proc Soc Exp Biol Med 162:295–298

39. Hippe E, Schonau JR, Olesen H (1977) Cobalamin binding proteins in stomach and serum from various animal species. Data for B_{12} binding capacities and molecular sizes of the binding proteins. Comp Biochem Physiol [B] 56:305–309

40. Hom BL (1967) Plasma turnover of ^{57} cobalt-vitamin B_{12} bound to transcobalamin I and II. Scand J Haematol 4:321–332

41. Jacob E, Baker SJ, Herbert V (1980) Vitamin B_{12}-binding proteins. Physiol Rev 60:918–960

42. Katz M, O'Brien R (1979) Vitamin B_{12} absorption studied by vascular perfusion of rat intestine. J Lab Clin Med 94:817–824

43. Kolhouse JF, Allen RH (1977) Absorption, plasma transport and cellular retention of cobalamin analogues in the rabbit. J Clin Invest 60:1381–1392

44. Lasser U, Kierat L, Grob PJ, Hitzig WH, Fráter-Schroder M (1985) Transcobalamin II, a serum protein reflecting autoimmune disease activity, its plasma dynamics, and the relationship to established serum parameters in systemic lupus erythematosus. Clin Immuno Immunogathol 36:345–357

45. Levine JS, Allen RH, Alpers DH, Seetharam B (1984) Immunocytochemical localization of the intrinsic factor-cobalamin receptor in dog-ileum: distribution of intracellular receptor during cell maturation. J Cell Biol 98:1111–1118

46. Magnus P, Magnus EM, Berg K (1984) Evidence for genetic effects on variation in plasma

unsaturated transcobalamin II and cobalamin (vitamin B_{12}). Scand J Haematol 33:180–186

47. Marcoullis G, Rothenberg SP (1983) Macromolecules in the assimilation and transport of cobalamin. In: Lindebaum J (ed) Contemporary issues in clinical nutrition. 5. nutrition in hematology. Churchill livingstone, New York, pp 59–119
48. Marcoullis G, Parmentier Y, Nicholas J-P (1980) Cobalamin malabsorption due to nondegradation of R proteins in the human intestins. J Clin Invest 66:430–440
49. Mellman I, Huntington WF, Rosenberg LE (1978) Cobalamin binding and cobalamin-dependent enzyme activity in normal and mutant human fibroblasts. J Clin Invest 62:952–960
50. Meyers RA, Carmel R (1984) Hereditary transcobalamin II deficiency with subnormal serum cobalamin levels. Pediatrics 74:866–871
51. Nexø E, Olesen H (1982) Intrinsic factor, transcobalamin, and haptocorrin. In: Dolphin D (ed) Vitamin B_{12}. Volume 2:Biochemistry and Medicine. Wiley, New York, pp 57–85
52. Porck HF (1985) Genetic and biochemical studies on human transcobalamin II. Thesis, University of Amsterdam, Holland
53. Porck HF, Fráter-Schröder M, Frants RR, Kierat L, Eriksson AW (1983) Genetic evidence for fetal origin of transcobalamin II in human cord blood. Blood 62:234–237
54. Rachmilewitz B, Rachmilewitz M, Chaouat M, Schlesinger M (1978) Production of TC II (vitamin B_{12} transport protein) by mouse mononuclear phagocytes. Blood 52:1089–1098
55. Rappazzo ME, Hall CA (1972) Transport function of transcobalamin II. J Clin Invest 51:1915–1918
56. Rappazzo ME, Salmi HA, Hall CA (1970) The content of vitamin B_{12} in adult and fetal tissue. Br J Haematol 18:425–433
57. Rothenberg SP, Weiss JP, Cotter R (1978) Formation of transcobalamin II-vitamin B_{12} complex by guinea-pig ileal mucosa in organ culture after in vivo incubation with intrinsic factor-vitamin B_{12}. Br J Haematol 40:401–414
58. Schneider RJ, Burger RL, Mehlman CS, Allen RH (1976) The role and fate of rabbit and human trancobalamin II in the plasma transport of vitamin B_{12} in the rabbit. J Clin Invest 57:27–38
59. Seetharam B, Alpers DH (1982) Absorption and transport of cobalamin (vitamin B_{12}). Annu Rev Nutr 2:343–369
60. Soda R, Tavassoli M, Jacobson DW (1985) Receptor distribution and the endothelial uptake of transcobalamin II in liver cell suspensions. Blood 65:795–802
61. Stenman U-H (1974) Studies on cobalamin. Vitamin B_{12} binding proteins of R-type. Thesis, The Minerva Foundation Institute for Medical Research, Helsinki
62. Youngdahl-Turner P, Rosenberg LE, Allen RH (1978) Binding and uptake of transcobalamin II by human fibroblasts. J Clin Invest 61:133–141
63. Zeitlin HC, Sheppard K, Baum JD, Bolton FH, Hall CA (1985) Homozygous transcobalamin II deficiency maintained on oral hydroxocobalamin. Blood 66:1022–1027
64. Zittoun J, Zittoun R, Marquet J, Sultan C (1975) The three transcobalamin in myeloproliferative disorders and acute leukaemia. Br J Haematol 31:287–298

Chapter 5

Cobalamin Absorption and Acquired Forms of Cobalamin Malabsorption

J. Belaïche and D. Cattan

Cobalamin (Cbl), or vitamin B_{12}, cannot be synthesized in the human body, and the only possible source for this vitamin is thus the diet. Daily requirements are 1–3 μg [31]. Dietary sources are primarily animal proteins. Liver, mammalian kidneys, and seafood are especially rich in this vitamin. Meat, fish, milk, and dairy foods also contain vitamin B_{12}, but in lesser amounts. Plants however, have no vitamin B_{12}, a finding which accounts for the megaloblastic anemias often observed in strict vegetarians. The principal dietary forms of this vitamin are adenosyl-Cbl and hydroxo-Cbl. Cyano-Cbl prescribed therapeutically and for physiologic properties is a synthetic form; its absorption is similar to that of of adenosylated and hydroxylated forms of the vitamin.

Cobalamin Absorption

Cbl supplied in the diet is combined with proteins. Salivary digestion puts it into contact with R proteins present in saliva and which possess a potential binding capacity of 40–80 ng/ml. Release of Cbl from its protein carrier, which begins during cooking, continues as the food passes into the stomach and is acted upon by pepsin and gastric acid [23]. In the stomach, Cbl may bind to two types of receptors: intrinsic factor (IF) and R proteins. IF is a glycoprotein secreted by human parietal cells contained in fundic glands. The daily secretion of IF is 70 000 U, but 500 U is adequate to absorb 0.5 μg dietary vitamin B_{12} [17]. IF secreted in gastric juice can bind 50–200 μg Cbl per day, which is 50 times higher than physiological requirements. Insulin [1], gastrin, pentagastrin [62, 74], food [21], histamine, and other H_2 (histamine) receptor agonists such as impromidine [61] stimulate IF secretion. Somatostatin inhibits IF secretion stimulated by pentagastrin [59]. While hydrochloric acid secretion remains high after stimulation by pentagastrin, IF secretion exhibits a peak 15–20 min after stimulation and then decreases uniformly even if the stimulatory agent is continued [17].

This rapid release indicates that IF is stored within the cells, and that stimulation has an effect on cell excretion only. Cyclic AMP might influence the mechanism of secretion [36]. The significant drop in the rate of IF secretion following vagotomy suggests the influence of the autonomic nervous system on the secretory mechanism [10]. R proteins, also known as cobalophilins or haptocorrins, may be defined as non-IF ligands of Cbl. They account for 10%–20% of Cbl ligands present in gastric juice [43]. This family of glycoprotein Cbl binders, without any IF activity, have molecular weights ranging from 120000 to 150000 daltons, as determined by gel filtration. The molecular weights range from 56000 to 64000 when measured by the equilibrium technique of sedimentation-ultra-centrifugation [2]. This apparently larger molecular weight as determined by gel filtration is due to the carbohydrate component, which accounts for 30%–40% of the molecule. Most of the R proteins present in gastric juice come from saliva [33, 34].

Secretion of R proteins by the stomach seems to be independent of parietal cell secretion [9]. Cbl affinity for R proteins in vitro is 50 times higher than for IF at pH 2 [3]. R proteins also have considerable affinity for Cbl analogues, while IF lacks this affinity. It thus is probable that while in the stomach, Cbl binds preferentially to R proteins present in gastric juice, which contradicts the conventional concept that Cbl binds to IF in the stomach, in the duodenum, the R-Cbl complex formed in the stomach is subjected to the action of pancreatic proteolytic enzymes. These enzymes degrade R proteins and reduce their affinity for Cbl by approximately 150 times. Cbl thus released can then bind to IF [3, 4].

The duodenum also receives an influx of Cbl from the liver, excreted in bile at concentrations of 1–4 ng/ml. Cbl present in the bile is bound to proteins which do not have any IF activity, and which belong to the class of R proteins [19, 29]. The bile excretes daily 3–9 µg Cbl, and only 1 µg is not reabsorbed in the ileum, which suggests the presence of an enterohepatic cycle. The presence of this enterohepatic Cbl cycle, although disputed, currently appears to be recognized. Cbl from the bile, initially bound to R proteins, crosses the intestine combined with IF. This thus assumes a transfer of Cbl from the bile to unbound IF during the passage between duodenum and ileum. In the duodenum lumen, release of biliary Cbl from its ligand occurs due to the action of pancreatic proteolytic enzymes [13]. In man, intestinal fluid contains an appreciable amount of free IF [35]. This IF might bind both biliary Cbl and Cbl synthesized by intestinal bacteria and thereby play the role of ileal permease. R proteins are almost entirely lacking from intestinal fluid in healthy individuals. However, a protein derived from R proteins has been demonstrated in jejunal fluid, known as corrinoid binder, which has an affinity for Cbl analogues, thereby conferring an antimicrobial action on it [47]. Indeed, unbound Cbl analogues might facilitate bacterial proliferation [47].

Once formed, the IF-Cbl complex passes through the small intestine until it reaches a specific receptor in the ileum. The first stage in crossing the ileum consists of attachment of the IF-Cbl complex to the microvilliform membrane of the enterocytes. This required the presence of the calcium ion and a pH above 5.6. What becomes of IF after its attachment to this receptor is not entirely clear; some investigators think that IF remains in the ileal lumen while Cbl enters the enterocyte. Other researchers feel that the IF-Cbl complex might be separated in the brush border [41]; still others believe that the IF-Cbl complex is completely absorbed by pinocytosis and is later broken up in the enterocyte [49]. Whatever the absorption mechanism for Cbl, IF does not enter the bloodstream in man [18]. Crossing the enterocyte requires several hours, and labeled Cbl is recovered 6–12 h after it is found in hepatic portal blood as 5'-deoxyadenosyl-Cbl and methyl-Cbl. These are bound to a plasma carrier transcobalamin II. Cyano-Cbl can cross the intestinal barrier without metabolic change. In addition to this active mechanism, there is a mechanism of passive diffusion of Cbl absorption throughout the small intestine, but only at supraphysiological doses of Cbl.

Measurement of Cobalamin Absorption

Serum assay of Cbl concentrations reflects the size of the pool of this vitamin in the body. The liver contains 2–5 mg Cbl, and prolonged and continuing disturbance of entry would require 4–6 years to exhaust hepatic stores. This explains why it is possible to have Cbl malabsorption with normal serum Cbl levels. An examination of Cbl absorption provides proof of the malabsorption and may guide the physician to detect gastric or intestinal dysfunction. The different techniques employed to measure Cbl absorption are: (a) test of fecal excretion, (b) test of serum absorption, (c) external hepatic assessment, (d) external whole body assessment, and (e) test of urinary excretion. Each method poses problems in methodology, and the test of urinary excretion, the Schilling test, remains the most widely used technique. This test consists of the oral administration of 0.5–2 µg unbound or crystalline cyano[^{57}Co]Cbl to a patient who has been fasting from the day before and measuring the excretion of administered radioactive material in the urine for 24 h after its administration. At 1 h after its ingestion, a 1-mg dose of unlabeled vitamin B_{12} is administered intramuscularly; this is intended to block all tissue receptors for vitamin B_{12} and induce excretion of most of the absorbed Cbl in the urine. Normally, more than 10% of ingested radioactivity is eliminated in a 24-h urine specimen. This test must be repeated 1 week later, with the concomitant administration of an excess of exogenously provided IF. A normal result obtained in this second Schilling test demonstrates vitamin B_{12} malabsorption due to IF deficiency. However, the performance of test both with and without IF can only replace assay of IF in gastric juice if repeated Schilling tests demonstrate a marked malabsorption of unbound Cbl, corrected by IF administration. Repetition of Schilling tests makes it possible to eliminate such causes of error as alcoholism, administration of drugs that interfere with Cbl absorption, e.g., colchicine, and the effects of Cbl deficiency itself on enterocyte structure and intestinal functions [39].

The two-step Schilling test without and then with IF is currently preferred to the dual-labeling test (Dicopac). The latter technique examines in one step the ratio between the percent of radioactive Cbl bound to IF-cyano[^{58}Co]Cbl and the percent of free Cbl cyano[^{57}Co]Cbl which is absorbed. The reliability of this test has been challenged. In a study with 71 patients suffering from pernicious anemia, one-third of patients had Schilling test results by this technique which were not those typically expected [54]. We confirmed the poor reliability and poor reproducibility of this test in diagnosing pernicious anemia [14]. The poor reliability of the Dicopac test is probably due to exchange of radioactivity between the two forms of Cbl.

Schilling tests employing cyano-Cbl bound to dietary proteins demonstrate the malabsorption of dietary Cbl in patients with achlorhydria who still show adequate gastric IF secretion. This may explain low serum levels of Cbl in the absence of a deficient external supply. These methods appear to be closer to physiological reality in so far as they take gastric acid secretion into account. Different protein carriers have been used to perform this test (Table 1), in particular chicken serum proteins because of the ease of their preparation. In normal subjects, the percentage of urinary exretion varies according to the proteins used, the highest percentage is obtained with sheep liver (9%) and the lowest with chicken serum (2%). These differences can probably be explained by differences in the strength of the Cbl-protein bound. Test of fecal excretion of Cbl have also been used to study the absorption of Cbl bound to proteins [23, 25, 26, 30].

Acquired Malabsorption of Cobalamins

The forms of acquired Cbl malabsorption discussed here are those whose principal causes are: (a) gastric, (b) impaired pancreatic functions, (c) intraluminal abnormalities of the small intestine, (d) parietal disorder of the small intestine, (e) ileal resection, (f) drug-induced, and (g) alcoholism.

Gastric Sources

Gastrectomy. Cbl deficiency, in the absence of replacement therapy, inevitably occurs at an average of 5 years after total gastrectomy. Partial gastrectomy with gastrojejunal anastomosis may also cause Cbl deficiency. This deficiency was observed in 18% of a series of 351 patients [55]. Patients with partial gastrectomy may have Cbl deficiency while absorption of unbound Cbl is normal. This was observed in 30% of cases [42, 55]. Cbl deficiency in these patients is attributed to malabsorption of dietary Cbl which cannot be separated from its protein carrier because of the loss of acid and peptic secretion. Such a mechanism was demonstrated in studies using Cbl bound to various protein carrier substances: chicken serum [20, 37], eggs [23] ovalbumin [24], chicken meat [25], and fish [26].

Pernicious Anemia. It is typically in later life that pernicious anemia occurs. It is very rare before age 30, and its occurrence increases with age. The mean age at

Table 1. Comparison of the percentage of urinary excretion of cobalamin bound to various protein carriers in normal subjects

Substance	Urinary excretion (%)	Reference
Sheep liver	9.1	[31]
Chicken	6.7	[25]
Fish	6.5	[21]
Eggs	3.3	[23]
Ovalbumin	3.01	[24]
Chicken serum	0.93–3.16	[15, 20, 37, 65, 66, 67]

onset is 60.5 years. Its incidence is 110–130 cases per 100000 inhabitants [17] in Scandinavia and in English-speaking countries; the rate is lower in countries of the Mediterranean area and in Africa. Incidence is difficult to assess in countries in which Cbl is widely overprescribed.

Diagnosis of this disorder in its typical form is based on a combination of the following: (a) megaloblastic macrocytic anemia, (b) low serum Cbl, (c) achlorhydria resistant to stimulation, (d) gastric IF secretion less than 200 U/h (normal, over 2000) or zero, following stimulation, and (e) marked malabsorption of unbound Cbl corrected by exogenously supplied IF. Pernicious anemia in its latent form is diagnosed by these same criteria except for megaloblastic macrocytic anemia. In addition, following manifestations may also be present: serum antibodies in the form of IF-blocking antibodies (54% of cases), IF-precipitating antibodies (39%), and antibodies to parietal cells (84%) and, furthermore, elevated serum gastrin levels (75% of cases). There may be intestinal Cbl malabsorption during pernicious anemia, i.e., not improved by administration of IF. This intestinal malabsorption is observed if the Schilling test with IF is performed early in the diagnostic work-up. In a prospective study with 28 patients, it was calculated to involve 75% of cases [39]. It may be combined with sugar malabsorption or with moderate steatorrhea. These disturbances are attributed to dystrophic lesions of the intestinal mucosa induced by the Cbl deficiency itself [6, 12, 39]. The malabsorption regresses in 1–8 weeks following administration of this vitamin; it may continue to be present in patients with very high levels of antibodies to IF [39].

Pernicious anemia may occur concomitantly with certain autoimmune disorders [17]. It is observed with Graves' disease (2.4% of cases), myxedema (9%), vitiligo (7%), Addisonian adrenal atrophy (4.2%), myasthenia gravis (1%), and rheumatoid arthritis (1.05%). Increased incidence of pernicious anemia in insulin-dependent diabetics has not been convincingly demonstrated [17]. It may be associated with hypogammaglobulinemia of late onset [73] and with isolated IgA deficiency [63]. The age at onset of pernicious anemia in the presence of these other disorders is likely to be younger (3rd or 4th decades of life); atrophy of the gastric mucosa affects both the fundus and the antrum; there is no increase in serum gastrin; and antibodies to parietal cells, to intrinsic factor, and to thyroid gland are generally absent. A positive family history of pernicious anemia is noted in 7.9%– 30% of cases, according to various studies [75]. Prevalence of this disorder in certain families is 20 times higher than that of a normal population. This disease was observed 18 times in homozygous twins [17]. Its genetic mode of transmission is not known.

Non pernicious Anemia Types of Achlorhydric Gastritis. In persons over 60 years of age non pernicious anemia such as achlorhydric gastritis occur ten times more frequently than does pernicious anemia. Patients with achlorhydric gastritis without pernicious anemia do not meet criteria for diagnosis of latent pernicious anemia, and only rarely have antibodies to IF. As for each of the other laboratory test results, it is difficult to make a clear distinction between these two clinical entities; 47% of patients with achlorhydria without pernicious anemia have antibodies to gastric parietal cells [75], and 75% have elevated serum gastrin

levels [27]. It is generally recognized that patients with atrophic gastritis successively lose their ability to secrete gastric acid, pepsin, and IF. The flow of IF
secretion following its stimulation is as low as that observed in pernicious
anemia in 10% of subjects with achlorhydria, who nevertheless absorb unbound
Cbl normally. A deficiency in serum Cbl is observed in 14%–22% of these subjects.

As in the case with patients who have undergone partial gastrectomy, this deficiency cannot be explained by inadequate dietary supply or by malabsorption
of free Cbl [15, 23, 37]. The deficiency has been ascribed to malabsorption of
dietary Cbl; this was demonstrated by examining the absorption of Cbl bound to
different protein carriers: chicken meat [25], fish [26], eggs [23], ovalbumin [24],
and chicken serum [15, 37]. Absorption rates of Cbl bound to proteins were
found in all these studies to be intermediate between that of control subjects and
that of patients with pernicious anemia. Malabsorption of Cbl was not improved
by exogenously supplied IF [15, 37] but by administering hydrochloric acid [15].
The mechanism for the malabsorption was related to the loss of acid secretion,
as in the case of partial gastrectomy, preventing the release of Cbl from its protein carriers. Hematologic manifestations of vitamin deficiency generally are
moderate, consisting of macrocytosis or very mild anemia. Longitudinal studies
conducted over more than 20 years have demonstrated that the transformation
of achlorhydric gastritis into pernicious anemia occurs very rarely [17]. Nevertheless, the demonstration of Cbl deficiency in a subject with achlorhydria, since it
does not conform to all of the criteria for diagnosis of pernicious anemia, does
justify the institution of replacement therapy which, due to the good absorption
of unbound Cbl, can be administered orally.

Pancreatic Insufficiency

Malabsorption of unbound Cbl was observed in 40% of patients with chronic
pancreatitis [40, 48, 71] and in 100% of patients with cystic fibrosis [22]. In most
cases, administration of pancreatic extracts corrects this malabsorption [48, 70].
However, Cbl deficiency and consequent macrocytic anemia are rare during the
course of pancreatic disorders. Various mechanisms to explain this malabsorption have been suggested. The first hypothesis suggested that the secretion of IF,
normally found in pancreatic juice, might be decreased in impaired pancreatic
function [71]. Further studies demonstrated that the binder of Cbl present in pancreatic juice, both of animals and of man, did not possess either the chemical or
the physical structure of IF, and that it did not increase the intestinal uptake of
Cbl [40]. Calcium concentration and the pH of ileal fluid, equivalent in control
subjects and in patients with chronic pancreatitis, do not appear to influence
malabsorption [71]. Correction of the malabsorption by administration of pancreatic extracts and activation of pig IF by trypsin have led to suggestions of
possible activation of IF by pancreatic proteases [72]. This hypothesis assumes
that pancreatic secretion facilitates Cbl absorption by activating gastric IF, resulting in the presence of IF in the small intestine with different physicochemical
properties. However, many studies have shown that the IF found in the intesti-

nal lumen is the same as that found in the stomach, both in animals and in man, and that pancreatic proteases do not change the chemical structure of IF [45, 53].

The mechanism currently proposed is that of Cbl sequestration. Pancreatic proteases might be essential for degradation of R protein-Cbl complexes formed in the stomach [52, 70]. The Cbl thus released could then bind to IF later. In cases of impaired pancreatic function, Cbl would remain sequestered by R proteins.

There are many findings both in vitro [3, 5, 28] and in vivo [4, 46] which support this hypothesis:

1. Partial degradation of R proteins by bovine pancreatic enzymes. Trypsin is more active than chymotrypsin, which in turn is more active than elastase.
2. A 50% reduction in the molecular weight of R proteins, combined with an affinity for Cbl reduced by 150 times. This allows the transfer of Cbl to IF, when a preparation of unbound IF and R proteins bound to cyano[^{57}Co]Cbl is incubated with bovine proteases.
3. Absence of transfer to IF of Cbl bound to R proteins when pancreatic proteases are not present.
4. A no change in molecular weight and in IF affinity after incubation with pancreatic proteases.
5. Demonstration of Cbl bound to R proteins in the jejunal fluid of patients with pancreatic insufficiency and not to IF as in control subjects.
6. Improvement in malabsorption by administration of cobinamides or structural analogues of Cbl which bind to R proteins and not to IF.

The practical application of the theory of Cbl sequestration has led to the development of tests to detect exocrine pancreatic insufficiency [11, 50]. Although attractive, this hypothesis does not explain the following findings observed in pancreatic insufficiency:

1. Malabsorption is not always present, and 60% of patients with chronic pancreatitis absorb Cbl normally [40, 48, 71]. For any given patient, there is no correlation between the severity of pancreatitis and the degree of malabsorption [69]. In addition, in two patients who underwent total pancreatectomy, Cbl absorption was normal [48].
2. Pseudo-vitamin B$_{12}$ compounds which, like cobinamides, bind preferentially to R proteins do not correct Cbl malabsorption [64].
3. Administration of bovine elastase corrects the Schilling test more easily than ingestion of trypsin or chymotrypsin, while elastase has a very low protease activity [70].
4. There is normal absorption of Cbl bound to R proteins in the saliva [32].
5. Pancreatic juice degrades R proteins in vitro and this degradation is not inhibited by aprotinin, a potent trypsin inhibitor [13].
6. Decreased protease activity in duodenal juice is not correlated with results of Cbl absorption [8].

Thus, it appears that despite much research, the exact mechanism of Cbl malabsorption in chronic pancreatitis is still not entirely clear.

Intraluminal Abnormalities of the Small Intestine

Bacterial Proliferation in the Small Intestine. The Cbl malabsorption which may result from bacterial proliferation in the small intestine is reversible by orally administered antibiotics [56]. Several mechanisms have been suggested to explain this malabsorption. The traditional mechanism involves the destruction of Cbl by bacteria. This mechanism has been demonstrated in vitro by incubation of bacteria from intestinal loops containing stagnant material. These bacteria can break the bond between IF and Cbl, for which they possess a high affinity. Bacteria able to cause such effects in vitro are primarily *Escherichia coli, Pseudomonas aeruginosa, Proteus mirabilis, Klebsiella pneumoniae, Bacteroides,* and *Bacillus subtilis; Streptococcus faecalis,* on the other hand, cannot produce this effect. However, it has not been established that this mechanism is important in vivo. Indeed, it is hardly probable that bacteria could block Cbl binding to IF because this binding occurs quickly in the jejunum, a short time after the vitamins have been ingested. Moreover, the amount of unbound IF in the small intestine is greater than the number of microorganisms present in the intestinal lumen [47].

Another possible mechanism suggests the uptake of the IF-Cbl complex by bacteria. This uptake would occur by means of endocytosis. However, most of the bacteria (except for *Bacteroides*) isolated from the jejunum and the ileum of subjects with diverticula of the small intestine and Cbl malabsorption are unable to bind this complex. Synthesis of Cbl analogues could also affect this mechanism for malabsorption. This synthesis, which occasionally is considerable, could induce an apparent malabsorption of radioactive Cbl in the Schilling test by dilution of the trace element. However, it has been demonstrated in subjects with bacterial proliferation of the jejunum with *E.coli, Bacteroides,* and *Lactobacillus* that 30% of the cyano-Cbl bound to IF is transformed into structural analogues, the cobamides. The IF-cobamide complex has the same affinity for the ileal receptor as the IF-Cbl complex, with which its compete. R proteins appear to be antimicrobial. It has recently been demonstrated in man that there is a protein, called corrinoid binder, whose molecular weight is 60000 daltons, and which reacts with anti-R-protein antibodies [47]. This protein can bind Cbl analogues but does not compete with IF to bind unbound Cbl.

Parasites. Intestinal giardiasis may sometimes cause Cbl malabsorption. The mechanism responsible for this malabsorption is perhaps similar to that of bacterial proliferation in the small intestine. Elimination of the parasite restores absorption to normal [51]. Anemia caused by *Diphyllobothrium latum* is rarely observed outside of Finland. Cbl deficiency occurring as the result of malabsorption of vitamin B_{12} is observed in 50% of carriers of this parasite. The mechanism of the malabsorption is not identical in all patients and might involve decreased secretion of IF as well as decreased uptake of Cbl-IF complexes in the ileum.

This type of malabsorption is reversible after treatment to eliminate the fish tapeworm.

Other Sources

Tropical Sprue. More than two-thirds of patients with tropical sprue exhibit Cbl malabsorption. The mechanism responsible is still not entirely clear [57]. The role of intraluminal bacteria appears to be a determining factor because the malabsorption is improved 24–72 h after orally administered antibiotic therapy. How bacterial contamination causes malabsorption is not known. Stagnant material from intestinal loops does not appear to modify Cbl-IF complexes. The bacteria may use the Cbl or bind it to their walls. In late-onset cases, malabsorption is sometimes corrected only following lengthy antibiotic treatment, which probably is evidence for the dominant role of changes in the ileal mucosa in this type of malabsorption.

Disorders of the Ileum and Ileal Resection. Cbl malabsorption has been noted during Crohn's disease, following the intestinal bypass procedure to treat severe obesity, in celiac disease, and following resection of the ileum [69]. The length of ileal resection necessary to cause malabsorption varies from one individual to another; it is a constant finding when resection involves more than 60 cm of the ileum [68].

Drugs. Pharmaceutical agents known to interfere with Cbl absorption include: colchicine, para-aminosalicylic acid, metformin, neomycin, cholestyramine, and H_2 (histamine) receptor antagonists. The mechanism responsible for malabsorption has not been completely elucidated for each drug [44].

Colchicine hinders Cbl absorption by a direct action on intestinal epithelium. In animals, a decrease in the number of IF receptors has been demonstrated following colchicine administration. A similar mechanism has been observed in patients receiving treatment with para-aminosalicylic acid or biguanides. Decreased enzymatic activity of brush border enterocytes has been observed following administration of metformin. IF uptake by the ion-exchange resin might explain Cbl malabsorption observed during cholestyramine therapy. H_2 (histamine) receptor antagonists can induce Cbl-bound protein malabsorption [7, 58, 65, 67]. We observed malabsorption of Cbl bound to chicken serum proteins in patients treated with ranitidine, while absorption of unbound Cbl was normal. This type of malabsorption was rapidly reversed when treatment was discontinued (Fig. 1). It was not related to a decrease in IF secretion but to the potent antacid secretion agent ranitidine, which prevented Cbl from being released from its protein carrier. Cbl deficiency is improbable in patients treated with H_2 receptor antagonists since the duration of this treatment is never long enough to exhaust body reserves of Cbl.

Alcohol. The consumption of alcohol in chronic alcoholic patients can lead to Cbl malabsorption [38]. Functional abnormalities in the intestine due to the toxic

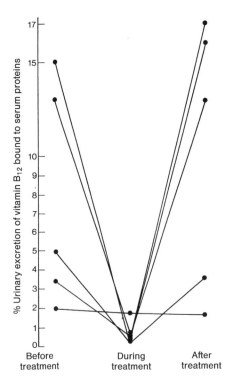

Fig. 1. Effect of ranitidine administration (300 mg/per day) on urinary excretion of cobalamin bound to chicken serum proteins in five patients with duodenal ulcer. (From [7])

effect of alcohol probably explain this malabsorption, as do anomalies in ultrastructure demonstrated in the ileum, dilatation of the endoplasmic reticulum, and cytoplasmic degeneration.

Diverse Causes. Cbl malabsorption, occasionally observed in the course of the Zollinger-Ellison syndrome has been ascribed to excessive acid secretion. The latter would decrease Cbl affinity in the intestine for IF; it would prevent the activation of pancreatic proteases [46] and inhibit binding of the Cbl-IF complex to the ileal receptor. Cbl malabsorption attributed to folate deficiency [60] seems to appear only when there is a disorder of the ileum or with excessive alcohol consumption [16].

Diagnostic Approach to Determine the Cause of Cobalamin Malabsorption

The three most frequent causes of Cbl malabsorption in adults are pernicious anemia, ileopathy or ileal resection, and bacterial proliferation in the small intestine. The Schilling test makes it possible for the physician to differentiate among these. This test is first performed without IF and then 1 week later with IF, thereby distinguishing gastric malabsorption from intestinal types of malabsorp-

Table 2. Results of Schilling test according to etiology

	Cbl	Cbl + IF	Cbl after antibiotic treatment
Pernicious anemia	−	+	−
Disorder of ileum or ileal resection	−	−	−
Bacterial colonization of small intestine	−	−	+

Cbl, cobalamin; IF, intrinsic factor; −, malabsorption; +, normal absorption.

tion. Table 2 summarizes Schilling test results in these three conditions. In impaired pancreatic function, analysis of Cbl absorption is unnecessary for the diagnosis, in as much as this is rarely responsible for megaloblastic anemia. In pernicious anemia, as noted above, there could be an enterocytic malabsorption explaining why the Schilling test is not corrected by the administration of exogenous IF, and why this malabsorption improves over time following therapy with vitamin B_{12}. In addition to inadequate dietary intake, there are cases of Cbl deficiency with normal absorption of free Cbl, in particular in chronic gastritis without pernicious anemia and following partial gastrectomy. In these patients, it may be useful in explaining the vitamin deficiency to investigate the absorption of Cbl bound to proteins. In these cases, Cbl bound to chicken serum proteins could be used in the Schilling test.

Acknowledgements. The authors wish to thank Mrs. V. Girard for her assistance in preparing the manuscript.

References

1. Ardeman S, Chanarin I (1965) Stimulation of gastric intrinsic factor. Br Med J 1:1417–1418
2. Allen RH (1975) Human vitamin B_{12} transport proteins. Prog Hematol 9:57–84
3. Allen RH, Seetharam B, Podell ER, Alpers DH (1978) Effect of proteolytic enzymes on the binding of cobalamin to R protein and intrinsic factor. In vitro evidence that a failure to partially degrade R protein is responsible for cobalamin malabsorption in pancreatic insufficiency. J Clin Invest 61:47–54
4. Allen RH, Seetharam B, Allen NC, Podell ER, Alpers DH (1978) Correction of cobalamin malabsorption in pancreatic insufficiency with a cobalamin analogue that binds with high affinity to R protein but not to intrinsic factor. In vivo evidence that a failure to partially degrade R protein is responsible for cobalamin malabsorption in pancreatic insufficiency. J Clin Invest 61:1628–1634
5. Andersen KJ, Lippe von der G (1978) The effect of proteolytic enzyme on the vitamin B_{12} binding proteins of human gastric juice and saliva. Scand J Gastroenterol 14:833–838
6. Arvanitakis C (1978) Functional and morphological abnormalities of the small intestinal mucosa in pernicious anemia. A prospective study. Acta Hepatogastroenterol (Stuttgart) 25:313–318
7. Belaiche J, Zittoun J, Marquet J, Nurit Y, Yvart J (1983) Effet de la ranitidine sur la sécrétion et sur l'absorption de la vitamin B_{12}. Gastroenterol Clin Biol 7:381–384
8. Belaiche J, Zittoun J, Marquet J, Yvart J, Cattan D (1987) In vitro effect of duodenal juice

on R binders cobalamin complexes in subjects with pancreatic insufficiency: correlation with cobalamin absorption. Gut 28:70–74

9. Bellanger J, Zittoun J, Marquet J. Belaiche J, Cattan D (1985) Etude du débit gastrique des ligands R au cours de diverses pathologies gastroduodénales. Gastroenterol Clin Biol 92 bis(res): 15A

10. Bitsch V, Christiansen PM, Faber V, Rodbro PV (1966) Gastric secretory patterns before and after vagotomy. Lancet 1:1288–1291

11. Brugge WR, Goff JS, Allen NC (1980) Development of a dual label Schilling test for pancreatic exocrine function based on the differential absorption of cobalamin bound to intrinsic factor and R protein. Gastroenterology 78:937–949

12. Carmel R, Herbert V (1967) Correctable intestinal effect of vitamin B_{12} absorption in pernicious anemia. Ann Intern Med 67:1201–1207

13. Carmel R, Abramson SB, Renner IG (1983) Characterization of pure human pancreatic juice: cobalamin content, cobalamin binding proteins and activity against human R binders of various secretions. Clin Sci 64:193–205

14. Cattan D, Belaiche J, Yvart et al. (1980) Faut -il utiliser le test de Schilling en double marquage? Gastroenterol Clin Biol 4(1 bis)(res):193 A

15. Cattan D, Belaiche J, Zittoun J, Yvart J, Chagnon JP, Nurit Y (1982) Rôle de la carence en facteur intrinsèque dans la malabsorption de la vitamin B_{12} liée aux protéines dans les achlorhydries. Gastroenterol Biol Clin 6:570–575

16. Cattan D, Belaiche J, Zittoun J, Yvart J (1982) Effect of folate deficiency on vitamin B_{12} absorption. Ann Nutr Metab 26:367–73

17. Chanarin I (1979) The megaloblastic anaemias. Blackwell, Oxford

18. Cooper BA, White JJ (1968) Absence of intrinsic factor from human portal plasma during 57CoB12 absorption in man. Br J Haematol 14:73–78

19. Cotter R, Rothenberg SP, Weiss JP (1979) Dissociation of the intrinsic factor vitamin B_{12} complex by bile: contributing factor to B_{12} malabsorption in pancreatic insufficiency. Scand J Gastroenterol 14:545–550

20. Dawson DW, Sawers AH, Sharma RK (1984) Malabsorption of protein bound vitamin B_{12} Br Med J 288:675–678

21. Deller DJ, Germar H, Witts LJ (1961) Effect of food on absorption of radioactive vitamin B_{12}. Lancet 1:574–577

22. Deren JJ, Arora B, Toskes PP, Hansell J, Sibinga MS (1973) Malabsorption of crystalline vitamin B_{12} in cystic fibrosis. N Engl J Med 288:627–632

23. Doscherholmen A, Swaim WR (1973) Impaired assimilation of egg Co57 vitamin B_{12} in patients with hypochlorhydria and achlorhydria and after gastric resection. Gastroenterology 64:913–919

24. Doscherholmen A, McMahon J, Ripley D (1976) Inhibitory effect of eggs on vitamin B_{12} absorption test. Br J Haematol 33:261–272

25. Doscherholmen A, McMahon J, Ripley D (1978) Vitamin B_{12} assimilation from chicken meat. Am J Clin Nutr 31:825–830

26. Doscherholmen A, McMahon J, Economom P (1981) Vitamin B_{12} absorption from fish Proc Soc Exp Biol Med 167:480–484

27. Ganguli PC, Cullen DR, Irvine WJ (1971) Radioimmunoassay of plasma gastrin in pernicious anemia, achlorhydria and in controls. Lancet 1:155–158

28. Gueant JL, Parmentier Y, Djali M, Bois F, Nicolas JP (1983) Sequestration of crystalline and endogenous cobalamin by R binder down to the distal ileum in exocrine pancreatic dysfunction. Clin Chim Acta 134:95–106

29. Gueant JL, Monin B, Gaucher P, Nicolas JP (1984) Biliary excretion of cobalamin analogues in humans. Digestion 30:151–157

30. Hamborg B, Kittang E, Schjonsby H (1985) The effect of ranitidine on the absorption of food cobalamins. Scand J Gastroenterol 20:756–758

31. Heyssel RM, Bozian RC, Darby WJ, Bell MC (1966) Vitamin B_{12} turnover in man. The assimilation of vitamin B_{12} from natural foodstuffs by man and estimates of minimal daily dietary requirements. Am J Clin Nutr; 18:176–184

32. Hofstad BE, Kittang E, Schjonsby H (1983) The effect of R protein on the absorption of vitamin B_{12} in chronic pancreatitis. Scand J Gastroenterol 18(Suppl 86):26A

33. Hurliman J, Zuber C (1969) Vitamin B_{12} binders in human body fluids. Antigenic and physicochemical characteristics. Clin Exp Immunol 4:125–140

34. Hurleman J, Zuber C (1969) Vitamin B_{12} binders in human body fluids II. Synthesis in vitro. Clin Exp Immunol 4:141–148

35. Kapadia GR, Mathan VI, Baker SI (1976) Free intrinsic factor in the small intestine in man. Gastroenterology 70:704–706

36. Kapadia GR, Schafer DE, Donalson RM, Ebersole ER (1979) Evidence for involvement of cyclic nucleotides in intrinsic factor secretion by isolated rabbit gastric mucosa. J Clin Invest 64:1044–1049

37. King CE, Liebach J, Toskes PP (1979) Clinically significant vitamin B_{12} deficiency secondary to malabsorption of protein-bound vitamin B_{12}. Dig Dis Sci 24:397–401

38. Lindenbaum J, Lieber CS (1976) Effects of chronic ethanol administration in man in the absence of nutritional deficiency. Ann NY Acad Sci 252:228–234

39. Lindenbaum J, Pezzimenti JF, Shea H (1974) Small intestinal function in vitamin B_{12} deficiency. Ann Intern Med 80:326–331

40. Lippe von der G, Andersen KJ, Schjonsby H (1976) Intestinal absorption of vitamin B_{12} in patients with chronic pancreatic insufficiency and the effect of human duodenal juice on the intestinal uptake of vitamin B_{12}. Scand J Gastroenterol 11:689–695

41. Mackenzie IJ, Donaldson RM (1972) Effect of divalent cations and pH on intrinsic factor mediated attachment of vitamin B_{12} to intestinal micro-villous membranes. J Clin Invest 51:2465–2471

42. Mahmud K, Ripley D. Doscherholmen A (1971) Vitamin B_{12} absorption test: their unreliability in postgastrectoma states. JAMA 261:1167–1171

43. Marcoullis G, Grasbeck R (1975) Vitamin B_{12}-binding proteins in human gastric mucosa – general pattern and demonstration of intrinsic factor isoproteins typical of mucosa. Scnad J Clin Lab 35:5–11

44. Marcoullis G, Rothenberg SP (1983) Macromolecules in the assimilation and transport of cobalamins. In: Lindenbaum J (ed) Contemporary issues in clinical nutrition. Churchill Livingston, New York pp 59–119

45. Marcoullis G, Merivuors H, Grasbeck R (1978) Comparative studies on intrinsic factor and cobalophilin in different parts of the gastrointestinal tract of the pig. Biochem J 173:705–712

46. Marcoullis G, Parmentier Y, Nicolas JP, Jimenez M, Gerard P (1980) Cobalamin malabsorption due to non-degradation of R proteins in the human intestine. Inhibited cobalamin absorption in exocrine pancreatic dysfunction. J Clin Invest 66:430–440

47. Marcoullis G, Nicolas JP, Parmentier Y, Jimenez M, Gerard P (1980) A derivative of R-cyanocobalamin binding protein, in the human intestine; a candidate antibacterial molecule. Biochim Biophys Acta 633:289–294

48. Matuchansky C, Rambaud JC, Modigliani R, Bernier JJ (1974) Vitamin B_{12} malabsorption in chronic pancreatitis. Gastroenterology 67:406–407

49. Nicolas JP, Jimenez M, Marcoullis G, Parmentier Y (1981) In vivo evidence that intrinsic factor cobalamin complex traverses the human intestine intact. Biochim Biophys Acta 675:328–333

50. Nicolas JP, Gueant JL, Gaucher P (1983) Malabsorption de la vitamine B_{12} et insuffisance pancréatique externe. Gastroenterol Clin Biol 7:307–314

51. Notis WM (1972) Giardiasis and vitamin B_{12} malabsorption. Gastroenterology 63:417–428

52. Okuda K, Kitazaki T, Takamatsu M (1971) Inactivation of vitamin B_{12} by a binder in rat intestine and the role of intrinsic factor. Digestion 4:35–48

53. Parmentier Y, Marcoullis G, Nicolas JP (1979) The intraluminal transport of vitamin B_{12} and the exovrine pancreatic insufficiency. Proc Soc Biol Med 160:396–400

54. Pathy MS, Kirkman S, Molloy MJ (1979) An evaluation of simultaneously administered free and intrinsic factor bound radioactive cyanocobalamin in the diagnosis of pernicious anaemia in the elderly. J Clin Pathol 32:244–250

55. Rygvold O (1974) Hypovitaminosis B_{12} following partial gastrectomy by the Billroth II method. Scand J Gastroenterol 9(Suppl 29):57–64

56. Salmeron M, Debure A, Rambaud JC (1982) Colonisation bactérienne chronique de l'intestin grêle et malabsorption. Gastroenterol Clin Biol 6:788–799
57. Salmeron M, Bories C, Rambaud JC (1982) Malabsorption tropicale. Gastroenterol Clin Biol 6:892–900
58. Salmon IL, Silvis SE, Doscherholmen A (1982) Effect of cimetidine on the absorption of vitamin B$_{12}$. Scand J Gastroenterol 17:129–131
59. Schtrumpf E, Vatn MH, Hanssen KF, Myren J (1978) A small dose of somatostatin inhipits the pentagastrin-stimulated gastric secretion of acid, pepsin and intrinsic factor in man. Clin Endocrinol (Coxf) 8:391–395
60. Scott RB, Kammer RB, Burger WF, Middletone FG (1968) Reduced absorption of vitamin B$_{12}$ in two patients with folic acid deficiency. Ann Intern Med 1:111–114
61. Sharpe PC, Mills JG, Horton MA, Hunt RH, Vincent SH, Milton-Thompson GJ (1980) Histamine H$_2$ receptors and intrinsic factor secretion. Scand J Gastroenterol 15:377–384
62. Shearman DJC, Finlayson NDC, Murray-Lyon IM, Samson RR, Girdwood RI (1967) Intrinsic factor secretion in response to pentagastrin. Lancet 2:192–194
63. Spector JI Juvenile achlorhydric pernicious anaemia with Ig A deficiency (a family study). JAMA 1974 228:334–336
64. Steinberg WH, Curington W, Toskes PP (1978) Evidence that failure to degrade R-binder is unimportant in the pathogenesis of cobalamin malabsorption in patients with chronic pancreatitis. Clin Res 26(6):794A
65. Steinberg WM, King CE, Toskes PP (1980) Malabsorption of protein-bound cobalamin during cimetidine administration. Dig Dis Sci 25:188–192
66. Streeter A, Duriappah B, Boyle R, O'Neill BJ, Pheils MJ (1974) Malabsorption of vitamin B$_{12}$ after vagotomy. Am J Surg 128:340–343
67. Streeter AM, Goulston KJ, Bathur FA, Hilmer RS, Crane GG, Pheils MT (1982) Cimetidine and malabsorption of cobalamin. Dig Dis Sci 27:13–16
68. Thompson WG, Wrathell E (1977) The relation between ileal resection and vitamin B$_{12}$ absorption. Can J Surg 20:461–464
69. Toskes PP (1980) Current concepts of cobalamin (vitamin B$_{12}$) absorption and malabsorption. J Clin Gastroenterol 2:287–297
70. Toskes PP, Deren JJ (1973) Vitamin B$_{12}$ absorption and malabsorption. Gastroenterology 65:662–683
71. Toskes PP, Hansell J, Cerda, Deren JJ (1971) Vitamin B$_{12}$ malabsorption in chronic pancreatic insufficiency: studies suggesting the presence of a pancreatic "intrinsic factor". N Engl J Med 284:627–632
72. Toskes PP, Deren JJ, Fruiterman J, Conrad M (1973) Specificity of the correction of vitamin B$_{12}$ malabsorption by pancreatic extract and its clinical significance. Gastroenterology 65:199–204
73. Twomey JJ, Jordan PH, Jarrold T, Trubowitz S, Ritz ND, Conn HIO (1969) Syndrome of immunoglobulin deficiency and pernicious anaemia. Am J Med 46:340–350
74. Wangel AG, Callender ST (1965) The effect of gastrin I and II on the secretion of intrinsic factor. Br J Med 1:1409–1411
75. Wintrobe MW (1981) Megaloblastic and non-megaloblastic anaemias. In: Clinical hematology. Lea and Febiger, Philadelphia, pp 559–561

Chapter 6

Current Gastroenterologic Aspects of Pernicious Anemia

D. CATTAN, A. M. ROUCAYROL, and J. BELAÏCHE

It is generally recognized that a patient suffering from atrophic gastritis of the fundic portion of the stomach becomes successively deficient in hydrochloric acid secretion, pepsin, and then intrinsic factor (IF). The atrophic gastritis observed in pernicious anemia causes failure of gastric secretion. The most direct consequences of this process include microbial overgrowth in the stomach (which is generally well tolerated), increased levels of serum gastrin secreted by the antrum (a relatively constant reaction to increased gastric pH when the antrum is spared), and vitamin B_{12} deficiency.

Pernicious anemia preoccupies the gastroenterologist for 4 reasons:

- The frequently deficient levels of B_{12} in serum draw attention to the diagnostic criteria used to characterize this disorder, the diagnosis of which is based on gastrointestinal investigations.
- Studies conducted with radioisotopic techniques, including radioactive vitamin B_{12} bound to different protein carriers, have confirmed the role of gastric acid secretion in the release of dietary vitamin B_{12} and have minimized the direct role of IF deficiency in the malabsorption of vitamin B_{12} from food in patients suffering from pernicious anemia. IF deficiency can seriously hinder reabsorption of vitamin B_{12} from bile.
- The true incidence of fundic adenocarcinoma, which is itself a well-known complication of atrophic gastritis, is currently under discussion.
- Hyperplasia of endocrine cells located in the gastric fundus (the incidence of which has recently been emphasized), microcarcinoidosis, and development of multiple carcinoid tumors of the fundus (which occur more rarely) might be consequences of the trophic influence of elevated serum gastrin.

Diagnostic Criteria for Pernicious Anemia

Pernicious anemia is characterized by a number of constant and a number of variable findings. The former include atrophic gastritis of the fundic region, achlorhydria resistant to stimulation, absent IF secretion (or secretion of less than 200 U/h after stimulation; $N \geqslant 2000$ U/h), and extensive malabsorption of crystalline vitamin B_{12} corrected by addition of IF. Megaloblastic macrocytic anemia characterizes this disease, while its absence, sometimes due to premature vitamin B_{12} administration, may suggest a latent form. Variable findings include:

elevated serum gastrin, serum anti-IF blocking and precipitating antibodies, bound and unbound anti-IF antibodies of gastric juice, antibodies to gastric parietal cells, normal or elevated levels of serum folates, and reduced levels of erythrocyte folates.

None of these findings is specific. Diagnosis of pernicious anemia is based, rather, on a collection of criteria. The number of criteria necessary and sufficient to establish the diagnosis of pernicious anemia cannot be clearly found. It is of interest to examine the elegibility criteria used in different published studies for selecting patients to participate in various investigations of this disorder. Among the 22 reports, with their varying aims, which have been published in the recent years, there have been the following reported findings:

Absence of free gastric acidity following stimulation: 16 cases
Malabsorption of crystalline vitamin B_{12}: 13 cases
Malabsorption of crystalline vitamin B_{12} corrected by administration of IF: 11 cases
Low levels of serum vitamin B_{12}: 14 cases
Low or no secretion of IF: 6 cases
Presence of antibodies to gastric parietal cells: 5 cases
Anti-IF antibodies: 7 cases
Bone marrow megaloblastosis: 7 cases
Macrocytic anemia: 13 cases
Normal levels of serum folates: 2 cases
Positive response to physiological doses of vitamin B_{12}: 8 cases

Combinations of criteria as proposed in certain published reports may be criticized, for example, on the following points: (a) malabsorption of crystalline vitamin B_{12} in combination with antibodies to parietal cells; (b) achlorhydria and megaloblastic macrocytic anemia responding favorably to physiological doses of vitamin B_{12}; and (c) diagnosis of pernicious anemia established in all cases by demonstrating the presence of vitamin B_{12} deficiency, after all other possible causes were eliminated. In actual medical practice these criteria are appropriate, at best, for elderly patients, in whom complicated investigations are impractical and the two-stage Schilling test (without and with IF) impossible. These criteria, however, are unacceptable for scientific studies.

It is currently recognized that only 70% of megaloblastic macrocytic anemias are due to pernicious anemia. Vitamin B_{12} deficiency was noted by routine laboratory assays in 7.7% of a test population of subjects about 60 years old; only 2.5% of these vitamin-deficient subjects were found to have pernicious anemia [84]. In 100 patients in our Department, suffering from various types of anemia and presenting with vitamin B_{12} deficiency determined by serum assays, pernicious anemia was diagnosed, according to strict diagnostic criteria, in 21 cases. In 13 cases, vitamin B_{12} deficiency was due to gastrectomy; in 5 cases to achlorhydria without pernicious anemia; in 13 cases to intestinal disorders; in 8 cases to dietary deficiency; and in 2 cases to drug treatment which interfere with vitamin B_{12} absorption. In 38 cases, the cause of the deficiency could not be determined; most of these were elderly or bed-ridden patients in whom key diagnostic procedures, in particular the Schilling test and gastric aspirates, could not be performed.

In addition, elevated levels of serum gastrin may be observed in all types of atrophic gastritis which spare the antrum; antibodies to gastric parietal cells are noted in 5%–10% of normal subjects and in 16% of women over 65 years of age. On the other hand, serum antibodies to IF have a relatively high specificity [35]. However, they may be observed in normal subjects (1%–5%) and in patients with insulin-dependent diabetes (2%), thyroid disorders without pernicious anemia (2–3%), and muscle diseases [25, 35].

Marked malabsorption of crystalline vitamin B_{12} corrected by the addition of IF is a basic criterion for diagnosis of pernicious anemia. However, this often utilizes the Schilling test with dual radioactive labeling (Dicopac); this test carries low reliability and reproducibility. Interpretation of its results must be made in terms of patient's diuresis, renal function, and any recently administered medications (or an interruption in alcoholism) between the two stages [63]; repetition of the test with and without instrinsic factor when marked vitamin B_{12} deficiency has led to intestinal malabsorption is necessary [64]. Pernicious anemia cannot be diagnosed without marked malabsorption of crystalline vitamin B_{12} corrected by administration of IF [64].

In patients with pernicious anemia, secretion of IF is always less than 200 U/h following stimulation. Subjects with atrophic gastritis and with achlorhydria resistant to stimulation may also show secretion of less than 200 U/h after stimulation and retain good absorption of crystalline vitamin B_{12} [3, 51, 52]. This discrepancy, which we noted in nine subjects, could be due to poor recovery of gastric secretion, despite correct gastric aspiration technique or to the presence in gastric secretion of active components of IF which are unmeasurable. Finally, as we shall see below, elderly subjects suffering from atrophic gastritis with achlorhydria resistant to stimulation may have low serum lelvels of vitamin B_{12} without greatly decreased IF secretion or malabsorption of crystalline vitamin B_{12}.

Table 1 provides the four most stringent combinations of criteria for diagnosing pernicious anemia. The most stringent, given by Chanarin and James [24], includes all constant features and excludes variable aspects, including serum antibodies to IF. Under these conditions, decreased IF secretion confirms the results of the Schilling test. The other three combinations each permit the interchangeability of three diagnostic criteria: (a) data obtained by gastric suction of only decreased IF secretion, (b) serum antibodies to IF, and (c) Schilling tests without or with addition of IF.

The specificity, sensitivity, and predictive value of these combinations are not known. Admission of a patient to a scientific study of pernicious anemia should require constant and reproducible finding of stimulation-resistant achlorhydria and malabsorption of crystalline vitamin B_{12} corrected by administration of IF observed on repeated occasions, regardless of hematological findings. In daily practice, the presence of antibodies to IF together with vitamin B_{12} deficiency, with achlorhydria or malabsorption of vitamin B_{12} corrected by administration of IF, could be sufficient for a positive diagnosis.

Table 1. Diagnostic criteria for pernicious anemia in various studies

Criterion	Study			
	Ganguli et al. [36]	Lindenbaum et al. [64]	Chanarin and James [24]	Carmel and Johnson [17]
Serum vitamin B_{12}	Low	Low	Low	Low
Gastric HCl following stimulation	0	0	0	
Gastric IF following stimulation (U/h)	<200	0	0 or low	0
Type I anti-IF antibodies in serum	+	+		+
Schilling test				
Without IF	−	−	−	− (Repeated)
With IF	+	+ (Performed early)	+	+
Serum folates		Normal or elevated		

Criteria within boxes are exchangeable; other criteria are required for positive diagnosis.
IF, intrinsic factor. Schilling test: − signifies poor absorption of radioactive vitamin B_{12}; + signifies good absorption of radioactive vitamin B_{12}.

Vitamin B_{12} Deficiency, Pernicious Anemia, and Chronic Atrophic Gastritis of the Fundus

A subject suffering from atrophic gastritis of the gastric fundus successively loses hydrochloric acid in gastric juice, then pepsin, and finally IF. An hourly poststimulation flow of IF equal to zero (or less than 200 U/h, such as is encountered in pernicious anemia) signifies a complete loss of gastric secretion. Marked malabsorption of crystalline vitamin B_{12}, a constant finding in pernicious anemia, is a manifestation of this cessation of secretory activity. Subjects with atrophic gastritis of the fundus, stimulation-resistant achlorhydria, and adequate dietary intake of vitamin B_{12} exhibit serum vitamin B_{12} deficiency without showing malabsorption of crystalline vitamin B_{12} [15, 25, 38, 56, 83, 93]. Most frequently, these are elderly subjects, who may have anemia due to dietary folate deficiency, an inflammatory type of anemia, or a primary disorder of erythropoiesis. Some of these patients may develop pernicious anemia, but this is probably uncommon [25, 83]. The vitamin B_{12} deficiency reflected by serum levels in these subjects is probably due to a defect in vitamin B_{12} release from dietary protein carriers, a consequence of the cessation of gastric hydrochloric acid secretion [70]. Indeed, such patients exhibit malabsorption of vitamin B_{12} bound to proteins [20, 56] uncorrected by the administration of IF [20, 56], and in most cases they still have adequate poststimulation IF secretory flow; this explains the good absorption of crystalline vitamin B_{12} evidenced by three of our patients (Table 2). Malabsorp-

Table 2. Characteristics observed in three patients with non pernicious anemia-type gastric achlorhydria with vitamin B_{12} deficiency

	Patient 1	Patient 2	Patient 3
HCl prior to/after stimulation	0/0	0/0	0/0
IF after stimulation (U/h; N > 2000)	1163	429	1277
Vitamin B_{12} (pg/ml; N > 200)	100	90	150
Serum folates/erythrocyte folates (ng/ml; N > 5/250)	11.2/505	8.9/185	5.8/255
Hb (g%)	9.1	13.6	13.4
MCV (μm^3)	94	89.7	98
Bone marrow	Normal	–	Abnormal, erythropoiesis
Schilling test			
Crystalline vitamin B_{12} (N ⩾ 10%)			
Without IF	10%	16%	16.8%
With IF	14%	17%	10%
Protein vitamin B_{12} (chicken; serum proteins N ⩾ 0.71%)			
With IF	0	0	0
Serum anti-IF antibodies	Absent	Absent	Absent

IF, intrinsic factor; MCV, mean corpuscular volume; N, normal.

a Urinary excretion as percentage of ingested radioactivity

tion of vitamin B_{12} bound to proteins in this syndrome has been demonstrated using vitamin B_{12} bound to chicken serum proteins [20, 56, 89]. Justification for the application of such a carrier for clinical investigations has been questioned [57] because of the tight binding and the artificial aspect of this "dietary" protein. Use of such a vector, besides the ease with which it may be prepared, however, is justified by the similarity of results obtained (in terms of good or poor absorption) with true dietary proteins, labeled in vitro or in vivo [20, 28, 32, 47, 56] both in patients suffering from pernicious anemia and in those with achlorhydria not caused by pernicious anemia.

Table 3 presents results obtained by the Schilling test with vitamin B_{12} incorporated into various proteins – in normal subjects, in patients with pernicious anemia, and in those with achlorhydria without pernicious anemia. In normal subjects, levels of urinary excretion of vitamin B_{12} decrease depending on whether the vitamin is added to sheep liver, chicken meat, fish, eggs, ovalbumin, or chicken serum (Table 3). In pernicious anemia, this urinary excretion is uniformly decreased. In subjects with achlorhydria not associated with pernicious anemia, i.e., subjects who well absorb crystalline vitamin B_{12}, levels of excretion of vitamin B_{12} bound to different protein carriers were intermediate between normal levels and those observed in patients with pernicious anemia; they were reduced by 50% for vitamin B_{12} added to chicken meat and to fish, 88% for vitamin B_{12} added to eggs and ovalbumin, and 95% for vitamin B_{12} bound to chicken serum proteins. Differences in this reduction, depending on the type of

Table 3. Percentages of urinary excretion of vitamin B_{12} bound to various protein-containing foods ingested in normal subjects, patients with pernicious anemia, and patients with achlorhydria

Substance (reference)	Normal IF−	Pernicious anemia			Achlorhydria		
		n	IF−	IF+	n	IF−	IF+
Sheep liver [71]	9.1	5	1.8	–	–	–	–
Chicken [30]	6.7	2	0.00–0.03	–	3	2.7–4	–
Fish [32]	6.5	2	0.2–0.03	–	2	3.3–3	–
Eggs [28, 29]	3.3	5	0.03	–	5	0.42	–
Ovalbumin [28, 29, 31]	3.01	5	0.00	0.3–1.1	6	0.37	–
Chicken serum [56]	2.32	–	–	–	5	0.06–0.34	NS
Chicken serum [20]	3.06	13	0.11	0.47	11	0.20	0.32

Results obtained by the Schilling test; IF+, with intrinsic factor; IF−, without intrinsic factor; figures represent means or ranges.

protein carrier employed, are thus inverse to the percent of urinary excretion of vitamin B_{12} bound to these carriers observed in normal subjects. Therefore, a subject with a non pernicious anemia – type achlorhydria should certainly eat fish or chicken rather than eggs, and although no radioisotope investigation has yet demonstrated it, this subject should certainly eat liver, because the rate of absorption of vitamin B_{12} bound to liver proteins (which are easily degraded) is in normal subjects about the same as that of crystalline vitamin B_{12}.

Studies conducted with vitamin B_{12} bound to chicken serum proteins demonstrate, in particular, that administration of IF does not correct malabsorption, either in patients with pernicious anemia or in subjects with achlorhydria from another cause. We observed that levels of urinary excretion of vitamin B_{12} bound to chicken serum proteins following administration of IF were not statistically different in patients with pernicious anemia and in subjects with achlorhydria without pernicious anemia (Table 4; Fig. 1).

Thus, in the light of these studies it is reasonable to assume that the role of IF deficiency in the malabsorption of dietary vitamin B_{12} in patients suffering from pernicious anemia is critical in foods from which vitamin B_{12} is easily separated

Table 4. Schilling test results; percentage urinary excretion of vitamin B_{12} bound to chicken serum proteins

	n	Without intrinsic factor		With intrinsic factor	
		Mean	Range	Mean	Range
Controls	11	3.06	0.71–7.8	–	–
Pernicious anemia	13	0.11	0–1.0	0.47	0.01–1.4
Achlorhydria	11	0.20	0–0.7	0.32	0.00–0.9
Total gastrectomy	4	–	–	0.19	0.0–0.3

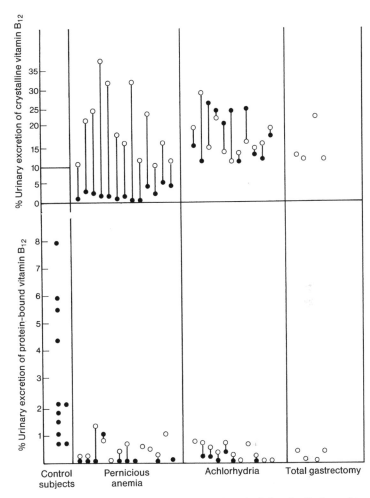

Fig. 1. Percent urinary excretion of crystalline vitamin B_{12} (*upper*) and of vitamin B_{12} bound to chicken serum proteins (*lower*) in control subjects and in patients with pernicious anemia, with achlorhydria, and with total gastrectomy. *Open circles*, with intrinsic factor; *filled circles*, without intrinsic factor

and is less important in foods from which vitamin B_{12} is difficult to separate. The direct role of IF deficiency alone in the malabsorption of dietary vitamin B_{12} in patients with pernicious anemia must be minor for protein carriers which, to release their vitamin B_{12} content, require normal amounts of gastric acidity and secretion.

In normal subjects, an excess of IF is secreted. The daily secretion of IF is approximately 70000 U, while an externally provided source, at a level 140 times less (500 U), is adequate to restore normal absorption of an amount of crystalline vitamin B_{12} equal to daily physiological requirements in patients who have undergone total gastrectomy [25]. Massive secretion of IF explains why it is found in unbound form throughout the small intestine [55], and this certainly results in the reabsorption of vitamin B_{12} from digestive secretions. The role played by the loss of endogenous vitamin B_{12} in the development of vitamin B_{12} deficiency in patients with pernicious anemia, however, is difficult to assess. The respective amounts of vitamin B_{12} and of vitamin B_{12} structural analogues found in bile and in digestive secretions are poorly known; the amount of vitamin B_{12} reabsorbed each day from digestive secretions might be at least equal to the amount supplied in the diet [4, 41, 42, 43, 54].

This would explain accelerated clearance of total amount of vitamin B_{12} in the body following parenteral administration in patients with pernicious anemia, vis-à-vis clearance in normal subjects and in strict vegetarians with the same degree of vitamin B_{12} deficiency as in patients with pernicious anemia [2]. The total cessation of IF secretion which characterizes pernicious anemia would thus lead to severe loss of endogenous vitamin B_{12} every day independently of the type of diet. A less severe decrease in IF secretion, insufficient to result in malabsorption of crystalline vitamin B_{12}, might have a similar effect, although to a lesser degree, in subjects with achlorhydria without pernicious anemia.

Gastric Cancer and Pernicious Anemia

Atrophic gastritis observed in pernicious anemia is considered as a precancerous lesion. Studies conducted some years ago have demonstrated that the risk of developing gastric carcinoma is three to five times higher in patients with pernicious anemia than in a control population, that this cancer is most frequently located to the fundus, and that it develops 7–10 years after the diagnosis of pernicious anemia has been established [7, 69, 97]. A correlation between the geographic prevalence of pernicious anemia and that of gastric cancer has been demonstrated [72]. In ordinary atrophic gastritis of the gastric fundus, with achlorhydria but without pernicious anemia, the risk of onset of gastric cancer is as high as in pernicious anemia, and it is recognized that the precancerous lesion in the former is the atrophic gastritis of the fundus itself [46].

The presence of a factor in the gastric lumen which may be carcinogenic in pernicious anemia has been postulated [90]. A correlation between achlorhydria, proliferation of nitrate-reductase bacteria and formation of nitrites and nitrosamines has been demonstrated [13, 27, 74, 79, 82, 88]. However, in man the carcinogenic action of nitrate compounds in the stomach is still hypothetical. Consumption of smoked meat and other foods that lead to nitrite formation might nonetheless explain the higher incidence of gastric cancer observed in Japan and Scandinavia.

Studies of cell reproduction and differentiation from the gastric mucosa conducted in vitro with [^{3}H]thymidine using cultures of biopsy explants and with [^{3}H]actinomycin using fixated biopsy specimens show a reduction in mitoses of regenerating cells in the neck of gastric glands in atrophic gastritis; this reduction is associated with increased cell turnover and with extension of the regenerative area towards the surface in dysplastic crypts, which are almost completely composed of poorly differentiated cells [26, 96]. This derepression is not observed in glands in which intestinal metaplasia occurs or in hyperplastic polyps [44]. Intestinal metaplasia, the incidence of which in fundic atrophic gastritis in pernicious anemia is estimated at 80% of cases [37], does not appear to be a stage in gastric cancer [91]. On the other hand, slight or, especially serious gastric dysplasia must be considered as a true precancerous lesion, when not already present with cancer [73].

Gastric polyps fall into two categories. Generally, adenomatous polyps account for 10–25% of gastric polyps and must be considered as true precancerous lesions since malignant transformation of these polyps is observed in 40% of cases [67, 73, 76]. On the other hand, hyperplastic polyps (hyperplasiogenic, regenerative, surface epithelial-type polyps), which generally account for 75%–90% of gastric polyps, have a very low or no malignant potential. Most probably, they occur as a result of secondary cell regeneration and do not possess characteristics of a true malignant process [44, 67, 73, 76]. In pernicious anemia, routine endoscopic examinations (Table 5) show that the prevalence of slight or severe dysplasia is about 10%, that of hyperplastic polyps 22 to 42% and that of adenomatous polyps is very low [8, 9, 87].

The fortuitous discovery of gastric polyps by digestive endoscopy should lead the physician to look for pernicious anemia and to monitor these patients, carefully, whatever the type of polyp. Thus, in a patient population of 357 subjects with gastric polyps, only 10% of which were of the adenomatous type. Laxen [61] detected pernicious anemia in 11% of patients, pernicious anemia – like atrophic gastritis of the fundus in 26%, and antibodies to IF in 16%. Among the 357 subjects, 31 gastric cancers were diagnosed, either by the first endoscopic examination or later, which represents 4.9 times the figure expected in a matched control population. In 17 cases (55%), gastric cancer developed in an area distant from polyps. A 5-year follow-up led to the detection of three gastric cancers in 147 patients, all of which were distant from hyperplastic polyps. This was seven times the expected risk.

Table 5. Results of endoscopic studies on pernicious anemia

	Stockbruegger et al. [87]	Borch [8], Borch et al. [9]
Number of patients with pernicious anemia	80	123
Mean age (years)	61.8	68
Duration of disease (years)	4.9	9
Adenocarcinoma	1	4
Dysplasia		
Moderate	6	8/80 (10%)
Severe	3	
Hyperplastic polyps	17	52
With severe dysplasia	1	
Adenomatous polyps	0	1
Carcinoid tumors	1	5

The question currently arising is whether the risk of gastric cancer in subjects with pernicious anemia had not previously been exaggerated, and whether the decreased incidence of gastric cancer observed in Western countries in recent decades also included subjects with pernicious anemia [14, 50, 92]. Many critical comments have been raised concerning studies that demonstrate an increased risk of gastric cancer in pernicious anemia [49, 81]. Diagnostic criteria for this disorder were evidently inadequate before the 1960s and relied upon poorly defined control populations, inadequate or different types of follow-up, disproportionately high proportions of cases in which the diagnoses of pernicious anemia and gastric cancer were made concomitantly, too many patients lost to follow-up, and retrospectively designed studies.

Analysis of published results up to 1977 by Chanarin [25] shows the wide variability of findings published since 1931. Autopsy investigations from 1936 to 1957 demonstrate that the incidence of gastric cancer in patients with pernicious anemia averaged 5.2% (range, 0.7%–12.9%) in 2115 autopsies. Radiological examinations conducted from 1945 to 1949 reveal 366 cases of pernicious anemia, with a mean incidence of cancer equal to 6.6% (range, 3%–20%). Clinical studies from 1931 to 1977 included 10509 subjects suffering from pernicious anemia, with a mean incidence of gastric cancer of 2.2% (range, 0%–12%). Currently, great interest is focused on results of epidemiological studies, including endoscopic examination, the aim of which is to establish the prevalence and incidence of gastric lesions that develop in atrophic gastritis with pernicious anemia.

Results of modern epidemiological studies are still highly variable. Schaffer et al. [81] made a retrospective investigation of 152 subjects with pernicious anemia observed between 1950 and 1979 in Minnesota. These patients represent 1555 years of case-reports. These investigators detected one type of gastric cancer for which 1.02 cases would be expected in a control population matched for age, sex, and geographical factors. This study is important because diagnostic criteria to affirm pernicious anemia were strict, patient matching was correctly made, relatively few patients were lost to follow-up (20 out of 152), and estimation of the number of cancer cases expected was based upon examination of a cancer register which took into account the differences in the incidence of cancer in this geographic area compared to previous years.

Elsborg and Mosbech [33] in 1972 studied the incidence of pernicious anemia over 1 year in a patient population with gastric cancer recorded in a national cancer register in Denmark. Out of 877 cases of gastric cancer, they observed 19 patients with pernicious anemia. The incidence of the latter in patients with gastric cancer (2.2%) was higher than that observed in patients with colon cancer (0.2%); nevertheless, it is possible that the incidence of pernicious anemia in the population of patients with colon cancer studied in this investigation was underestimated [34]. Among 19 cases of gastric cancer with pernicious anemia, 12 patients were over 70 years of age. The investigators estimated that in the same year, the number of patients with pernicious anemia over 70 years old was 3240 in Denmark. The calculated incidence of gastric cancer in pernicious anemia over age 70 years would thus be 3.7 per 1000 versus 1.3 per 1000 in the general population.

Using the same technique, Eriksson et al. [34] studied the prevalence of pernicious anemia in patients with gastric cancer, later verified at autopsy, in a population of 240 000 inhabitants between 1958 and 1976. Among the 917 patients with gastric cancer selected, the investigators detected pernicious anemia in 19. This figure was not statistically different from the number of cases of this disorder detected in a series of control subjects free from gastric cancer and matched for sex and age. The difference became significant only when the patients with pernicious anemia for less than 5 years were excluded: 16 cases of pernicious anemia in the gastric cancer group and 8 cases in the control group. The investigators concluded that the incidence of gastric cancer observed during pernicious anemia was at most twice the normal rate, and that the incidence had been overestimated in previous studies, most probably because of the inadequate nature of the control populations employed.

Only prospective endoscopic studies of subjects with pernicious anemia will be able to determine the prevalence and incidence of gastric cancer in the course of this disorder. The prevalence of gastric lesions observed by endoscopic examination in pernicious anemia was recently provided by Stockbruegger et al. [87] and by Borch et al. [8, 9] (Table 5). The prevalence of gastric cancer was 1.25% in the former series and 3.25 in the latter. This difference may most likely be explained by the younger age of the patients and the shorter duration of pernicious anemia in the former series; the percentage in the latter, which includes endoscopic examination of subjects over 70 years of age, is doubtlessly closer to reality. Moreover, 61 patients in the series of Borch et al. were re-examined later 18–60 months (mean 32) after their first endoscopy. No new gastric lesion was detected during this follow-up [9].

It is possible that future endoscopic examinations will reveal pernicious anemia to be a disorder during which the prevalence of gastric cancer is high but its incidence low. Borch et al. estimate that in Sweden the first endoscopic examination performed routinely in patients under than 75 years of age with pernicious anemia would enable detection of 330 malignant gastric neoplasms [9].

In practice, it now appears that every subject in whom the diagnosis of pernicious anemia is established should have a gastric endoscopy to detect cancer as well as hyperplastic or adenomatous polyps or dysplasia. Repetition of this procedure at a later date should depend on the subject's age and clinical context, and upon results of the first endoscopy. When there are no contraindications to possible surgical procedures, subjects having hyperplastic or adenomatous polyps, even if these have been removed by endoscopic procedure, and those with moderate to severe areas of gastric dysplasia, should be followed up [73]. If no lesion is detected during the first fiberendoscopy, it seems adequate to repeat this procedure every 3 years.

Multiple Carcinoid Tumors of the Gastric Fundus and Pernicious Anemia

The combination of carcinoid tumors of the fundus and atrophic gastritis of this area of the stomach, initially reported in 1952 [66], was described in 1955 [23] in patients with pernicious anemia. Current interest in carcinoid tumors may be a result of their resembling carcinoid tumors in rats receiving high doses of omeprazole for a prolonged period; this agent is a potent inhibitor of gastric acid secretion. Common features of these two types of carcinoid tumors include their location to the fundus, their enterochromaffin – like cell (ECL) cell, the presence of lesional continuity between carcinoid tumors, microcarcinoidosis and nontumoral hyperplasia of ECL cells, achlorhydria, and elevated levels of serum gastrin [59].

Routine gastric endoscopic studies have determined the prevalence of carcinoid tumors in patients with pernicious anemia: 5 cases out of 123 in the series of Borch [8], 3 cases out of 44 with Lehtola et al. [62], and 1 in 80 cases with Stockbruegger et al. [87]. According to Borch et al., the prevalence of gastric carcinoid tumors in pernicious anemia (4.1%) is as high as that of adenocarcinoma (3.2%). In a population of subjects in Sweden with pernicious anemia estimated at 16 000, 660 might have gastric carcinoids tumors and 128 metastasized gastric carcinoid tumors [9]. This relatively high prevalence of carcinoid tumors in patients with pernicious anemia is in contrast to their rare occurrence in the general population. Following 11 000 gastroscopies conducted by Lehtola et al. over a 10-year period, only one case of carcinoid tumor was detected [62]. Traditionally, gastric carcinoid tumors account for only 2%–4% of digestive carcinoid tumors and 0.3% of gastric tumors [39, 60, 65]. In the opinion of Rogers and Murphy [75], however, 7% of malignant gastric tumors classified as adenocarcinomas should actually be considered as carcinoid tumors or mixed tumors. Many gastric cancers associated with pernicious anemia, as in the case with carcinoid tumors, are located in the gastric fundus and at multiple sites. It is possible that the true incidence of carcinoid tumors in the course of chronic gastritis, and particularly in pernicious anemia, have been hitherto underestimated due to lack of histological staining of gastric endocrine cells.

Whatever the true incidence of gastric carcinoid tumors may be, it is currently recognized that more than half of cases develop in the course of atrophic gastritis of the fundic region, with or without pernicious anemia. In 42 cases of gastric carcinoid tumors observed by Wilander et al. [95], 24 were situated in the fundus, including 6 with pernicious anemia and 3 with achlorhydria. Carney et al. [18] classified 30 cases of gastric carcinoids according to anatomical site, number, and condition of the underlying mucosa: 6 cases involved carcinoid tumors in a normal fundic mucosa, 6 cases antral carcinoids in a normal mucosa, 2 cases fundic carcinoids combined with Zollinger-Ellison syndrome, and 16 cases involved multiple fundic carcinoids in the fundic mucosa during atrophic gastritis, 12 of which occurred in patients with pernicious anemia.

Among the 55 cases of gastric carcinoids tumors with atrophic gastritis recorded in the literature at the end of 1985, 29 involved subjects with pernicious anemia tumor. The patient's age at the time when pernicious anemia was diagnosed was on average 10 years lower than the usual age, and the diagnosis of pernicious anemia preceded that of carcinoid tumors by an average of 10 years. The majority of cases were in women; carcinoid tumors were latent; and most were detected during routine fiberendoscopies or at autopsy.

A pseudo-peptic ulcer clinical picture rarely occurs [18]. In a few cases, carci-

noid tumors were revealed by iron-deficiency anemia [19]. These tumors, located
to the gastric fundus, were polypoid, sessile or pedunculate, and rarely ulcer-
ated; their size ranged from 2 mm to 60 mm, but most were less than 10 mm in
diameter. They were likely to be multiple, with a patient reported by Goldman et
al. who had 59 tumors [40]. These were trabecular-type tumors, composed of
small cells, with a clear and regular nucleus, not mucus-secreting, but argyrophilic and not argentaffin. Infiltration of the tumors extended to the lamina pro-
pria and to the submucosa, rarely to the muscularis mucosae. Metastatic adeno-
pathies and hepatic metastases were noted in 23% of cases.

These carcinoid tumors were based in severe atrophic gastritis of the fundus.
Within this atrophic gastritis, hyperplasia of argyrophilic cells was observed,
sometimes endotubular, forming crescents within the tubes, and at other times
extratubular, forming clusters of cells located deep in the mucosa, above the
muscularis mucosae. When these clusters had a diameter greater than or equal to
that of an adjacent tube, they formed microcarcinoids, and sometimes microcarcinoidosis extended to the whole fundus [62, 80] (Fig. 2). The antrum, generally
normal, was the site of hyperplasia of gastrin-secreting cells [18].

The latent and poorly invasive characteristics of these carcinoid tumors paral-
leled the remarkably slow development; death occurred from causes other than
carcinoid tumors [18].

The principal characteristic of these carcinoid tumors was to present with very

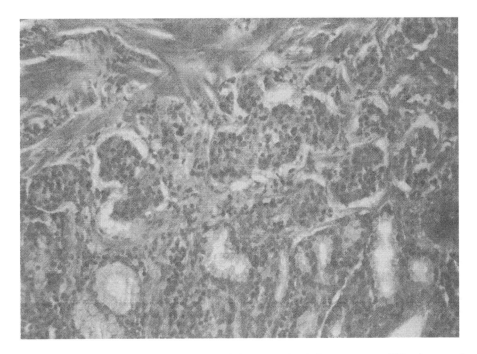

Fig. 2. Microcarcinoidosis (by Grimelius staining procedure). Larger argyrophilic clusters of
cells, often confluent ($\times 160$)

high serum gastrin levels. Hypergastrinemia was from gastric origin, as evidenced by its regression following total gastrectomy [40, 58, 80], and antral, as demonstrated by hyperplasia of G cells in the antrum and the lack of fixation of antigastrin serum to the fundic mucosa, carcinoid tumors, and fundic endocrine hyperplasia. Secretion by carcinoid tumors is unclear. Immunohistochemical studies carried out with 12 antisera in 31 cases of carcinoid tumor with atrophic gastritis were negative [1, 8, 18, 40, 53, 58, 68, 80, 94]. Serum and urinary peptide levels or hormonal-type compounds were generally normal, in particular regarding serotonin, its metabolites, and histamine.

It is recognized that these carcinoid tumors develop from ECL cells [8, 16, 18, 48, 58, 80]. These argyrophilic (and non-argentaffin) cells are characterized among the gastric argyrophilic endocrine cells by a lack of known secretion and by their ultrastructure. Electron microscopy demonstrates round, densely centered endosecretory granules in the cytoplasm, delineated by a membrane; some of these are peculiar due to the presence of a very large, clear halo, often lateral and irregularly shaped, which confers a vacuolar aspect.

Currently, these carcinoid tumors of the gastric fundus developed in atrophic gastritis are considered as the consequence of endocrine hyperplasia of the fundus. This hyperplasia is a virtually constant event; it was observed in 14 out of 16 cases by Carney et al. [18], in 5 out of 5 by Borch [8], and in 9 out of 11 in various other published studies. Using standard histological techniques, histochemical staining, and electron microscopy, these hyperplastic endocrine cells have been shown to be identical to carcinoid tumor cells; immunohistochemical studies that were carried out were negative with all the antisera tested [6, 8, 18, 68, 80]. Thus, according to Bordi et al. [11, 12], it is acknowledged that this endocrine hyperplasia is formed by ECL cells.

In rats, ECL cells constitute 66% of gastric endocrine cells; they secrete and store histamine and contain histidine decarboxylase; their histidine decarboxylase activity is regulated by gastrin [5, 45]. All conditions which produce increased serum gastrin lead to ECL cell hyperplasia. In rats treated with high doses of omeprazole for prolonged periods, there is a lesional continuity between isolated carcinoid tumors, microcarcinoidosis, and ECL cell hyperplasia. In this model, there is a correlation between levels of circulating gastrin, on the one hand, and levels of histamine and histidine decarboxylase in the fundic mucosa, on the other hand [59].

In normal humans, ECL cells of the gastric fundus have a lower density than those of the rat. They do not contain histamine. ECL cell hyperplasia combined with high serum gastrin, observed in the syndrome of gastric carcinoid tumor in combination with atrophic gastritis, has also been observed in the Zollinger-Ellison syndrome and in the course of atrophic gastritis of the fundus with or without pernicious anemia [10, 11, 78, 86]. In pernicious anemia, it is recognized that diffuse hyperplasia of fundic argyrophilic cells, composed of ECL cells, is observed in 69% of cases and that nodular hyperplasia is observed in 13% of cases [8]. The hypothesis that a correlation exists between this hyperplasia and increased serum gastrin level has been proposed [10, 11, 78]. In a series of 30 patients, with serious atrophic gastritis and achlorhydria resistant to pentagastrin stimulation, including 20 subjects with pernicious anemia, we investigated correlations between fundic endocrine hyperplasia, on the one hand, and hourly output of IF following stimulation, serum levels of vitamin B_{12}, absorption of crystalline vitamin B_{12} with the Schilling test, presence of antibodies to IF, and serum levels of gastrin, on the other hand. Of all the possible histological and secretory parameters to describe atrophic gastritis, only increased serum gastrin was correlated with the presence of argyrophilic hyperplasia of the gastric fundus [77]. An increase in histamine and in histidine decarboxylase activity of the fundic mucosa was observed only in atrophic gastritis with achlorhydria, increased serum gastrin, and ECL cell hyperplasia [21, 22].

Finally, carcinoid tumors which develop as a complication of atrophic gastritis of the fundus in man, especially in pernicious anemia, appear as the final outcome of endocrine hyperplasia of the fundus, most probably due to the trophic action of gastrin.

These recent advances should have practical value for the monitoring subjects with pernicious anemia. The frequency of endoscopic examinations should depend on the presence of severe endocrine hyperplasia observed after Grimelius staining procedure and/or high serum gastrin levels. The proper therapeutic management of carcinoid tumors of the fundus during pernicious anemia is still under discussion. Given their slow development, their poorly invasive nature, and the relatively rare occurrence of metastases, many clinicians recommend endoscopic excision of these tumors, followed by regular check-ups when they are small and few in number, and surgical excision, or even total gastrectomy, in those cases in which tumors are large (over 1.5 cm), numerous, and combined with severe fundic hyperplasia and with high serum gastrin levels [9].

Because of the painless course of even advanced carcinoids, others investigators feel that agressive surgical resection should be strongly considered [48]. Finally, if these tumors are truly gastrin-induced and gastrin-dependent, endoscopic excision might ideally be supplemented by antrectomy, which would render the mutilation of total gastrectomy unnecessary. However, the outcome of endocrine hyperplasia in the portion of the stomach remaining after antrectomy is still uncertain.

References

1. Alumets J, Sundler F, Falkmer J, Jungberg O, Hakanson R, Martensson H et al. (1983) Neurohormonal peptides in endocrine tumors of the pancreas, stomach and upper small intestine. Immunohistochemical study of 24 cases. Ultrastruct Pathol 5:55–72
2. Amin S, Spinks T, Ranicar A, Short MD, Hoffbrand AB (1980) Long-term clearance of (57 Co)-cyanocobalamin in vegans and pernicious anaemia. Clin Sci 58:101–103
3. Ardeman S, Chanarin I (1986) Intrinsic factor secretion in gastric atrophy. Gut 7:99–101
4. Ardeman S, Chanarin I, Berry V (1965) Studies on human gastric intrinsic factor. Observations on its possible absorption and enterohepatic circulation. Br J Haematol 11:11–14
5. Aures D, Akanson R, Schauer A (1968) Histidine decarboxylase and dopa decarboxylase in the rat stomach. Properties and cellular localization. Eur. J. Pharmacol 3:217–234
6. Black WC, Haffner HE (1968) Diffuse hyperplasia of gastric argyrophil cells and multiple carcinoid tumors. Cancer 21:1080–1099
7. Blackburn EK, Callender ST, Dacie JV et al (1968) Possible association between anemia

and leukemia: a prospective study of 1,625 patients with a note on the very high incidence of stomach cancer. Int. J. Cancer 3:163–170

8. Borch K (1986) Epidemiologic, clinicopathologic and economic aspects of gastroscopic screening of patients with pernicious anemia. Scand J. Gastroenterol 21:21–30

9. Borch K, Renvall H, Liedberg G (1985) Gastric endocrine cell hyperplasia and carcinoid tumors in pernicious anemia. Gastroenterology 88:638–648

10. Bordi C, Costa A, Missale G (1975) ECL cell proliferation and gastrin levels. Gastroenterology 68:205–206

11. Bordi C, Gabrielli M, Missale G (1978) Pathological changes of endocrine cells in chronic atrophic gastritis. An ultrastructural study on peroral gastric biopsy specimens. Arch. Pathol. Lab. Med 102:129–135

12. Bordi C, Ravazzola M, De Vita O (1983) Pathology of endocrine cells in gastric mucosa. Ann. Pathol 3:19–28

13. Boriello SP, Reed PJ, Dolby JM, Barclay FE, Webster ADB (1985) Microbiologic and metabolic profile of achlorhydric stomach: comparison of pernicious anaemia and hypogammaglobulinaemai. J. Clin. Pathol 38:946–953

14. Brunet M, Berlie J, Janin ML, Zimmerman B, Hucher M, Ducourneau R (1979) Mortalité par cancer de l'estomac de 1954 à 1974. Nouv. Presse Méd. 21:1743–1744

15. Callender ST, Retief FP, Witts LJ (1960) The augmented histamine test with special reference to achlorhydria. Gut 1:326–336

16. Capella C, Polak JM, Timson CM, Frigerio B, Solcia E (1980) Gastric carcinoids of argyrophil cells. Ultrastruct Pathol 1:411–418

17. Carmel R, Johnson CS (1978) Racial patterns in pernicious anemia. N. Engl. J. Med 298:647–650

18. Carney JA, Go VL, Fairbanks VF, Moore SB, Alport EC, Nora FE (1983) The syndrome of gastric argyrophil carcinoid tumors and non-antral gastric atrophy. Ann. Intern Med 99:761–766

19. Caruso MP, Beardwoodd, Maxwell R, Wagner C (1984) Gastric atrophy and carcinoid tumors. Ann. Intern Med 100:459

20. Cattan D, Belaïche J, Zittoun J. Yvart J, Chagnon P, Nurit Y (1982) Rôle de la carence en facteur intrinsèque dans la malabsorption de la vitamine B_{12} liée aux protéines dans les achlorhydries. Gastroentérol. Clin Biol 6:570–575

21. Cattan D, Roucayrol AM, Launay JM, Callebert J (1986) Histamine et activité histidine décarboxylase de la muqueuse fundique dans les gastrites atrophiques fundiques avec et sans hyperplasie des cellules ECL. Gastroentérol. Clin Biol 10(2bis):138A

22. Cattan D, Roucayrol AM, Launay JM, Callebert T (1989) Circulating gastrin endocrine cells hystamine content and histidine decarboxylase activity in atrophic gastritis. Gastroenterology, to be published.

23. Cattan R, Nallet J, Libeskind M, Habib R, Pariente P (1955) Carcinoide de l'estomac et anémie de Biermer. Arch. Mal. Appar Dig. 1955; 44:922–8

24. Charnarin I, James D (1974) Humoral and cell-mediated intrinsic factor. Lancet I:1078–1083

25. Chanarin I (1979) Megaloblastic anaemia, 2nd edn. Blackwell, Oxford

26. Deschner EE, Winawer SJ, Lipkin M (1972) Patterns of nucleic acid and protein synthesis in normal human gastric mucosa and atrophic gastritis. J Natl. Cancer Inst 48:1567–1574

27. Dolby JM, Webster ADB, Boriello SP, Barclay FE, Bartholomew BA, Hill MJ (1984) Bacterial colonization and nitrite concentration in the achlorhydric stomachs of patients with primary hypogammaglobulinaemia or classical pernicious anaemia. Scand. J. Gastroenterol 19:105–110

28. Doscherholmen A, Swaim WR (1973) Impaired assimilation of egg 57 Co-vitamin B_{12} in patients with hypochlorhydria and achlorhydria, and after gastric resection. Gastroenterology 64:913–919

29. Doscherholmen A, McMahon J, Ripley D (1976) Inhibitory effect of eggs on vitamin B_{12} absorption: description of a simple ovalbumin 57 Co-vitamin B_{12} absorption test. Br. J. Haematol 33:261–272

30. Doscherholmen A, McMahon J, Ripley D (1978) Vitamin B_{12} assimilation from chicken meat. Am. J. Clin. Nutr 31:825–830

31. Doscherholmen A, McMahon J, Ripley D (1978) Physiologic role of gastric acid and pepsin in the assimilation of vitamin B_{12} from food. Clin. Res. 26:661A
32. Doscherholmen A, McMahon J, Economon P (1981) Vitamin B_{12} absorption from fish. Proc. Soc. Exp. Biol. Med 167:480–484
33. Elsborg L, Mosbech J (1979) Pernicious anaemia as a risk factor in gastric cancer. Acta. Med Scand 208:315–318
34. Eriksson S, Clase L, Moquist-Olsson I (1981) Pernicious anaemia as a risk factor in gastric cancer. Acta. Med. Scand 210:481–484
35. Fairbanks VF, Lennon VA, Kokmen E, Howard FM (1983) Test for pernicious anemia: serum intrinsic factor blocking antibody. Mayo Clin. Proc. 58:203–204
36. Ganguli PC, Cullen DR, Irvine NJ (1971) Radioimmunoassay of plasma gastrin in pernicious anemia achlorhydria, without pernicious anemia-hypochlorhydria and in controls. Lancet I:155–158
37. Glass GB, Pirchumoni CS (1975) Atrophic gastritis. Hum. Pathol 6:219–250
38. Glass GB, Speer FD, Nierburgs HE et al. (1960) Gastric atrophy, atrophic gastritis and gastric secretory failure. Correlative study by suction biopsy and exfoliative cytology of gastric mucosa, paper electrophoretic and secretory assay of gastric secretion and measurements of intestinal absorption and blood levels of vitamin B_{12}. Gastroenterology 39:429–453
39. Godwin J (1975) Carcinoid tumors. An analysis of 2,837 cases. Cancer 36:560–569
40. Goldman H, French S, Burbridge E (1981) Cell hyperplasia and multiple metastasizing carcinoids of the stomach. Cancer 47:2620–2626
41. Grasbeck R, Nyberg W, Reinzenstein P (1958) Biliary and fecal vitamin B_{12} excretion in man. An isotope study. Proc. Soc. Exp. Biol. Med 97:780–784
42. Green R, Jacobsen DW, Van Tonder SV, Kew MC, Metz J (1982) Enterohepatic circulation of cobalamin in the non-human primate. Gastroenterology 61:773–776
43. Gueant JL, Monin B, Gaucher P, Nicolas JP (1984) Biliary excretion of cobalamin and cobalamin analogues in the human. Digestion 30:151–157
44. Haentjens P. Williams G (1984) Precancerous lesions in the stomach. Acta. Clin. Belg 84:277–282
45. Hakanson R, Larsson JI, Liedberg G (1976) Effects of antrectomy and porto-caval shunting on the histamine-storing endocrine-like cells in oxyntic mucosa of the rat stomach. J Physiol (Lond) 259:785–800
46. Hastrup-Svendsen J, Dahl C, Bosvendsen L, Christiansen (1986) Gastric cancer risk in achlorhydric patients. A long term follow-up study. Scand. J. Gastroenterol 21:16–20
47. Heysell RM, Bozian RC, Darby WJ, Bell MC (1966) Vitamin B_{12} turnover in man. The assimilation of vitamin B_{12} from natural foodstuff by man and estimates of minimal daily dietary requirements. Am. J. Clin. Nutr. 18:176–184
48. Hodges JR, Isaacson P, Wright R (1981) Diffuse enterochromaffin-like (ECL) cell hyperplasia and multiple gastric carcinoids: a complication of pernicious anemia. Gut 22:237–241
49. Hoffman NR (1970) The relationship between pernicious anemia and cancer of the stomach. Geriatrics 25:90–95
50. Imai T, Kubo T, Watanabe H (1971) Chronic gastritis in Japanese with reference to high incidence of gastric carcinoma. J. Natl. Cancer. Inst. 47:179–195
51. Irvine WJ, Davies SH, Haynes RC (1965) Secretion of intrinsic factor in response to histamine and to gastrin in the diagnosis of addisonian pernicious anaemia. Lancet I:397–401
52. Irvine WJ, Cullen DR, Scarth L, Simpson JD (1968) Intrinsic factor secretion assessed by direct radioimmunoassay and by total body counting in patients with achlorhydria and in acid secretors. Lancet I:184–188
53. Itsumo M, Watanabe H, Iwafuchi M, Yanahihara N, Ito S, Uwo J (1983) Multiple gastric carcinoids occurring in atrophic fundic mucosa. Gan No Rinsho 29:915–922
54. Kanazawa S, Herbert V, Herzlich B, Brivas G, Manusselis C (1983) Removal of cobalamin analogue in bile by enterohepatic circulation of vitamin B_{12}. Lancet I:707–708
55. Kapadia CR, Mathan VI, Baker SJ (1976) Free intrinsic factor in the small intestine in man. Gastroenterology 70:704–706
56. King CE, Liebach J, Toskes PP (1979) Clinically significant vitamin B_{12} deficiency secondary to malabsorption of protein-bound vitamin B_{12} Dig. Dis. Sci 24:397–402

57. Kittang E, Hamborg B, Schonsby H (1985) Absorption of food cobalamin assessed by the double isotope method in healthy volunteers and in patients with chronic diarrhea. Scand. J. Gastroenterol 20:500–507

58. Larsson LI, Rehfeld JF, Stockbruegger R et al (1978) Mixed endocrine gastric tumors associated with hypergastrinemia of antral origin. Am. J. Pathol 93:53–68

59. Larsson H, Carlsson E, Mattson H et al (1986) Plasma gastrin enterochromaffin-like cell activation and proliferation: studies with omeprazole and ranitidine in intact and antrectomized rats. Gastroenterology 90:391–399

60. Lattes R, Grossi C (1956) Carcinoid tumors of the stomach. Cancer 9:698–711

61. Laxen F (1984) Gastric carcinoma and pernicious anemia in long-term endoscopic follow-up of subjects with gastric polyps. Scand. J Gastroenterol 19:515–540

62. Lehtola J, Karttunen T, Krekela I, Niemela S, Rasanen O (1985) Gastric carcinoids with minimal or no macroscopic lesion in patients with pernicious anemia. Hepatogastroenterology 32:72–76

63. Lindenbaum J (1983) Status of laboratory testing in the diagnosis of megaloblastic anemia. Blood 61:624–627

64. Lindenbaum J, Pezzimenti JF, Shea N (1974) Small intestinal function in vitamin B₁₂ deficiency. An Intern Med 80:326–331

65. MacDonald RA (1956) Study of 356 carcinoids of the gastrointestinal tract: report of four cases of the carcinoid syndrome. Am J. Med. 21:867–878

66. Martin JD, Atkins EL (1952) Carcinoid of the stomach. Surgery 27:698–704

67. Ming SC (1977) The classification and significance of gastric polyps. In: Yardley JH, Morson BS, Abell MD (eds) Intenational Academy of Pathology Monographs Williams and Wilkins Baltimore, pp 149–175

68. Morgan JE, Kaiser CW, Johnson W et al. (1983) Gastric carcinoid (gastrinoma) associated with achlorhydria (pernicious anemia). Cancer 51:2332–2340

69. Mosbech J, Videbaek A (1950) Mortality from and risk of gastric carcinoma among patients with pernicious anemia. Br J 2:390–394

70. Nicolas JP (1982) Le protocole classique du double test de Schilling, est-il toujours bien adapté au diagnostic des malabsorptions intestinales de la vitamine B₁₂? Importance de la secrétion chlorhydro-peptidique de la forme d'administration de la vitamine B₁₂ Gastroentérol. Clin. Biol 6:554–556

71. Okuda K, Takara I, Fujii T (1968) Absorption of liver-bound vitamin B₁₂ in relation to intrinsic factor. Blood 32:313–323

72. Payne RW (1961) Pernicious anemia and gastric cancer in England and Wales. Br. Med. J 1807–1808

73. Potet E, Camilleri JP (1982) Populations à haut risque et dysplasies précancereuses de l'estomac, définition et attitude pratique. Gastroentérol. Clin. Biol 6:454–461

74. Reed PI, Smith PLR, Haines K, House FR, Walters CL (1981) Gastric juice. N-nitrosamines in health and gastroduodenal disease. Lancet II:550–552

75. Rogers LW, Murphy RC (1979) Gastric carcinoid and gastric carcinoma morphological correlates of survival. Am J pathol 3:195–202

76. Rotterdam H, Sommer SC (1981) Biopsy diagnosis of the digestive tract. Raven, New York. pp 96–117

77. Roucayrol AM, Cattan D (1985) Corrélation entre l'hypergastrinémie et l'hyperplasie des cellules endocrines argyrophiles non-gastrino-secrétantes du fundus gastrique au cours des gastrites chroniques atrophiques. Gastroentérol Clin Biol 9(2 bis):28A

78. Rubin W (1969) Proliferation of endocrine-like (enterochromaffin) cells in atrophic gastric mucosa. Gastroenterology 57:641–649

79. Rudell WS, Bone ES, Hill M, Walters CL (1978) Pathogenesis of gastric cancer in pernicious anemia. Lancet I:521–523

80. Saint-Andre JP, D'Aubigny N, Ben Gouali A, Goethals G, Simard C (1983) Gastrite chronique atrophique avec tumeur carcinoïde et "microcarcinoïdes". An Pathol 3:235–240

81. Schafer LW, Larson DE, Melton LJ, Higgins JA, Zinsmeister AR (1985) Risk of development of gastric carcinoma in patients with PA; a population-based study in Rochester. Mayo Clin Proc 60:444–448

82. Schlag P, Bockler R, Peter M (1982) Nitrite and nitrosamines in gastric juice: risk factors for gastric cancer. Scand J Gastroenterol 17:145–150

83. Siurala M, Eramaa E, Nyberg W (1960) pernicious anemia and atrophic gastritis Acta Med Scand 166:213–223

84. Slingerland DW, Cardarelli JA, Belton AB, Miller A (1984) The utility of serum gastrin levels in assessing the significance of low serum vitamin B_{12} levels. Arch. Intern. Med 144:1167–1168

85. Solcia E, Capella C, Vasallo G (1971) Endocrine cells of the stomach and pancreas in states of gastric hypersecretion. Rendic. Gastroentereology 2:147–158

86. Solcia E, Capella C, Vassalo G, Buffa R (1975) Endocrine cells of the gastric mucosa. Int Rev. Cytol 42:223–286

87. Stockbruegger RW, Menon GG, Beilby JOW, Mason RR, Cotton PB (1983) Gastroscopic screening in 80 patients with pernicious anemia. Gut 24:1141–1147

88. Stockbruegger RW, Cotton PB, Menon GG et al. (1984) Pernicious anemia, intragastric bacterial overgrowth and possible consequences. Scand J Gastroenterol 19:355–364

89. Streeter AM, Shum HY, Duncombe VM, Hewson JW, Thrope MEC (1976) Vitamin B_{12} malabsorption associated with a normal Schilling test result. Med J Aust 1:54–55

90. Tannenbaum SR (1983) N-nitrosocompounds: a perspective on human exposure. Lancet I:629–631

91. Walker IR, Strickland RG, Ungar B, Mackay IR (1971) Simple atrophic gastritis and gastric carcinoma. Gut 12:906–911

92. Waterhouse J, Muir C, Shanmuguratnam K, Powell J (1982) Cancer incidence in four continents, vol IV. International Agency for Research on Cancer Lyons IARC scientific publication no 42

93. Whiteside MG, Mollin DL, Cognill NF, Williams AW, Anderson B (1964) The absorption of radioactive vitamin B_{12} and the secretion of hydrochloric acid in patients with atrophic gastritis. Gut 5:385–399

94. Wilander E, Grimelius L, Lundquist G, Skoog V (1979) Polypeptide hormones in argentaffin and argyrophil gastroduodenal endocrine tumors. Am J Pathol 96:519–530

95. Wilander E, El-Salhy M, Pitakanen P (1984) Histopathology of gastric carcinoids: a survey of 42 cases. Histopathology 8:183–193

96. Willems G (1979) Cell population kinetics in the mucosa in the gastrointestinal tract. In: Duthie HL, Wormsley KG (eds) Scientific basis of gastroenterology. Churchill Livingstone, Edinbourg

97. Zamchek N, Grable E, Ley A, Norman L (1955) Occurrence of gastric cancer among patients with pernicious anemia at the Boston City Hospital. N Engl. J. Med 252:1103–1110

Cobalamin Malabsorption in Exocrine Pancreatic Insufficiency in Adults and Children

J. L. GUÉANT, B. MONIN, and J. P. NICOLAS

Many publications have suggested that pancreatic function may affect the utilization of vitamin B_{12}. In this review, the evidence for such function and the possible mechanisms for it are reviewed and evaluated.

Vitamin B_{12} in the Gastrointestinal Tract

Vitamin B_{12} is a micronutrient present at low concentrations in most foods of animal origin and derived exclusively in nature from the biosynthetic activities of microorganisms. It is soluble in water, belongs to the cobalamins (Cbl), which are corrinoid compounds, and consists of a planar tetrapyrrol structure surrounding a cobalt atom to which are attached a benzimidazole base in one plane and a variety of ligand in the other. Corrinoids lacking Cbl structure do not have vitamin B_{12} activity. Some of these are vitamin B_{12} analogues.

Vitamin B_{12} in food is protein bound, either to an enzyme using the vitamin as cofactor or to a transport protein. Dietary intake in most Western populations exceeds the estimated daily requirements of 1–4 µg. Ingested vitamin B_{12} is released from its binder by sequential exposure to cooking and acid-peptic digestion in the stomach [8, 42, 63, 68]. The released vitamin is then exposed to two binders in gastric juice: haptocorrin (Hc; R protein), derived from saliva, and intrinsic factor (IF), the physical properties of which are summarized in Table 1. The affinity of vitamin B_{12} for Hc is 50 times that for IF at pH 2 [1] so that most of the vitamin B_{12} in gastric contents binds to Hc [50], which is not digested by pepsin at low pH unless incubated for 10 h [22, 60]. Pancreatic enzymes digest Hc, so that intragastric digestion may occur during duodenogastric reflux [34].

In the duodenum, gastric contents mix with the bile, which contains 0.52 ± 0.45 nmol/l Cbl and 1.62 ± 0.48 nmol/l total corrinoids [24]. The concentration of total corrinoids in bile has been reported to be negatively correlated with that in plasma, supporting the hypothesis that the liver functions to clear the plasma of analogues of vitamin B_{12} [24]. Bile also contains a variable concentration of Hc (0–4.33 nmol/1) [24]. Although transcobalamin II has been reported to be present in bile [11], we have not succeeded in confirming this [24].

In the duodenum, saturated and unsaturated Hc is degraded. Such degradation has been demonstrated in vitro using bovine proteases and intestinal fluid [1, 22] and in vivo using jejunal aspiration after intragastric instillation of labelled vitamin B_{12} bound to salivary Hc [22]. Transfer of labelled vitamin B_{12} from Hc to IF in the duodenum has been described both in vitro and in vivo [1, 3, 28, 50]. This transfer depends on partial digestion of Hc, resistance of IF to proteolysis, and neutralization of acid chyme to pH 8, which reduces by 70–150 times the affinity of Hc for vitamin B_{12} [1, 22], leaving the affinity of Hc for vitamin B_{12} only four fold that of IF. Partial degradation of Hc results in a molecule with a smaller Stockes radius and higher isoelectric points than intact Hc [22, 29] (Table 1). Under these conditions, IF is not degraded, although its Stokes radius and isoelectric point have been reported to change slightly (from 3.28 to 3.30 nm and 5.20 to 5.09, respectively) [50, 57] with relative increase of carbohydrate content. This is consistent with greater resistance to digestion by IF molecules with greater sialic acid content. Figure 1 illustrates a hypothesis which predicts that the carbohydrate core of these molecules determines their resistance to such digestion. Biliary vitamin B_{12} is completely bound to Hc and is susceptible to the same modifications and metabolic changes in the duodenum as is dietary Cbl, thus producing an enterohepatic circulation of vitamin B_{12} [18]. Bile salts inhibit the binding of Cbl to IF [66, 70], and analogues of vitamin B_{12} bind weakly to IF compared with their considerable affinity for Hc. Duodenal Hc has been calculated to have a binding capacity for B_{12} analogues of 13.5 nmol/l [21, 29]. The enterohepatic circulation appears selectively to excrete vitamin B_{12} analogues in the feces, while conserving vitamin B_{12} [23, 24, 37].

In the ileum, vitamin B_{12} analogues may be synthesized by intestinal bacteria [4]. Some of these may bind to IF [39, 43, 53], and the IF-analogue complex may compete with IF-B_{12} for binding to receptor sites on ileal cells [53]. Whether these complexes are absorbed is unclear, but analogues of vitamin B_{12} bound to Hc have been reported in human plasma [27]. These are probably cleared from plasma by a mechanism which utilizes the galactose receptor on hepatocytes [43].

The IF-vitamin B_{12} complex binds to specific receptors on ileal mucosa. These have an affinity for the complex of about 10 nM^{-1}, and binding requires a pH greater than 5.5 and divalent cations, especially calcium and magnesium [41, 44]. The binding of IF-B_{12} to the receptor involves calcium ions, and such binding may be enhanced by bile salts [38, 66].

The mechanism of transfer of vitamin B_{12} from the surface of the ileal cell into the portal venous blood is unknown [65]. Most studies of this phenomenon have utilized intestinal preparations from mices, guinea pigs, or pigs [17, 40, 61]. Data have been presented suggesting selective transport of vitamin B_{12} after its release from the IF at or near the intestinal surface [36, 46] and of cytosolic transport by a carrier with some of the characteristics of transcobalamin II [40]. We have recently completed studies of radioautography using purified, labeled, undenatured IF [26, 30] which retained its biological reactivity with guinea pig ileal receptor [30]. Radioautographic electron microscopy has revealed binding of the Cbl-IF complex to a receptor site at the base of the microvilli, entry of the complex into the cytoplasm of the enterocyte, possible localization in endocytic or lysosomal vesicles, and localization of radioactivity to subapical mitochondria [30]. The function of pH [65] and enzymatic digestion [17] in this process remains unclear.

Table 1. Physicochemical characteristics of digestive carrier proteins of vitamin B_{12}

	Molecular weight[a]	Stokes radius[b] (nm)	Isoelectric points		Unsaturated binding capacity (nM/l)	Association constant (nM^{-1})
			Extremes	Mean		
Intrinsic factor	57 500	3.30	4.5–5.7	5.20	15–80	7.1
Haptocorrin saliva[c]	120 500	4.48	2.8–4.7	3.78	7–45	32.0
Haptocorrin bile[c]	128 000	4.65	2.8–4.6	3.74	1–4	28.9
Partially degraded haptocorrin[c]	54 500	3.11	2.7–3.8	3.36	5–20	0.1

[a] Determined by Sephadex G 200 gel filtration using iodinated human albumin serum and iodinated IgG as reference molecules.
[b] Association constant for cyanocobalamin at pH 7.4 and at 25 °C.
[c] Also known as R binder or R protein.

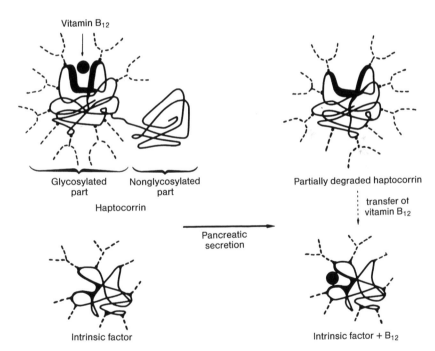

Fig. 1. Only a portion of the haptocorrin is glycosylated. This portion of the molecule makes up the site of binding with vitamin B_{12}; it is wholly resistant with regard to proteases, whereas the nonglycolyzed portion can be degraded. Intrinsic factor is wholly protected from pancreatic enzymes by its carbohydrate core. The partially degraded haptocorrin has an affinity for vitamin B_{12} two times weaker than that of intrinsic factor. Pancreatic enzymatic secretion is therefore responsible for the transfer of vitamin B_{12} from haptocorrin onto intrinsic factor

Cobalamin Malabsorption and Exocrine Pancreatic Insufficiency in the Adult

Malabsorption of crystalline Cbl (measured by Schilling test) is observed in 30%–70% of cases of exocrine pancreatic insufficiency (EPI) [29, 47, 54, 72, 78], whereas deficiency of vitamin B_{12} in such patients is less common [32]. EPI as a possible cause of megaloblastic anemia secondary to deficiency of vitamin B_{12} was suggested more than 50 years ago [9]. The mechanism by which absorption of vitamin B_{12} was affected by EPI has been investigated by many investigators.

In one study of five patients, the abnormal Schilling test observed in four was not corrected by either exogenous IF or pancreatic extracts; one patient studied appeared to be deficient in vitamin B_{12}; and the one who did not have an abnormal Schilling test has pancreatectomy [47]. An association between chronic pancreatic insufficiency and abnormal Schilling test was suggested, based on a study of four patients [33].

In 1962, Veeger et al. described malabsorption of vitamin B_{12} in three of seven patients with EPI and showed that malabsorption was corrected by feeding of bicarbonate or pancreatic extract [78], whereas calcium was inactive. Gräsbeck et al [15] described decreased urinary excretion of vitamin B_{12} in normal subjects when this was fed with ethylene-diamine-tetraacetate (EDTA); calcium did not interfere with absorption, and calcium lactate corrected abnormal Schilling tests in four of six patients with steatorrhea [19]. In one of these four, steatorrhea was secondary to celiac disease. Calcium supplements have been reported to improve Schilling tests in patients with hypoparathyroidism, supporting studies demonstrating a role for calcium in Cbl-IF receptor interaction [48]. Toskes et al. showed that malabsorption of vitamin B_{12} in EPI could not be due to calcium lack or to low pH since the calcium concentration and the pH in ileal contents of such patients was normal [74]. Malabsorption of vitamin B_{12} in EPI appeared not due to bacterial overgrowth in the intestine since malabsorption was not improved by broad-spectrum antibiotics [47]. Lebauer et al. [45] demonstrated that malabsorption of vitamin B_{12} in EPI appeared due to deficiency of pancreatic enzymes. Abnormal Schilling tests were corrected by feeding of pancreatic enzymes, and the effect was potentiated by bicarbonate and calcium. They suggested that pancreatic enzymes promoted absorption of vitamin B_{12} in three ways: (a) aiding formation of the IF-Cbl complex, (b) affecting binding of the complex to the ileal receptor, and (c) partially degrading IF-Cbl to promote absorption.

Correction of malabsorption of vitamin B_{12} in EPI by pancreatic extracts was confirmed by several authors [54, 76, 81] despite absence of a clear relationship between pancreatic protease secretion in the intestine and absorption of vitamin B_{12} [57, 76, 81]. A Cbl binder in pancreatic juice was demonstrated, and its role in B_{12} absorption was postulated [6, 74]. This binder was subsequently shown not to affect absorption of vitamin B_{12} [6, 80], and it was identified as Hcd. An additional low molecular weight peptide in pancreatic juice was also suggested to function in Cbl absorption [73]. Additional studies revealed the following: the physicochemical characteristics of Cbl-IF and IF alone appeared identical in gastric and ileal contents [50, 57]; Cbl absorption in EPI was improved by feeding cobinamide, a Cbl derivate which saturates Hc without significant binding to IF; and the pancreatic factor postulated has neither been purified nor characterized [71].

In other studies, Toskes et al. have demonstrated that IF in human and porcine gastric juice promotes absorption of vitamin B_{12} more effectively if it is first treated with trypsin or insoluble chymotrypsin [77]. This treatment did not alter the molecular weight, mobility, of affinity for the ileal receptor of IF. Although the methods used could not exclude proteolytic alteration of the IF molecule [56], the results could also be explained by digestion of competing Hc by the proteases. Support for the role of pancreatic secretion on absorption of vitamin B_{12} includes evidence that Cbl association with isolated loops of guinea pig ileum was increased by pancreatic juice, and inhibited by trypsin inhibitors [7] in the absence of added IF. The only Cbl binder described in pancreatic juice is Hc.

As indicated above, it would appear that Cbl is first bound to Hc and is transferred to IF only after digestion of the Hc. Such a sequence was postulated in 1962 by Highly and Ellenbogen [35] without identification of the initial binder as Hc. The potential interference with absorption of Hc in the gastrointestinal lumen was suggested by Gräsbeck [16], and this has been confirmed by careful studies [1, 3] which demonstrated that bovine proteases degrade Hc and reduce its affinity for Cbl by 150-fold. Aspirated duodenal juice obtained from patients with EPI did not effectively degrade human salivary Hc [5, 22, 28, 29, 52], but pancreatic juice from such patients was effective [6]. The differences may be due to the much greater concentration of proteolytic enzymes in the latter preparation, whereas physiological conditions probably were better duplicated by experiments using duodenal juice. In other studies, retention of labeled Cbl by undigested Hc in intestinal fluid of patients with EPI and its failure to be transferred to IF was demonstrated [29, 50].

Other studies in support of the hypothesis that malabsorption of vitamin B_{12} in EPI patients is due to retention by undigested Hc include correction of abnormal Schilling test in three EPI patients by feeding cobinamide to saturate the Hc [2] and demonstration of undigested unsaturated Hc in intestinal fluid from a patients with EPI [59]. Toskes et al., however, [67, 69] did not observe correction of Schilling tests in such patients when Cbl-adenine (pseudo-Cbl) was used to block Hc binding and observed correction of the Schilling test by cobinamide in only four of nine EPI patients. The latter may be due to binding of pseudo-Cbl to IF, for which its inhibition constant is 9.9 nM^{-1} compared with that for cobinamide, which is 10 μM^{-1}. The pseudo-Cbl may thus have inhibited the IF in these studies. Disparity among the studies with cobinamide might be due to clinical differences between patients studied: differences in unsaturated Hc binding capacity in intestinal fluid or presence of Hcd in gastric juice, which has 1000 times greater affinity for Cbl than for cobinamide [34].

In three EPI patients in whom Hc-bound labeled Cbl was instilled into the gastric antrum, the complex remained undegraded in the ileum [22]. In the jejunum of eight EPI patients fed labeled Cbl, either free or bound to IF, 13%–85% of the Cbl was bound to Hc, whereas in eight control subjects, more than 95%

was bound to IF [20, 50]. Transit from gastric antrum to jejunum did not change the characteristics of the IF, when analyzed by gel filtration, immunological reactivity, and isoelectric focusing, suggesting that IF does not require activation by a pancreatic factor. Brugge et al. [5] have developed a modified Schilling test in which $[^{57}Co]B_{12}$ is bound to IF and $[^{58}Co]B_{12}$ is bound to porcine Hc. The ratio of $^{58}Co/^{57}Co$ excretion ranged from 0.5 to 1.0 in control subjects, and was less than 0.15 in EPI patients. These results correlated with steatorrhea ($p < 0.01$) and trypsin output ($p < .01$).

These studies demonstrate that failure to degrade Hc appears an important factor in causing malabsorption of vitamin B_{12} in EPI. That other factors also contribute to the malabsorption is demonstrated by the absence of a demonstrable correlation between pancreatic protease output, measured by intubation after hormone stimulation, and Schilling test [29, 54, 76] and by the presence of Cbl malabsorption in patients with chronic pancreatitis without exocrine deficiency [54, 74].

Elastase may be more effective than trypsin or chymotrypsin in correcting Cbl malabsorption in EPI patients [69], but in vitro, elastase appeared less effective that chymotrypsin or trypsin in digesting Hc [1]. The effectiveness of different pancreatic extracts in treatment of EPI is variable; contact with acid chyme may alter the relative activity of different proteases; and different preparations from different manufactures have different potency. For these reasons, variable effects of feeding pancreatic extracts in EPI patients is not surprising.

In more than 55% of EPI patients, Schilling test is normal [29, 54] despite severe pancreatic exocrine abnormality in some. We have recently studied 22 cases of EPI and 22 control subjects to compare Hc, Hc degradation, IF, and other factors in intestinal juice (Table 2). In four cases of EPI, normal Schilling test was associated with deficient trypsin, and chymotrypsin secretion with HDT (Hc degradation test) less than 10% (normal 80%). In these, most of the Cbl was bound to Hc [29], demonstrating that failure to release Cbl from Hc is not invariably associated with malabsorption of Cbl. Allen et al. [1, 2] have reported that normal Schilling tests are often seen in EPI patients with achlorhydria, possibly because at neutral pH the affinity of Hc and that of IF for Cbl are almost equivalent. This could explain the reports that abnormal Schilling test in EPI patients may be corrected by bicarbonate administration [54, 71, 78], which has been confirmed by Marcoullis et al. [50], but confounded by a report that in a patient with pernicious anemia and abnormal Schilling test absorption of vitamin B_{12} was improved by bicarbonate and pancreatic extract [75]. In the study described above [29] on 22 patients with EPI, a correlation was observed between Schilling test and HDT + ($10 \times pH$), demonstrating the interrelationship between gastric pH, Hc digestion, and absorption of vitamin B_{12}.

Factors in addition to gastric pH may improve absorption of vitamin B_{12} in EPI patients. In our study of 22 EPI patients [29], HDT was measured with different concentrations of basal intestinal juice. In 50% of cases, Hc digestion increased with increasing concentration of intestinal juice, with a significant hyperbolic relationship found between HDT and trypsin-chymotrypsin activity ($p < .01$), the proportion of vitamin B_{12} bound of IF, and the Schilling test result ($p < .01$). Because, however, Cbl transport from upper intestine through the gut

Table 2. Unsaturated binding capacity (UBC) and mode of transport of vitamin B_{12} in the intestinal fluid of nine pancreatic insufficiency patients (EPI) and four control subjects compared on the haptocorrin degradation test (HDT) and the Schilling test

	UBC vitamin B_{12} (nmol)			Combination of vitamin B_{12}			Schilling test	
	Intrinsic factor	Haptocorrin	Partially degraded haptocorrin	Intrinsic factor	Haptocorrin	HDT (%)	B_{12}	B_{12} intrinsic factor
EPI patients	5.2±5.3	2.1±3.8	8.7±4.9	37.5±26.6	62.5±26.6	29.6±32.5	16.1±6.7	16.5±8.9
Control subjects	6.8±5.3	0.0	12.7±8.0	80.4±11.7	19.6±11.7	95.2±7.2	19.5±11.4	20.9±9.8

From [29].

Table 3. Concentration of vitamin B_{12}, total corrinoids, and digestive carrier proteins of corrinoids in the meconium and the feces of children with cystic fibrosis (CF) and child control subjects

	Unsaturated corrinoid binding capacity			Vitamin B_{12} (pmol/g)	Total corrinoids (pmol/g)
	Haptocorrin	Partially degraded haptocorrin	Intrinsic factor (pmol/g)		
Meconium					
CF (n=4)	405.7±176.1	0.0	<0.01	19.0±12.6	34.7±22.4
Premature (n=4)	0.0	0.4±0.1	<0.01	2.0±2.2	3.0±2.9
Control subjects (n=13)	<0.3	1.0±0.2	<0.01	5.3±3.4	7.3±7.1
Feces					
CF (n=9)	0.0	14.3±15.2	5.3±7.6	85.3 (34.2)[a]	151.5 (75.6)[a]
Control subjects (n=5)	0.0	0.2±0.1	Traces	2.1±0.7	18.3±5.2

[a] The median of distribution is indicated in parentheses; the standard deviation is not shown owing to the wide dispersion of values obtained. From [23, 79].

lumen and into the portal venous blood is affected by many factors, one would not expect a linear relationship between pancreatic protease output and Schilling test. One such factor is the bile. Cotter et al. [11] demonstrated that biliary Hc could compete for Cbl with intestinal IF. The proportion of Cbl bound to Hc in the jejunum was greater than that in the gastric contents in five of eight cases described by Marcoullis et al., while in the case of EPI with obstructed jaundice, less Cbl was bound to Hc in the jejunum than in the stomach [20, 50]. In EPI, biliary Cbl coupled to Hc may not be digested, since such digestion is not observed in vitro with EPI intestinal fluid [21]. Of the Cbl in basal intestinal fluid from EPI patients, $62 \pm 26.6\%$ was bound to Hc [29] (Table 2). Bile salts may also affect binding of Cbl to IF [66, 70]. Even partial interruption in reutilization (enterohepatic circulation) of Cbl would be expected to accelerate the appearance of Cbl deficiency. It is thus surprising that EPI does not produce Cbl deficiency more frequently.

Macrocytosis in patients with EPI is difficult to interpret, since many have ingested alcohol and have low level of folate [64]. Of 90 cases of EPI studied by Evans and Wolleger, Cbl deficiency causing macrocytic anemia was present in only two [14]. The patients described by Lebauer et al. [45] was alcoholic, and folate levels were not measured.

Malabsorption of dietary Cbl may not be identical with abnormal Schilling test, since dietary constituent may buffer gastric and jejunal pH and alter absorption of Cbl. Absorption of dietary Cbl in EPI may depend on several factors: residual pancreatic enzyme secretion, gastric and duodenal pH, biliary Hc, and a possible role of bacterial proteases in the ileum (see following section).

Vitamin B$_{12}$ Malabsorption in Cystic Fibrosis

The congenital exocrinopathy of cystic fibrosis is the most common cause of pancreatic exocrine deficiency in children and is the most common congenital disease in Europe. Its incidence in France is 1:2400. Deren et al. in 1973 described malabsorption of labeled Cbl in 11 cases aged 8–20 years [13]. None of these children was deficient in Cbl, This malabsorption was confirmed by Harms et al. [31] using whole body counting. Malabsorption of labeled Cbl and of the endogenous Cbl in a test meal was corrected by pancreatic extracts or sodium bicarbonate. The more frequent Cbl malabsorption in such children than in EPI patients may be due to lower gastric pH in the children [12], in whom gastric hyperacidity after stimulation was demonstrated [12]. Hypersecretion of gastric IF was also present. In the presence of gastric hyperacidity, the isoelectric distribution of gastric IF is unchanged, whereas that of salivary Hc alters with predominance of molecules with higher pI [55].

We have purified IF from gastric juice of children with cystic fibrosis and observed normal biological activity (by Schilling test [26]) but increase of fucose/sialic acid ratio. Hc concentration in meconium of infants with cystic fibrosis is increased over that in intestinal contents of normal newborns (Fig. 2) [79], representing more than 95% of Cbl binders on the former. The meconium excretion of Cbl and total corrinoids also is greater than in normal infants (Table 3). In cystic

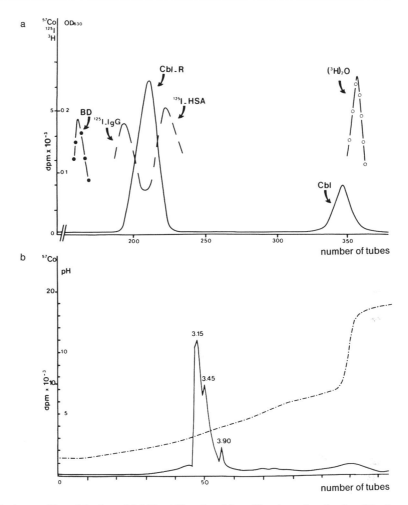

Fig. 2 a, b. Elution profiles of Sephacryl S 300 gel filtration (**a**) and isoelectrofocalization (**b**) of unsaturated binders of vitamin B_{12} contained in the cystic fibrosis meconium. In gel filtration, a single binder is eluted with a molecular weight estimated at $110\,100 \pm 10\,100$ daltons and an unsaturated vitamin binding capacity of about 405.7 ± 176.1 pmol/g. This binder corresponds to saturated and nondegraded haptocorrin (Cbl-R) because it displays in isoelectrofocalization three principal isoelectric isoproteins between pH 3.15 and 3.90 with a mean isoelectric point of 3.67 ± 0.20

fibrosis, Cbl is sequestered to Hc in the proximal intestine, which is partially degraded by bacterial proteases to molecules with MW of 44 300 and 20 300 daltons. Such bacterial decomposition of Hc may prevent malabsorption of Cbl in such patients and may explain the greater proportion of Cbl bound to IF in the ileum than elsewhere in the small gut in two patients with EPI with no detectable secretion of trypsin or chymotrypsin [67]. We have shown that IF represents 58%–72% of Cbl binders in feces in cystic fibrosis patients, whereas it represents less than 2% in feces of normal children [23], supporting the suggestion [82] that bacterial proteases of colonic anaerobes can degrade IF.

As in EPI, Cbl malabsorption in cystic fibrosis may be affected by many factors: failure of Hc degradation by pancreatic proteases, gastric hypersecretion, and digestion of binders by bacterial proteases.

No patient with Cbl deficiency caused by cystic fibrosis has been described. One suspect case was published [62]: a 2-year-old child with macrocytic anemia and abnormal Schilling test in whom anemia was corrected by injection of Cbl. The rarity of such deficiency may be due to buffering of the acid chyme or by early treatment with pancreatic extracts.

References

1. Allen RH, Seetharam B, Podell E, Alpers DH (1978) Effect of proteolytic enzymes on the binding of cobalamin to R protein and intrinsic factor. J Clin Invest 61:47–54
2. Allen RH, Seetharam B, Allen NC, Podell E, Alpers DH (1978) Correction of cobalamin malabsorption in pancreatic insufficiency with cobalamin analogue that binds with high affinity to R protein but not to intrinsic factor. J Clin Invest 61:1628–1634
3. Andersen KJ, Von der Lippe G (1978) The effect of proteolytic enzymes on the vitamin B_{12} binding proteins of human gastric juice and saliva. Scand J Gastroenterol 14:833–838
4. Brandt LJ, Bernstein LH, Wagle A (1977) Production of vitamin B_{12} analogues in patients with small-bowel bacterial overgrowth. Ann Intern Med 87:546–551
5. Brugge WR, Goff JS, Allen NC, Podell ER, Allen RH (1980) Development of a dual label Schilling test for exocrine pancreatic function based on the differential absorption of cobalamin bound to intrinsic factor and R proteins. Gastroenterology 78:937–949
6. Carmel R, Abramson SB, Renner IG (1983) Characterization of pure human pancreatic juice: cobalamin content, cobalamin binding proteins and activity against human R binders of various secretions. Clin Sci 64:193–205
7. Carmel R, Hollander D, Gergely HM, Abramson SB, Renner IG (1984) Pure human pancreatic juice directly enhances cobalamin uptake by guinea pig ileum. Clin Res 32:489 A
8. Cattan D, Belaiche J, Zittoun Y, Yvart J, Chagom JP, Murit Y (1982) Rôle de la carence en facteur intrinsèque dans la malabsorption de la vitamine B_{12} liée aux protéines, dans les achlorhydries. Gastroenterol Clin Biol 6:570–575
9. Cheney G, Niemand F (1932) A possible relationship of pancreatic insufficiency to Addison-Biermer (pernicious) anemia. Arch Intern Med 49:925–933
10. Clarkson R, Kowlessan OD, Horwith M, Sleisenger MH (1960) Clinical and metabolic study of a patient with malabsorption and hypoparathyroidism Metabolism 9:1093–1106
11. Cotter R, Rothenberg SP, Weiss JP (1979) Dissociation of the intrinsic factor-vitamin B_{12} complex by bile: contributing factor to B_{12} malabsorption in pancreatic insufficiency. Scand J Gastroenterol 14:545–550
12. Cox KL, Isenberg JN (1978) Hypersecretion of gastric acid in patients with pancreatic exocrine insufficiency due to cystic fibrosis. Gastroenterology 74:1022
13. Deren JJ, Arora B, Toskes PP, Hansell J, Sibinga MS (1973) Malabsorption of crystalline vitamin B_{12} in cystic fibrosis. N Engl J Med 288:949–950
14. Evans WB, Wollaeger EE (1966) Incidence and severity of nutritional deficiency states in chronic exocrine pancreatic insufficiency: comparison with non-tropical sprue. Am J Dig Dis 11:594–606
15. Gräsbeck R (1962) Applications of radioactive vitamin B_{12} in clinical diagnosis. In: Heinrich HC (ed) Vitamin B_{12} and intrinsic factor. 2nd European symposium, Hamburg, 1961 Enke, Stuttgart, p 138
16. Gräsbeck R (1975) Absorption and transport of vitamin B_{12}. Br J Haematol 31 (Suppl):103–110

17. Gräsbeck R, Kouvonen I (1983) The intrinsic factor and its receptor: are all membrane transport system related? TIBS 203–205
18. Gräsbeck R, Nyberg W, Reizenstein PG (1958) Biliary and fecal vitamin B_{12} excretion in man. An isotropic study. Proc Soc Exp Biol Med 97:780–784
19. Gräsbeck R, Kantero I, Siurala M (1959) Influence of calcium ions on vitamin B_{12} absorption in steatorrhea and pernicious anaemia. Lancet 1:234
20. Guéant JL (1980) Pathogénie de la malabsorption de la vitamine B_{12} dans l'insuffisance pancréatique externe. Thèse de Médecine Expérimentale, University of Nancy I, pp 134, 218
21. Guéant JL (1982) Contribution à l'étude du rôle et des propriétés physico chimiques des protéines R dans le transit digestif de la vitamine B_{12}. Thèse de 3 ème Cycle en Nutrition, Sciences de l'Alimentation et Biochirnie Appliquée. University of Nancy I, p 169
22. Guéant JL, Parmentier Y, Djalali M, Bois F, Nicolas JP (1983) Sequestration of crystalline and endogenous cobalamin by R binders down to the distal ileum in exocrine pancreatic dysfunction. Clin Chim Acta 134:107–127
23. Guéant JL, Vidailhet M, Pasquet C, Djalali M, Nicolas JP (1984) Effect of pancreatic extracts on the faecal excretion and on the seric concentration of cobalamin and cobalamin analogues cystic fibrosis. Clin Chim Acta 137:33–41
24. Guéant JL, Monin B, Gaucher P, Nicolas JP (1984) Biliary excretion of cobalamin and cobalamin analogues in the human. Digestion 30:151–157
25. Guéant JL, Vidailhet M, Djalali M, Michalski JC, Nicolas JP (1984) Cobalamin R binder as a possible model molecule for glycoprotein study in cystic fibrosis. Clin Chim Acta 143:217–223
26. Guéant JL, Kouvonen I, Michalski JC, Masson C, Gräsbeck R, Nicolas JP (1985) Purification of human intrinsic factor using HPLC as a final step. FEBS Letters 184:14–19
27. Guéant JL, Kamel L, Champigneulle B, Parmentier Y, Monin B, Boissel P, Gaucher P, Nicolas JP (1985) Etapes portale et hépatique du cycle entérohépatique de la vitamine B_{12} et de ses analogues chez l'homme. Gastroenterol Clin Biol 9 (s bis):98
28. Guéant JL, Champigneulle B, Djalali M, Bigard MA, Gaucher P, Hassouni A, Nicolas JP (1986) An in vitro test of degradation of haptocorrin (TDH) for the biological diagnosis of exocrine pancreatic dysfunction using duodenal juice collected during endoscopy. Lancet 2:709–712
29. Guéant JL, Djalali M, Aouadj R, Gaucher P, Monin B, Nicolas JP (1986) In vitro and in vivo evidence that the malabsorption of cobalamin is related to its binding on haptocorrin (R binder) in chronic pancreatitis. Am J Clin Nutr 44:265–377
30. Guéant JL, Gérard A, Monin B, Gérard H, Khanfri J, Nicolas JP (1986) Autoradiographic localization of iodinated human intrinsic factor (IF) in the guinea-pig ileum using electron microscopy. Gastroenterology 90:1442
31. Harms HK, Kennel O, Bertele RM, Bildlingmeier F, Bohne A (1981) Vitamin B_{12} absorption and exocrine pancreatic insufficiency in childhood. Eur J Pediatr 136:75–79
32. Henderson JR, Warwick RG, Simpson JD, Shearman DJC (1972) Does malabsorption of vitamin B_{12} occur in chronic pancreatitis? Lancet 2:241–243
33. Herzlich B, Herbert V (1984) The role of the pancreas in cobalamin (vitamin B_{12}) absorption. Am J Gastroenterol 79:6
34. Herzlich B, Herbert V (1985) Cobalamin-specific R binder in pernicious anemia gastric juice: production by digestive enzyme action on salive R binder. Am J Gastroenterol 80:841–852
35. Highley DR, Ellenbogen L (1962) Studies on the mechanism of vitamin B_{12} absorption: in vivo transfer of vitamin B_{12} to intrinsic factor. Arch Biochem Biophys 99:126–131
36. Hines JD, Rosenberg A, Harris JW (1968) Intrinsic factor-mediated radio-B_{12} uptake in sequential incubation studies using everted sacs of guinea pig small intestine: evidence that IF is not absorbed into the intestinal cell. Proc Soc Biol Med 129:653–658
37. Kanazawa S, Herbert QV, Herzlich B (1983) Removal of cobalamin analogues in bile by enterohepatic circulation of vitamin B_{12}. Lancet 1:707–708
38. Kanazawa S, Herzlich B, Herbert V (1985) Enhancement by human bile of the free and intrinsic factor-bound cobalamin to small bowel epithelial cell receptors. Am J Gastroenterol 80:964–969

39. Kapadia CR, Mathan VI, Baker SI (1976) Free intrinsic factor in the small intestine in man. Gastroenterology 70:704-706

40. Kapadia CR, Del Serfilippi KV, Donaldson RM (1983) Intrinsic factor-mediated absorption of cobalamin by guinea pig ileal cells. J Clin Invest 71:440-448

41. Katz M, Cooper BA (1974) Solubilized receptor for intrinsic factor-vitamin B_{12} complex from guinea-pig intestinal mucosa. J Clin Invest 54:733-739

42. King CE, Leibach J, Toskes PP (1979) Clinically significant vitamin B_{12} deficiency secondary to malabsorption of protein-bound vitamin B_{12}. Dig Dis Sci 24:397-402

43. Kolhouse JF, Allen RH (1977) Absorption, plasma transport and cellular retention of cobalamin analogues in the rabbit. J Clin Invest 60:1381-1392

44. Kouvonen I, Gräsbeck R (1981) Topology of the hog intrinsic factor receptor in the intestine. J Biol Chem 256:154-158

45. Lebauer EK, Smith K, Greenberger JJ (1968) Pancreatic insufficiency and vitamin B_{12} malabsorption. Arch Intern Med 122:423-425

46. Levine JS, Nakane PK, Allen RH (1982) Immunocytochemical localization of intrinsic factor-cobalamin bound to the guinea pig ileum in vivo. Gastroenterology 82:284-290

47. MacIntyre PA, Sachs MV, Krevans JR (1956) Pathogenesis and treatment of macrocytic anemia. Arch Intern Med 98:541-549

48. Mackenzie IJ, Donaldson RM (1972) Effect of divalent cations and pH on intrinsic factor mediated attachment of vitamin B_{12} to intestinal microvillous membranes. J Clin Invest 51:2465-2471

49. Marcoullis G, Gräsbeck R (1975) Vitamin B_{12} binding proteins in human gastric mucosa. General pattern and demonstration on intrinsic factor isoproteins of mucosa. Scand J Clin Lab 35:5-11

50. Marcoullis G, Parmentier Y, Nicolas JP, Jimenez M, Gérard P (1980) Cobalamin malabsorption due to non-degradation of R proteins in the human intestine. J Clin Invest 66:430-440

51. Marcoullis G, Nicolas JP, Parmentier Y, Jimenez M, Gérard P (1980) Derivative of R-type cyanocobalamin binding proteins in the human intestine. A candidate antibacterial molecule. Biochim Biophys Acta 633:289-294

52. Marcoullis G, Guéant JL, Nicolas JP (1986) Development of a radioimmunoassay specific for exocrine pancreatic insufficiency. Clin Chem 32:453-460

53. Mathan VI, Babior BM, Donaldson RM (1974) Kinetics of the attachment of intrinsic factor bound cobinamide to ileal receptor. J Clin Invest 54:598-608

54. Matuchansky C, Rambaud JC, Modigliani R (1974) Vitamin B_{12} malabsorption in chronic pancreatitis. Gastroenterology 67:406-407

55. Naimi D, Guéant JL, Hambaba L, Vidailhet M, Monin B, Nicolas JP (1984) Gastric intrinsic factor hypersecretion in cystic fibrosis. J Pediatr Gastroenterol Nutr. 6:899-903

56. Nicolas JP (1982) Le protocole classique du double test de Schilling est-il toujours bien adapté au diagnostic des malabsorptions intestinales de la vitamine B_{12}? Importance de la sécrétion chlorhydro-peptique et de la forme d'administration de la vitamine. Gastroentero Clin Biol 6:554-556

57. Nicolas JP, Marcoullis G, Jimenes M, Parmentier Y, Gérard P (1981) In vivo evidence that intrinsic factor cobalamin-complex traverses the human intestine intact. Biochim Biophys Acta 675:328-333

58. Nicolas JP, Guéant JL, Gaucher P (1983) Malabsorption de la vitamine B_{12} et insuffisance pancréatique exocrine. Gastroenterol Clin Ciol 7:307-314

59. Parmentier Y, Marcoullis G, Nicolas JP (1979) The intraluminal transport of vitamin B_1 and exocrine pancreatic insufficiency. Proc Soc Exp Biol Med 160:396-400

60. Reizenstein PG (1959) Excretion of non-labeled vitamin B_{12} in man. Acta Med Scand 165:313-320

61. Robertson JA, Gallagher ND (1985) In vivo evidence that cobalamin is absorbed by receptor-mediated endocytoses in the mouse. Gastroenterology 88:908-912

62. Rucker RV, Harrison GM (1973) Vitamin B_{12} deficiency in cystic fibrosis. N Ebgl J Med 289:329

63. Schade SG, Schilling RF (1967) Effect of pepsin on the absorption of food vitamin B_{12} and iron. Am J Clin Nutr 20:636-640

64. Schilling RF (1983) The role of the pancreas in vitamin B_{12} absorption. Am J Hematol 14:197–199
65. Seetharam B, Alpers DH (1985) Cellular uptake of cobalamin. Nutr Rev 43:97–102
66. Seetharam B, Jimenez M, Alpers DH (1983) Effect of bile and bile acids on binding of intrinsic factor to cobalamin and intrinsic factor-cobalamin complex to ileal receptor. Am J Physiol 245:72–77
67. Steinberg W, Curington CW, Toskes PP (1979) Evidence that failure to degrade R binder is unimportant in the pathogenesis of cobalamin malabsorption in patients with chronic pancreatitis. Gastroenterology 76:1255 A
68. Steinberg GW, King CE, Toskes PP (1980) Malabsorption of protein-bound cobalamin but not unbound cobalamin during cimetidine administration. Dig Dis Sci 25:188–192
69. Steinberg W, Toskes P, Curingtin C, Shan R (1980) Further studies which cast doubt upon a role for R proteins in the cobalamin malabsorption of chronic pancreatitis. Gastroenterology 78:1270 A
70. Teo NH, Scott JM, Reed B, Neale G, Weir DG (1981) Bile and acid inhibition of vitamin B_{12} binding by intrinsic factor in vitro. Gut 22:270–276
71. Toskes PP (1980) Current concepts of cobalamin absorption and malabsorption. J Clin Gastroenerol 2:287–297
72. Toskes PP, Deren JJ (1973) Vitamin B_{12} absorption and malabsorption. Gastroenterology 65:662–683
73. Toskes PP, Smith G (1978) Isolation of a low molecular weight vitamin B_{12} promoting protein from preparations of trypsin and chymotrypsin Gastroenterology 74:1106
74. Toskes PP, Hansell S, Cerda S, Deren JJ (1971) Vitamin B_{12} malabsorption in chronic pancreatic insufficiency. Studies suggesting the presence of a pancreatic "intrinsic factor". N Engl J Med 284:627–632
75. Toskes PP, Deren JJ, Conrad ME, Smith GW (1973) Trypsin-like nature of the pancreatic factor that corrects vitamin B_{12} malabsorption associated with pancreatic dysfunction. J Clin Invest 52:1660–1664
76. Toskes PP, Deren JJ, Fruiterman J, Conrad M (1973) Specificity of the correction of vitamin B_{12} malabsorption by pancreatic extract and its clinical significance. Gastroenterology 63:199–204
77. Toskes PP, Smith GW, Francis GM, Sander EG (1977) Evidence that pancreatic proteases enhance vitamin B_{12} absorption by acting on crude preparation of hog gastric intrinsic factor and human gastric juice. Gastroenterology 72:31–36
78. Veeger W, Abels J, Hellemans N, Nieweg HO (1962) Effect of sodium bicarbonate and pancreatin on the absorption of vitamin B_{12} and fat in pancreatic insufficiency. N Engl J Med 267:1341–1344
79. Vidailhet M, Guéant JL, Monin B, Pasquet C, Djalali M, Morali A, Nicolas JP (1984) Unsaturated Cbl R binder in cystic fibrosis meconium In: Lawson M (ed) Proceedings of the 9th international cystic fibrosis congress, 1984 Wiley, New York, p 132
80. Von der Lippe G (1977) The absorption of vitamin B_{12} in chronic pancreatic insufficiency. Scand J Gastroenterol 12:257
81. Von der Lippe G, Andersen JJ, Schjonsby H (1976) Intestinal absorption of vitamin B_{12} in patients with chronic pancreatic insufficiency and the effect of human duodenal juice on the intestinal uptake of vitamin B_{12}. Scand J Gastroenterol 11:689–695
82. Welkos SL, Toskes PP, Bear H (1981) Importance of anaerobic bacterial in the cobalamin malabsorption of the experimental rat blind loop syndrome. Gastroenterology 80:313–320

Making Sense of Laboratory Tests of Folate Status: Folate Requirements to Sustain Normality

V. Herbert

Introduction

The first event in the sequence leading to nutrient deficiency is development of negative nutrient balance, i.e., nutrient utilization, requirement, catabolism, and excretion exceeds nutrient ingestion, absorption, and anabolism. Persistent negative nutrient balance produces a sequence of stages or degrees along a continuum of nutrient depletion which are not always clearly demarcated one from the next but show an increasing deficiency, usually marked by sequential appearance of various biochemical and clinical deficits. In the case of folate, the continuum from normality to folate-deficiency megaloblastic anemia begins with negative folate balance and proceeds through three sequential stages, or degrees, of folate depletion, each defined by the appearance of specific biochemical and/or hematologic markers (Table 1, Fig. 1).

The first evidence of negative folate balance is a low serum folate level (below 3 ng folate/ml serum), determined either microbiologically or by radioassay. A single low serum folate value, or even several such values, does not diagnose folate deficiency. A low serum folate diagnoses only negative balance at the time the serum sample was drawn, i.e., the amount of folate being taken up by tissues, catabolized and excreted at the point in time when the serum sample was drawn was greater than the amount being absorbed. Serum folate stabilizes at a level below 3 ng/ml after only 2-3 weeks of negative folate balance [1-4]. If folate utilization does not continue to exceed folate absorption, this stage may never lead to tissue folate depletion; therefore low serum folate by itself does not mean that treatment with folate is indicated, but low serial folates over more than a month may indicate that stores are low [5, 6].

Treatment with folate should be considered when there is increased demand coupled with folate depletion (Fig. 1). Folate depletion is a state of reduced folate stores characterized by a fall in the red cell folate level to below 160 ng folate/ml erythrocytes [1-4, 7], a level at which there is little reserve of folate to be called on in times of increased demand, such as in pregnancy. There is no biochemical or clinical functional deficit at this stage. Red cell and liver folate stores fall together as folate deficiency progresses [1, 4, 7].

Folate depletion may be primary (global), i.e., lack of all forms of folate, as occurs with inadequate folate ingestion and/or absorption, or it may be secondary, as occurs with vitamin B_{12} deficiency, in which, due to a lack of holome-

Table 1. Sequence of events in the development of folate-deficiency megaloblastic anemia in subjects on an experimental regimen with very little folate (less than 5 µg/day) [1, 3, 28]

Sequence of changes	Time after start of regimen (weeks)
Decrease in serum folate (<3 ng/ml); modest increase in mean diameter of medullary erythroblasts	3
Hypersegmentation of polynuclear neutrophils in the bone marrow (mean no. of lobes 3.5); dU suppression test abnormal in medullary cells	5
Hypersegmentation of polynuclear neutrophils in the peripheral blood; in the marrow, increase in the rate and no. of abnormalities of mitosis and presence of basophil megaloblasts of intermediate size; dU suppression test abnormal in peripheral blood lymphocytes	7
Appearance in the marrow of giant metamyelocytes and a certain no. of polychromatophilic megaloblasts of intermediate size	10
Increase in urinary formiminoglutamic acid	13
Acidophilic megaloblasts of intermediate size in the marrow	14
Decrease in red blood cell folate	17
Macroovalocytosis; presence of numerous giant metamyelocytes in the marrow	18
Obvious medullary megaloblastosis	19
Anemia	20

Folate deficiency in persons consuming <5 µg folate daily

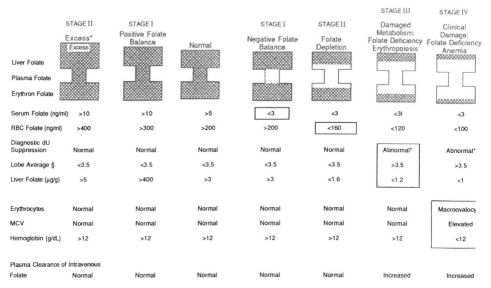

* Dietary excess of folate reduces zinc absorption.

Fig. 1

thionine synthetase (5-methyltetrahydrofolate homocysteine methyltransferase; E.C. 2.1.1.13), there is a decrease in all tetrahydro-folate coenzymes except methylfolate. Due to lack of B_{12}, folate is trapped as methylfolate [8–11], which is a poor substrate for polyglutamate synthesis [12]. This results in decreased production of tissue folate stores, which are mainlyl polyglutamates, and leakage of short-chain folates from cells, with resultant tissue folate depletion [12]. On this subject, much evidence existed in the literature of the late 1950s and early 1960s for the concept, stated two decades ago [7] that "vitamin B_{12} may play a role in the transport of folate into all cells and/or retention of folate by cells as well as in utilization by cells." There are abnormal folate polyglutamate ratios in B_{12} deficiency, corrected by B_{12} therapy [13].

In addition to primary and secondary folate deficiency, there is also selective folate deficiency in one cell line but not another [1, 3, 14, 15]. There is evidence that folate therapy in vitamin B_{12} deficient subjects, because of the folate trap, may exacerbate nerve damage [16].

The second stage of folate depletion is folate-deficient erythropoiesis (Fig. 1) characterized first by an abnormal diagnostic deoxyuridine (dU) suppression test on bone marrow cells, i.e., inadequate suppression by nonradioactive dU of the incorporation into DNA of subsequently added radioactive thymidine, with the abnormality corrected by either methylfolate or PGA (Pteroylglutamic acid) or any other folate form, but not by hydroxobalamin [3, 17, 18]. This is shortly followed by an abnormal diagnostic dU suppression test on peripheral blood lymphocytes [3, 19,20]. When there is coincident iron deficiency, dU suppression becomes abnormal first in the lymphocytes and only subsequently in bone marrow cells [21]. This second stage has been referred to as "subclinical deficiency" or "subtle megaloblastic anemia" [3, 17]. The diagnostic dU suppression test becomes clearly abnormal when intracellular folate levels fall to < 200 ng folate/ 10^9 cells, as reported by Colman and Herbert [27] using normal human lymphocytes, and confirmed by Steinberg et al. [23] using Friend malignant erythroleukemia cells. Treatment with folate is always appropriate when folate depletion reaches the second stage, since there is now a biochemical function deficit requiring folate for correction.

The third stage of folate depletion, folate deficiency anemia (Fig. 1), is characterized by the appearance of macroovalocytic erythrocytes and a hemoglobin level below 130 g/l in men, below 120 g/l in women, and below 110 g/l during pregnancy. Folate deficiency without anemia (i.e., folate depletion and folate-deficient erythropoiesis) is more prevalent than folate-deficiency anemia in many countries, just as iron deficiency without anemia (i.e., iron depletion and iron deficient erythropoiesis) is more prevalent than iron-deficiency anemia [24, 25].

Newer potential diagnostic approaches to assessing folate status have recently been reviewed [3], including noncompetitive radioassay and percent saturation of folate-binding proteins. Hidden folate deficiency is common in patients with defects in hemoglobin synthesis, such as iron deficiency and thalassemia [3]. The morphologic clue to this hidden folate deficiency is the presence of hypersegmentation in the peripheral blood neutrophils [3, 21]. In these situations, for unknown reasons, serum and red cell folate are artificially elevated despite lymphocyte and bone marrow folate depletion, demonstrable by dU suppression tests diagnostic for folate deficiency [21, 26]. the diagnosis is confirmed by giving low-dose (100–200 μg/day) folate therapy and observing the ensuing reticulocy-tosis and rise in hemoglobin. To avoid confusion, the folate dose should not exceed 200 μg/day. Larger doses can produce hematologic response in patients who have only vitamin B_{12} deficiency [27], because the megaloblastosis of B_{12} deficiency results from trapping of folate as methylfolate [10, 12], and nonme-thylfolate by-passes the trap [18]. When the iron deficiency is worse than the folate deficiency, all the red cells may be small and pale, with none of the large »hyperchromic red« cells characteristic of folate deficiency.

Folate Requirements to Sustain Normality

The minimal daily adult folate requirement to sustain normality (i.e., to sustain normal DNA synthesis) in the absence of increased metabolic need is approxi-mately 50 μg for adults (about 1 μg/kg body weight). This figure is based on observations that such orally administered amounts sustain the serum folate of normal young women [1, 2], that daily parenteral administration of this amount successfully treats uncomplicated folate deficiency anemia [28, 29], and on other studies in human volunteers [30]. Complicated cases may fail to respond to such doses [31, 32]. Hyperthyroidism, pregnancy, hemolytic anemia [3, 4, 15], need for intensive care [33, 34], or any other sustained metabolic drain [35], may increase folate need up to six- to eightfold. Loss of folate from the liver, on an intake of essentially no folate (i.e., 2 μg folate daily), as assessed by liver biopsies, varies from 35 to 47 μg daily [36]. Assuming that extrahepatic stores are approximately half those in the liver, total daily folate loss in an adult eating essentially no folate would average about 60 μg. In such adults, morphologic evidence of folate deficiency in bone marrow and peripheral blood does not appear until liver fo-late falls below 1 μg/g [35].

In Canada, the mean national daily folate intake at ages 12–65 is approxi-mately 3 μg/kg body weight or about 205 μg/day for men and 149 μg/day for women [36]; this diet permits maintenance of normal and similar liver folate levels in both sexes [37]. Of 560 assayed livers from autopsies of children and adults in Canada, only 2 had folate content below 3 μg/g liver [37]. From this one can conclude that a typical North American daily dietary intake of approx-imately 1 μg folate/kg body weight keeps liver folate above 1 μg/g liver in more than 99% of cases and represents approximately the minimal daily requirement to sustain DNA normality but low stores. Daily dietary folate intake correlates significantly with red cell folate [38], and red cell folate reflects liver folate fairly closely [4, 14, 39]. From these correlations, one can predict that less than 1% of adults would have less than 1 μg folate/g liver on daily dietary intakes of ap-proximately 1 μg/kg body weight, i.e., 1 μg food folate per kilogram body weight would approximate the minimal daily adult requirement to sustain biochemical normality and low stores, that is no folate deficiency but stores inadequate to protect against situations of sustained increased demand, such as pregnancy.

On the typical United Kingdom diet, which contains about 190 μg folate/day

[38, 40, 41], a daily oral supplement of 100 µg PGA prevented any fall in mean red cell folate during pregnancy [42]. In women with poor folate stores who received essentially no other dietary folate, the progression of folate deficiency was as effectively prevented by administering 300 µg PGA daily in a food that impaired folate availability by 44% (thereby reducing the effective dose to 168 µg PGA daily) as it was by higher doses or more efficient vehicles [43]. Maternal milk folate content may be as high as 50–60 µg/l [43], suggesting a need for a daily supplement in this range for lactating women with minimal stores. However, supplementation may be unnecessary in lactating women in the middle socioeconomic class [45].

An appropriate maintenance dose to prevent folate deficiency in a premature infant is 50–100 µg/day, which is adequate to prevent the folate deficiency that commonly accompanies childhood hemolytic anemias [46]. In a study of 20 infants aged 2–11 months, Asfour et al. [47] demonstrated the nutritional adequacy of diets providing 3.6 µg folate per kg body weight daily over 6- to 9-month periods. In full-term infants, liver stores are about 224 µg [48]. By 2 weeks of age, serum and red cell folate levels fall below adult values and remain there during the entire 1st year of life [45, 48]. The needs of infants are adequately met by human or cow milk, which contain 50–60 µg folate/l [44, 49], but not by goat milk, which has a much lower folate content [47]. Milk from humans, cows, and goats contains a factor essentially unaffected by pasteurization which facilitates folate uptake by gut cells [50, 51]. This factor facilitates both absorption of dietary folate and reabsorption of bile folate. In Canadians, on a diet of approximately 3 µg folate/kg body weight, liver folate stores appear to be satisfactory (> 3 µg/g liver) in children as well as in adults [36].

The elderly are in the same category as younger adults with respect to folate needs [52]. On diets estimated to contain 135 µg folate/day, 21 elderly men and women living at home sustained red cell folate greater than 100 ng/ml and were hematologically normal; but 9 had red cell folate below 150 ng/ml [38]. Russell et al. [53] presented data showing that loss of gastric acid reduces PGA absorption in the elderly. There are more enteric bacteria in the achlorhydric elderly ($9 \cdot 10^7$ bacteria/ml of succus entericus) as compared to the normal elderly ($4 \cdot 10^3$/ml) [55, 56]. Red cell folate was not lower in achlorhydrics [54], suggesting that their reduced *percent* absorption of folate from food may be compensated for by the increased *quantities* of folate supplied by the increased colonies of enteric bacteria in the achylorhydric stomach and upper intestine [55]. Enteric bacteria can make more folate than there is in the average United States daily diet. there were 57–577 µg folate/24-h stool in six United States subjects [56].

Ettinger and Colman [56] found a higher serum folate:erythrocyte folate ratio in the elderly than in young adults in the United States, raising the possibility of less efficient cell uptake of folate and/or increased folate leakage from cells in the elderly. This is in keeping with the peaking of liver folate levels in the fourth decade of life in Canadians and a fall thereafter [37]. Vitamin B_{12} deficiency is more common in the elderly and is accompanied by an increased serum folate:erythrocyte folate ratio [3]; whether or not this explains the above findings remains to be determined.

This is a very active field of investigation. Current deliberations on folate requirements, in which the present author participated, will appear in the proceedings of a 13–22 March 1985 Geneva meeting of a Joint FAO/WHO expert Group on Requirements of Vitamin A, Iron, Folate, and Vitamin B_{12}, anticipated to be published in the WHO Technical Report Series. The underlying chemistry and biochemistry of the folates were reviewed in the last years [57–60]. The next review will be the published proceedings of the Eighth International Symposium on Pteridines and Folic Acid Derivatives which was held at McGill University in Montreal on 8–13 June 1986.

Conclusion

This review indicates that the ground-breaking model created by Finch and his colleagues [24] to stage the sequence of events in developing iron deficiency by using specific laboratory tests to identify specific stages can also be applied to folate deficiency (adding a "negative balance" stage) [61]. We have also applied this model to vitamin B_{12} deficiency [61] and suggest that it can and should be applied to any essential nutrient.

When staging models are used, it is important to recognize that variance of any laboratory test from the individuel's norm may signal developing deficiency before the test result exceeds the laboratory norm, which is derived from a group of persons [64].

The use of staging models in the B_{12} and folate field can go a long way toward clarifying the status of laboratory testing in the diagnosis of megaloblastic anemia [62]. This subject was reviewed in *Blood* in 1983 [63], prior to the preparation of models for the staging of degrees of vitamin depletion as measured by specific tests.

References

1. Herbert V (1962) Experimental nutritional folate deficiency in man. Trans Assoc Am Physicians 75:307–320
2. Herbert V (1962) Minimal daily adult folate requirement. Arch Intern Med 110:649–652
3. Herbert V (1985) Biology of disease: megaloblastic anemia. Lab Invest 52:3–19
4. Chanarin I (1979) The megaloblastic anemias, 2nd ed. Blackwell, Oxford
5. Herbert V, Colman N, Drivas G (1985) A proposed model of sequential stages in the development of folate deficiency anemia. Blood 66 (Suppl 1):30A
6. bis. Herbert V (1986) Folate deficiency, Book of abstracts. XXI congress, International Society of Hematology Sydney, Australia, 11–16 May 1986, p 216
7. Herbert V (1964) Studies of folate deficiency in man. Proc R Soc Med (Lond) 57:377–384
8. Weir DG, McGing PG, Scott JM (1985) Commentary: folate metabolism, the enterohepatic circulation and alcohol. Biochem Pharmacol 34:1–7
9. Herbert V, Das KC (1976) The role of vitamin B12 and folic acid in hemato- and other cell-poesis. Vitam Horm 34:1–30
10. Herbert V, Zalusky R (1962) Interrelations of vitamin B12 and folic acide metabolism: folic acid clearance studies. J Clin Invest 1263–1276
11. Noronha JM, Silverman M (1962) On folic acid, vitamin B12, methionine, and formiminunoglutamate metabolism. In: Heinrich HC (ed) Vitamin B12 and intrinsic factor, 2nd European symposium Enke, Stuttgart, pp 728–736

12. Shane B, Stokstad ELR (1985) Vitamin B12-folate interrelationships. vol. 5. Ann Rev Nutr 5:115–141
13. Perry J, Chanarin I (1977) Abnormal folate polyglumate ratios in untreated pernicious anemia corrected by therapy. Br J Haematol 35:397–402
14. Herbert V (1977) Folic acid requirement in adults (including pregnant and lactating females) Summary of the workshop. In: Folic acid. Biochemistry and physiology in relation to the human nutrition requirement. National Research Council. Food and Nutrition Board, Washington DC. National Academy of Sciences Ed., pp 247–55, 277–293
15. Lindenbaum J. (1977) Folic acid requirement in situations of increased need. In: Folic acid. Biochemistry and physiology in relation to the human nutrition requirement. National Research Council. Food and Nutrition Board, National Academy of (Ed) Sciences, pp 256–276
16. Scott JM, Weir DG (1981) The methyl trap hypothesis. A physiological response in man to prevent methyl group deficiency in kwashiorkor (methionine deficiency) and an explanation for folic acid-induced exacerbation of subacute combined degeneration in pernicious anemia. Lancet 2:337–340
17. Carmel R, Karnaze DS (1985) The deoxyuridine suppression test identifies subtle cobalamin deficiency in patients without typical megaloblastic anemia. JAMA 253:1284–1287
18. Metz J, Kelly A, Swett VC et al. (1968) Deranged DNA synthesis by bone marrow from vitamin B12 deficient humans. Br J Haematol 14:575
19. Das KC, Herbert V (1978) The lymphocyte as a marker of past nutritional status: Persistence of abnormal lymphocyte deoxyuridine (dU) suppression test and chromosomes in patients with past deficiency of folate and vitamin B12. Br J Haematol 38:219–233
20. Das KC, Manusselis C, Herbert V (1980) Simplifying lymphocyte culture and the deoxyuridine suppression test by using whole blood (0.1 ml) instead of separated lymphocytes. Clin Chem 26:72
21. Das KC, Herbert V, Colman N, Longo D (1978) Unmasking covert folate deficiency in iron-deficient subjects with neutrophil hypersegmentation: dU suppression tests on lymphocytes and bone marrow. Br J Haematol 39:357–375
22. Colman H, Herbert V (1980) Abnormal lymphocyte deoxyuridine suppression test: a reliable indicator of decreased lymphocyte folate levels. Am J Hematol 8:169–174
23. Steinberg SE, Fonda S, Campbell CL, Hillman RS (1983) Cellular abnormalities of folate deficiency. Br J Haematol 54:605–612
24. Bothwell TH, Charlton RW, Cook JD, Finch CA (1979) Iron metabolism in man. Blackwell, St-Louis
25. English E, Finch CA (1984) Iron deficiency: A systematic approach. Drug Ther Bull 14(4):45–46, 51–53
26. Green R, Kuhl W, Jacobson R et al. (1982) Masking of macrocytosis ba α-thalassemia in blacks with pernicious anemia. N Engl J Med 307:1322
27. Herbert V (1963) Current concepts in therapy: megaloblastic anemia. N Engl J Med 268:201–203, 368–371
28. Herbert V, Colman N, Jacob E (1980) Folic acid and vitamin B12. In: Goddhart RS, Shils ME (eds) Modern nutrition in health and disease. Lea and Febiger, Philadelphia, pp 229–259
29. Zalusky R, Herbert V (1961) Megaloblastic anemia in scurvy with response to 50 µg folic acid daily. N Engl J Med 265:1033–1038
30. Herbert V (1968) Nutritional requirements for vitamin B12 and folic acid. Am J Clin Nutr 21:743–752
31. Hoogstraten B, Cutner J, Natovitz B (1984) Sequence of recovery from multiple manifestations of folic acid deficiency. Mt Sinai J Med 31:10–16
32. Marshall RA, Jandl JH (1960) Response to "physiologic" doses of folic acid on megaloblastic anemia. Arch Intern Med 105:352
33. Amos RJ, Amess JAL, Hind CJ, Mollin DL (1982) Incidence and pathogenesis of acute megaloblastic bone marrow change in patients reveicing intensive care. Lancet 2:835–839
34. Amos RJ, Amess JAL, Nancekievill DG, Rees GM (1984) Prevention of nitrous oxide-induced megaloblastic changes in bone marrow using folinic acid. Br J Anaesth 56:103–107

35. Gailani SD, Carey RW, Holland JF, O'Malley JA (1970) Studies of folate deficiency in patients with neoplastic diseases. Cancer Res 30:327-333
36. Canada Bureau of Nutritional Sciences Dept. of Health and Welfare Nutrition Canada. Ottawa (1977) Food consumtpion report, 1977.
37. Hoppner K, Lampi B (1980) Folate levels in human liver from autopsies in Canada. Am J Clin Nutr 33:862-864
38. Bates CJ, Fleming M, Paul AA et al. (1980) Folate status and its relation to vitamin C in healthy elderly men and women. Age Ageing 9:241-248
39. Wu AI, Chanarin I, Slavin G, Levi AJ (1975) Folate deficiency in the alcoholic - its relationship to clinical and hematological abnormalities, liver disease, and folate stores. Br J Haematol 29:469-478
40. National Research Council. Folic acid: biochemistry and physiology in relation to the human nutrition requirement. DC: Food and Nutrition Board, National Academy of Sciences, Washington
41. Spring JA, Robertson J, Buss DH (1979) Trace nutrients. 3. Magnesium, copper, zinc, vitamin B6, vitamin B12, and folic acid in the British national household food supply. Br J Nutr 41:487-493
42. Chanarin I, Rothman D, Ward A, Perry J (1968) Folate status and requirement in pregnancy. Br Med J 2:390-394
43. Colman N, Green R, Metz J (1975) Prevention of folate deficiency by food fortification. II. Absorption of folic acid from fortified staple foods. Am J Clin Nutr 28:459-464
44. World Health Organization (WHO) (1968) Nutritional anemias: Report of a WHO scientific group. WHO technical report series 405. WHO, Geneva, p 37
45. Ek J (1983) Plasma, red cell, and breast milk folacin concentrations in lactating women. Am J Clin Nutr 38:929
46. Herbert V (1981) Nutritional anemias of childhood - folate, B12: The megaloblastic anemias. In: Suskind RM (ed) Textbook of pediatric nutrition. Raven, New York, pp d133-144
47. Asfour R, Wahbea N, Waslien C et al. (1977) Folacin requirements of children. III. Normal infants. Am J Clin Nutr 30:1098-1105
48. Dallman P (1974) Iron deficiency and related nutritional anemias. In: Nathan DG, Oski F (eds) Hematology of infancy and childhood. Saunders, Philadelphia, pp 126-134
49. FAO/WHO (Food and Agriculture Organization/World Health Organization) (1970) Expert Group: requirements of ascorbic acid, vitamin D, vitamin B12, folate and iron. WHO technical report series 452. WHO, Geneva, p 75
50. Colman N, Hettiarachchy N, Herbert V (1980) Detection of a milk factor that facilitates folate uptake by intestinal cells. Science 211:1427-1429
51. Colman N, Chen J-F, Gavin W, Herbert V (1981) Factors affecting enhancement by milk of folate uptake into intestinal cells. Blood 58(Suppl 1):26A
52. Rosenberg IH, Bowman BB, Cooper BA et al. (1982) Folate nutrition in the elderly. In: Symposium on the evidence relating selected vitamins and minerals to health and disease in the elderly population in the United States. Am J Clin Nutr 36 (Suppl):1060-1066
53. Russell RM, Krasinski SD, Samloff IM (1984) Correction of impaired folic acid (PteGlu) absorption by orally administered HCI in subjects with gastric atrophy. Clin Res 32:633A
54. Drasar BS, Hill MJ (1974) Human intestinal flora. Academic, New York
55. Herbert V, Drivas G, Manusselis C et al. (1984) Are colon bacteria a major source of cobalamin analogues in human tissues? 24-hour human stool contains only about 5 µg cobalamin but about 100 µg apparent analogue (and 200 µg folate). Trans Assoc Am Physicians 97:161-171
56. Ettinger S, Colman N (1985) Altered relationship between red cell and serum folate in the aged, suggesting impaired erythrocyte folate transport. Fed Proc 44(4):1283
57. Blair JA (1983) Chemistry and biology of pteridines: pteridines and folic acid derivatives. de Gruyter, New York
58. Goldman ID, Chabner BA, Bertino JR (1983) Folyl and antifolyl polyglutamates. Plenum, New York
59. Blakley RL, Benkovic SJ (1984) Folates and pterins: chemistry and biochemistry of folates, vol I. Wiley, New York
60. Hillman RS, Finch C (1985) The red cell manual, edn 5. Davis, Philadelphia

61. Herbert V (1987) The Herman Award Lecture. Nutrition science as a continually unfolding story: the folate and vitamin B12 paradigm. Am J Clin Nutr 46:387–402

62. Lindenbaum J, Nath B (1980) Megaloblastic anemia and neutrophil hypersegmentation. Br J Haematol 44:551

63. Lindenbaum J (1983) Status of laboratory testing in the diagnosis of megaloblstic anemia. Blood 61:624–627

64. Herbert V, Memoli D, McAleer E, Colman N (1986) What is normal? Variation from the individuals norm for granulocyte "lobe average" and holo-transcobalamin II (holo-TC II) diagnoses vitamin B12 deficiency before variation from the laboratory norm. Clin Res 34:718A.

Chapter 9

Prevalence of Folic Acid Deficiency in the French Population

A. LEMOINE, G. POTIER DE COURCY, S. HERCBERG, and
C. LE DEVEHAT

Introduction

The most recent French studies on vitamins and public health have identified folates as one of the primary high-risk vitamins, that is, for which there is the greatest risk of deficiency. While we have only recently become aware of this situation, it is actually not new. As in other industrialized countries, changes in life-style and dietary habits in France have been particularly rapid, and our current diet is very different from traditional models in terms of composition, scheduling, distribution, etc. These dietary changes and the transformation in our life-style have had an influence not only on the composition of meals but perhaps also upon nutritional requirements themselves.

In view of the current opulence in our diet, while the folates must be given particular priority, the other micronutrients must not be neglected: "abundance" is not synonymous with "balance." Furthermore, the decrease in requirements and, thus, in caloric intake requires that foodstuffs have a higher relative concentration of vitamins. Unfortunately, we find that there is an increase in consumption of "empty calories" in the form of highly concentrated or refined foodstuffs (pure sugars, alcohols, fats, etc.). The nutrient concentration in the diet is thus decreasing, and the decrease in vitamin intake is worrisome, especially when caloric intake is low. Unlike other micronutrients, vitamins are fragile and are sensitive to various types of physical and chemical aggression, such as those seen during storage, transformation, and preparation.

Unlike the specific vitamin-deficiency diseases such as beri-beri, pellagra, and scurvy, which are exceedingly rare in France, effects of the far more common subclinical vitamin deficiencies are very difficult to assess. Clinical repercussions are more discrete, occur with far greater latency, and are less specific since they are frequently multifactorial. Rather than a specific pathological entity the primary short-term and medium-term effect of vitamin deficiency is a generalized increase in morbidity. Over the long term, the role of chronic vitamin deficiency in multifactorial pathologies such as cancer, arteriosclerosis, congenital malformations, and immune disorders is currently under study.

Folate deficiency is the second most common cause of nutritional anemia and the most common of the vitamin deficiencies [11]. If it is diagnosed only after the onset of anemia, it is detected 3–4 months after the onset of the first symptoms (e.g., granulocyte hypersegmentation [14]). It is far preferable to diagnose deficiencies as early as possible, before the onset of patent clinical pathology so that appropriate prophylactic and hygienic measures can be instituted. An evaluation of the risk of vitamin deficiencies in the overall population is thus indispensable. Systematic screening and study of groups at risk for vitamin deficiency should be maintained so that appropriate public health measures can be taken.

Dietary Folate Sources

Only recently studies of folate intake by the French population have been carried out, and the results of few such studies are available as of yet. The ESVI-TAF survey [19] was carried out between 1979 and 1982 to evaluate vitamin intake in a representative sample of the French population, compared with groups of obese subjects and excessive drinkers. A control group of 362 subjects in good health, aged 18–44 years, and living in metropolitan France was selected; they had no specific medical history, were not receiving medications or hormonal contraceptives, and had never received dietary counseling. Body weight was normal (less than 116% of the ideal weight; see Table 1). Men in this group consumed less than 44 g alcohol per day and women less than 22 g per day. Diet was determined over 7 days using a notebook, and patients were also interviewed concerning their diet. Vitamin composition of the diet was calculated using the computerized equivalence table of the Hospices Civils de Lyon, representing a compilation of analytical data and other published tables.

In parallel, intake of two groups considered to be at risk of dietary deficiency was calculated: 57 obese subjects and 107 heavy drinkers. These groups differed from the control population only in terms of body weight and alcohol consumption, respectively. Obese subjects were selected on the basis of their weight index ($\geqslant 1.47$; see Table 1), and heavy drinkers on the basis of their daily alcohol consumption ($\geqslant 88$ g for men; $\geqslant 44$ g fork women).

Mean folate ration in the control group was significantly higher in men than in women, while the opposite was the case for folate concentration in the diet (folate ration calculated per 1000 kcal [12] (Table 1).

These results, comparable to those seen in other European countries [25] are highly divergent from the recommended daily intake for the French population [8]: as a function of sex, between 71% and 77% of subjects consumed less than 50% of the recommended folate intake. Some authors have pointed out deficiencies of certain tables of folate concentrations [24], thus explaining the discordance between recommended and observed daily intakes [1]. Determination of dietary folates should take into account both free folates and polyglutamates, and this poses certain methodological difficulties. Depending upon the assay technique used for determination of folates, results can vary between 105 and 242 µg/24 h for the same dietary composition [20], and the folate intake in the typical American diet can range between 200 and 600 µg [3]; according to Spring et al. [25], the typical British diet provides between 130 and 300 µg folates/24 h.

A current recalculation of subgroups of the ESVITAF study using an updated table of dietary composition could show the current situation to be more in keeping with current French recommendations, and there appears to be a ten-

Table 1. Daily folate ration (DFR) and dietary folates concentration (DFC) expressed in µg/24 h

	Controls	Obese	Drinkers
Men			
n	189	11	92
DFR	182 ± 77	286 ± 76[b]	230 ± 120[b]
DFC	59.4 ± 24.9	80.9 ± 17.6[a]	59.3 ± 30.1
Women			
n	173	46	15
DFR	160 ± 77	169 ± 137	195 ± 99
DFC	68.4 ± 41.1	73.6 ± 43.9	61.9 ± 37.9

Data are from the ESVITAF study (see [19]).

"Obese" defined as having a weight index $\geqslant 1.47$; weight index was calculated as follows:

$$\text{Weight index} = \frac{\text{Weight (kg)}}{\text{Height (cm)} - 100 - \left(\dfrac{\text{Height (cm)} - 150}{x} \right)}$$

where $x = 4$ for men, 2 for women.

"Drinkers" defined as having daily alcoholic consumption of $\geqslant 88$ g for men, $\geqslant 44$ g for women.

Figures are mean $\pm$ SD; differences between men and women significant at $p < 0.05$.
[a] Difference from controls, $p < 0.01$.
[b] Difference from controls, $p < 0.001$.

dency for experts at the WHO to recommend less than the currently recommended level in France; these studies have not yet been published.

Compared with the control groups, folate intake is significantly higher in the populations of male drinkers and obese subjects. Among women intake is higher in the group with high alcohol consumption, however the difference from the control group is not significant, probably because of the limited population size.

Folate intake is thus more satisfactory overall in obese subjects than in controls, due to a greater dietary intake and a positive correlation between folate intake and intake of calories and macronutrients shown in this study $(p < 0.001)$.

The relatively high dietary folate consumption by drinkers is more surprising; some alcoholic drinks, in particular beer, contain nonnegligible quantities of folates, the bioavailability of which is questionable [18]. We should also note that the group of excessive drinkers, meeting the same selection criteria as the controls except for their alcohol intake, did not present the pathological symptomatology characteristic of chronic alcoholics, and this population is thus different from severe alcoholics who generally have a very poor diet and thus require a separate study.

In view of the high frequency of folate deficiency in pregnant women [7] several dietary surveys have been carried out on this subgroup, however they are still few and are difficult to compare. Among pregnant women living in Western countries, mean intakes of 82 µg/24 h [9] or even lower [23] have been described, however these are calculated for free folates, and a value of 190 µg total folates recently calculated for the English population appears to give a more realistic picture [4].

A recent French study [26] was carried out on 215 pregnant women living in the Paris region, either born in France or abroad; folate intake varied as a function of the stage of pregnancy and ethnic origin between 280 and 380 µg/24 h. At the beginning of pregnancy, in the immigrant subgroup folate consumption was 100 µg/24 h lower than that in the population of French origin. Calculating folate intake as a function of category of foodstuffs, 60% of dietary folates were provided by fruits and vegetables. Comparing the folate intake with dietary recommendations, the pregnant women here studied were in a situation comparable to that of women in the ESVITAF group; only 23% consumed more than half of the recommended ration (Fig. 1), 800 µg/24 h in pregnant women [8].

The criteria chosen as a threshold for vitamin deficiency are purely arbitrary, and the level of 2.5 µg/l for plasma folates is more rigorous than that chosen in the majority of international studies. It is generally agreed that the hematologic disorders characteristic of folate deficiency are significantly more frequent when levels are below this level, while subjects presenting plasma levels between 2.5 and 5 µg/l do not appear to show substantial clinical or laboratory abnormalities. Nevertheless, we consider this group to be at risk of deficiency in case of a momentary decrease in dietary consumption and/or an increase in requirements (pregnancy, disease, etc.).

Laboratory findings confirm results of the dietary survey for obese subjects: circulating folate levels are higher than in controls, with a significant difference in erythrocyte folate levels (Table 3). On the other hand, the folate status appears to be worse in the drinkers than in the other groups (Table 3). The subjects studied were not severe alcoholics and had no clear clinical signs of chronic alcoholism, and this level of alcohol consumption is relatively common in France. De-

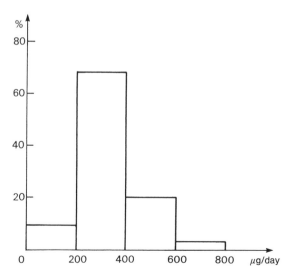

Fig. 1. Distribution of dietary folate intake in a group of pregnant women in the Paris region ($n = 215$) [26]. Daily recommendations in France for folacin intake: adults, 400 µg; lactating women, 500 µg; pregnant women, 800 µg

spite folate intake levels comparable to those in corresponding controls, abnormally low folate levels were significantly more frequent (Fig. 2). Severe alcoholics probably constitute the largest high-risk group in France, however at present we do not have a representative study of this population, which is difficult both to define and to study. The very high frequency of folate deficiency (sometimes severe) in alcoholic patients is well known [5, 6, 13, 17].

Folate status in 47 women from Paris at the beginning of pregnancy [26] was very similar to that in the ESVITAF reference female population (plasma folates 7.1 ± 3.1 µg/l, erythrocyte folates 314 ± 98 µg/l). Despite equivalent folate intakes, these values were significantly lower in the group of 96 women in the 9th month of pregnancy (respectively, 3.9 ± 1.3 and 213 ± 74 µg/l). Despite the fact that folate intake was higher here than in other European surveys [4, 9, 23], it was inadequate to maintain initial circulating folate levels at the end of pregnancy.

Correlations between dietary intake and folate status were fairly poor, similar to those reported by other authors [2]. Alongside the positive correlation between folate and caloric intake indicated above, the ESVITAF survey showed a high correlation between circulating folates and vitamin C intake in women, as previously reported by Bates et al. [2] in a population of elderly subjects. On the basis of information for the ESVITAF group, study of sensitivity, specificity, and predictive value of the dietary survey confirms the relatively poor correlation between laboratory values and dietary intakes [15]. Nevertheless, it is important to note that disturbances in laboratory values develop only after a long period of severe deficiency [14], and a better assessment of diet along with more precise tables would probably provide a better correlation with folate levels. This is indeed the case for the study of pregnant women in the Paris region [26], which showed a positive correlation between circulating and dietary folate levels, in particular in terms of consumption of fruits, vegetables, and cereals. A significant negative correlation was seen in the study between folate status and consumption of tea, high in French women at the end of pregnancy.

Folate status in elderly subjects has long been the source of considerable concern, and studies on this group have recently been carried out. Folate deficiency in this population, vitamin B_{12} and iron deficiencies are also very frequent, is probably responsible for the majority of dietary anemias [27]. Institutionalized elderly subjects have been given particular attention since dietary surveys of subjects living at home are very difficult for methodological reasons. Dietary intake has been inadequately evaluated in France, and there are no specific recommendations for the elderly. Nevertheless, if we consider that the recommended intake is identical to that in younger adults (400 µg/24 h [8]) and in view of recom-

Table 2. Folate ration (µg/24 h) in elderly subjects living at home

Study		Men	Women
Bates et al.	[1]	136 ± 44	116 ± 29
Nutrition Canada	[22]	151	130
Jagerstad	[16]	157	129

mendations in other Western countries (Table 2), we can assume that dietary intake is clearly insufficient in this population. Dietary habits are quite divergent between regions, and it is likely that the prevalence of this type of deficiency is highly variable in France.

Folate Status

Results concerning plasma and erythrocyte folate levels in various groups studied in the ESVITAF survey are summarized in Table 3. A positive correlation ($p < 0.001$) was seen between plasma and erythrocyte folate levels. While the folate status in the control group would initially appear to be satisfactory, the pre-

Table 3. Plasma and erythrocyte folate levels in the various groups studied, expressed in µg/l

	Controls	Obese	Drinkers
Men			
n	189	11	92
Plasma	6.8 ± 2.9*	7.1 ± 2.8	7.0 ± 3.2
Erythrocyte	299 ± 96	408 ± 177[b]	293 ± 134
Women			
n	173	46	15
Plasma	6.8 ± 3.1	7.4 ± 3.9	6.1 ± 2.6
Erythrocyte	282 ± 100	341 ± 130[b]	230 ± 77[a]

*: mean ± SD
[a] Difference from controls, $p < 0.05$.
[b] Difference from controls, $p < 0.01$.

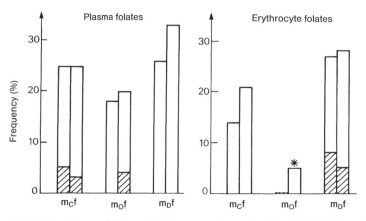

Fig. 2. Prevalence of folate deficiency, as determined by ESVITAF study [19]. *C*, controls; *O*, obese; *D*, drinkers; *m*, male subjects; *f*, female subjects. *Hatched bars*, deficient; *unhatched bars*, marginal; *asterisk*, difference from controls, $\alpha < 0.01$ (χ^2 test). Levels of plasma folates (folacin status) defined as follows (µg/l): normal, >5; marginal, 2.5–5; deficient, <2.5. Erythrocyte folates: normal, >200; marginal, 100–200; deficient, <100

valence of folate deficiency as defined by the thresholds set in Fig. 2 ranges between 14% and 25%, a high proportion for a reference population selected as a function of rigorous inclusion criteria and thus actually representative of a small proportion of the overall French population.

Conclusions

Biological signs of folate deficiency are among the most common vitamin deficiency in industrialized countries; requirements are not being adequately met. Furthermore, there are certain subgroups of the population whose significantly higher folate requirements cannot be met with conventional dietary measures; the recommended intake of 800 µg/24 h in pregnant women [8] is a case in point. The recommendations are thus met only rarely, and questions must be posed as to whether they are realistic; the recommended daily intake is defined as the level meeting requirements of 95% of a given population [21]. This is a public health objective, and for this reason the American Academy of Sciences has recently decided not to revise downward the recommended intake of vitamins and mineral oligo-elements, levels which some authors currently consider to be excessive.

At present, we are poorly informed concerning folate requirements and intake levels in the French population due to the insufficient information provided by dietary studies, as well as to the poor precision of the food tables. Concerning the latter, a dietary data bank is now being constituted in France; this will constitute an invaluable aid to specific and rigorous dietary studies [10].

Furthermore, studies of dietary folate bioavailability are essential since at least 50% of dietary folates are in the form of polyglutamates, and the mean percentage of absorption of these compounds is not known. In view of this uncertainty, as well as of the substantial dietary requirements in pregnant women, we consider it imperative that the recommended folate intake in adult women be maintained at 400 µg/24 h, and systematic supplementation is recommended in high-risk groups.

References

1. Bates CJ, Black AE, Philipps DR, Wright AJA, Southgate DAT (1982) The discrepancy between normal folate intakes and the folate RDA. Hum Nutr Appl Nutr 36A:422–429
2. Bates CJ, Fleming M, Paul AA, Black AE, Mandal AR (1980) Folate status and its relation to vitamin C in healthy elderly men and women. Age Ageing 9:241–248
3. Beck WS (1977) Folic acid deficiency. In: Williams WJ et al. (eds) Haematology, 2nd edn. McGraw-Hill, New York, pp 334–335
4. Black AE, Paul AA, Hall C (1985) Footnotes to food tables 2. The underestimations of in take of lesser B vitamins by pregnant and lactating women as calculated using the fourth edition of Mc Cance and Widdowsons's "The composition of foods". Hum Nutr Appl Nutr 39A:19–22
5. Bonjour JP (1980) Vitamins and alcoholism. II. Folate and vitamin B12. Int J Vitam Nutr Res 50:96–121

6. Celada A, Rudolf H, Donath A (1979) Effect of experimental chronic alcohol ingestion and folic acid deficiency on iron absorption. Blood 54:906–915
7. Chanarin I (1985) Folate and cobalamin. Clin Haematol 14:629–641
8. Dupin H (1981) Apports nutritionnels conseillés dans la population française. Technique et Documentation Lavoisier. CNRS, CNERNA, Paris
9. Elsborg L, Rosenquist A (1979) Folate intake by teenage girls and by pregnant women. Int J Vitam Nutr Res 49:70–76
10. Favier JC (1983) Elaboration d'une banque de données sur la composition des aliments. Cah Nutr Diet 18:137–143
11. Hall CA, Bardwelle SA, Allen ES, Rapazzo ME (1975) Variations in plasma folate levels among a group of healthy persons. Am J Clin Nutr 28:854–857
12. Hansen RG, Wyse BW (1980) Expression of nutrient allowances per 1000 kilocalories. J Am Diet Assoc 76:223–227
13. Heilmann E, Koschatzki J (1978) Folsäure und vitamin B12 bei chronischen alkoholikern. Schweiz Med Wochenschr 108:1920–1923
14. Herbert V (1962) Experimental nutritional folate deficiency in man. Trans Assoc Am Physicians 75:307–320
15. Herberth B, Potier de Courcy G, Sancho J et al. (1985) ESVITAF survey on the vitamin status of the french: relationship between nutrient intake and biochemical indicators. Acta Vitaminol Enzymol (Milano) 7:207–215
16. Jagerstad M, Westesson EK (1979) Folate. Scand J Gastroenterol 14 (Suppl 52):196–202
17. Lederer J (1981) L'anémie mégalocytaire des alcooliques et le rôle de l'acide folique. Sem Hop Paris 29:1259–1265
18. Lederer J, Bataille JP (1980) Influence respective de la bière, du vin et des spiritueux sur le taux d'acide folique et l'érythrocyte. Rev Alcool 26:135–143
19. Lemoine A, Le Devehat C, Herbeth B et al. (1986) ESVITAF: Enquête sur le statut vitaminique de trois groupes d'adultes français (témoins, obèses, buveurs excessifs). Ann Nutr Metab 30 (Suppl 1):1–94
20. Moscovitch LF, Cooper BA (1973) Folate content of diets in pregnancy: comparison of diets collected at home and diets prepared from dietary recors. Am J Clin Nutr 26:707–714
21. WHO (1972) Les anémies nutritionnelles. Série de rapports techniques no. 503. WHO, Genève
22. Nutrition Canada (1973) National Survey Canada. Information Canada, Ottawa
23. Papoz L, Eschwege E, Cubeau J, Pequignot G, Barrat J, Le Lorier G (1980) Comportement alimentaire au cours de la grossesse. Rev Epidemiol Santé Publique 28:155–167
24. Phillips DR, Wright AJA (1983) Studies on the response of *Lactobacillus casei* to folate vitamin in foods. Br J Nutr 49:181–186
25. Spring JA, Robertson J, Uss DH (1979) Trace nutrients. 3. Magnesium, copper, zinc, vitamin B6, Vitamin B12 and folic acid in the British house hold food supply. Br J Nutr 41:487–490
26. Tchernia G, Blot I, Zittoun J, Potier de Courcy G (1986) Folates: pour une supplémentation? In: L'alimentation des femmes enceintes. Colloque International, Cidil, Paris
27. Zittoun J, Potier de Courcy G (1985) Acide folique: une carence sans conséquences? In: L'alimentation des personnes âgées. Colloque International Cidil, Paris

Chapter 10

Effects of Some Drugs and Alcohol on Folate and Cobalamin Metabolism

J. ZITTOUN

Defects in DNA biosynthesis resulting in megaloblastic anemia often are due to folate and/or cobalamin deficiency. This vitamin deficiency can be due to deficient or inadequate diet, to impairment of absorption or utilization, or to increased needs of these vitamins.

Cobalamin and Folate Metabolism

Metabolism of cobalamin is extensively discussed in Chaps. 4 and 5 of this volume. Folic acid (pteroylmonoglutamic acid) is present in food. Almost all dietary folates are pteroylpolyglutamates, with three or more glutamate residues added in γ carboxyl linkage to the glutamic acid found in pteroylglutamate. These polyglutamates are reduced (at 5-8 positions of the pteridine moiety) and a methyl or formyl group is attached at the N5 or N10 position (Fig. 1).

Folate absorption at low concentrations is a carrier-mediated, saturable, pH-dependent process that occurs mainly through the proximal gut [1, 2]. At higher concentrations, transport of folate across the small intestine is passive. Dietary folates are first hydrolyzed to monoglutamate forms by an intestinal glutamyl-carboxypeptidase (conjugase) present in the intestinal juice and on the mucosal brush border.

Hydrolysis of polyglutamates appears to be an obligatory step in the absorption of folates. Methyltetrahydrofolate monoglutamates are absorbed rapidly and remain unchanged while other monoglutamates are converted by the mucosal cell to methyltetrahydrofolate and, to a lesser degree, to formyltetrahydrofo-

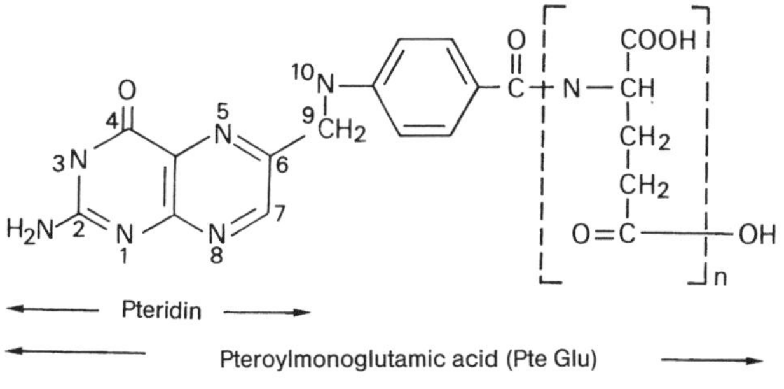

Fig. 1. Pteroyl mono and polyglutamates. (n: number of glutamate residues)

late. Some 10%–20% of methyltetrahydrofolate in the portal circulation is taken up by the liver, and the remainder is cleared by peripheral tissues [3]. Uptake of methyltetrahydrofolate by cells is an energy-dependent, carrier-mediated process (see Chap. 19, this volume). Inside the cells, methyltetrahydrofolate monogluta-mate is converted, after demethylation and possibly formylation, (see Chap. 3) into polyglutamates by folylpolyglutamylsynthetase [4]. The formation of poly-glutamates induces better retention of folate inside the cells; in addition, poly-glutamates are better substrates for folate-dependent enzymes than their corre-sponding monoglutamates [5]. Most methyltetrahydrofolate in the liver is poly-glutamate and rapidly released to peripheral tissues via a folate enterohepatic cycle. In a rat model, it has been shown that folates are excreted into bile by the liver and undergo reabsorption. The amount of folate in bile is much higher than that in plasma. The liver exerts a regulatory effect to maintain folate homeostasis [6].

Drug-Induced Megaloblastic Anemia

Drugs which affect DNA biosynthesis can induce megaloblastic anemia [7] (Table 1).

Table 1. Drugs inducing megaloblastic anemias

Drugs	Effects
Purine and pyrimidine analogs	
Purine (6-mercaptopurine, 6-thioguanine, azathioprine)	Inhibition of purine biosynthesis
Pyrimidine (5-fluorouracil, arabinosylcytosine)	Incorporation into RNA and DNA
Drugs inhibiting crucial enzymes in DNA biosynthesis	
Hydroxyurea	Inhibition of ribonucleotide reductase
5-fluorouracil	Inhibition of thymidylate synthase
Arabinosylcytosine	Inhibition of DNA polymerase
Drugs interfering with folate metabolism	
Methotrexate, pyrimethamine, triamterene, trimethoprim	Inhibition of dihydrofolate reductase
Sulfasalazine	Inhibition of absorption
Cholestyramine	Inhibition of absorption
Anticonvulsivants	Increase of folate catabolism
Drugs impairing cobalamin metabolism	
Neomycin, para-aminosalicylic acid, colchicine, biguanides	Inhibition of absorption
Histamine H_2 receptor antagonists	Inhibition of absorption
Nitrous oxide	Inactivation of methionine synthase

Pyrimidine and Purine Analogs

The pyrimidine and purine analog incorporate into DNA instead of the natural substrates and induce abnormalities of DNA replication (see Chap. 1).

The purine analog, 6-mercaptopurine, 6-thioguanine, and azathioprine have been reported to cause macrocytosis and megaloblastic changes in some cases. These drugs are known to exert an inhibitory effect at different points during the early stages of purine biosynthesis, but megaloblastic anemia caused by these drugs is less frequent than that induced by pyrimidine analogs.

The main pyrimidine analogs capable of inducing megaloblastic anemia are antineoplastic agents. 5-Fluorouracil (5FU) incorporates into RNA; it is also a potent inhibitor of thymidylate synthase (TS) and thus inhibits folate metabolism (see Chap. 20). Arabinosylcytosine (Ara-C), currently used in the treatment of acute myeloid leukemia, competes with the natural substrates, cytidine and deoxycytidine.

Inhibitors of Crucial Enzymes in DNA Biosynthesis

Inhibitors of Ribonucleotide Reductase. Hydroxyurea, a drug usually utilized in the treatment of myeloproliferative disorders, is a potent inhibitor of ribonucleotide reductase, the enzyme converting nucleotide diphosphates to deoxynucleotide diphosphates [8]. Patients treated with hydroxyurea, present with important macrocytosis and megaloblastic changes which disappear when treatment is stopped.

Inhibitors of Thymidylate Synthase. 5FU, the analog of the natural substrate uracil, is a potent inhibitor of TS and thus inhibits thymidylate biosynthesis. Normally, this synthesis involves methylation of deoxyuridylate (dUMP) by TS associated with the folate coenzyme N5N10 methylene tetrahydrofolate. 5FU after conversion into 5-fluorodeoxyuridylate (5FdUMP) binds TS with a higher affinity than the natural substrate dUMP. The stability of the covalent ternary complex, 5FdUMP-TS-folate coenzyme leads to the antineoplastic effects of 5FU (see Chap. 20).

Inhibitors of DNA Polymerase. The active form of Ara-C, the triphosphate form Ara-CTP, has been shown to inhibit DNA polymerase, the enzyme which catalyzes the assembly of the four deoxyribonucleotide triphosphates (see Chap. 1).

Drugs that Interfere with Folate Metabolism

Inhibitors of Dihydrofolate Reductase. Methotrexate, pyrimethamine, triamterene and trimethoprim inhibit dihydrofolate reductase (DHFR) but with marked differences in their propensity to induce megaloblastic changes in humans. Methotrexate has a very high affinity for mammalian DHFR, whereas the three others have a much decreased affinity for this enzyme [9]. (The metabolism and mechanism of action of methotrexate is reviewed in Chap. 20.) Pyrimethamine, used in combination with dapsone for malarial treatment, may induce severe megaloblastic anemia especially at high dosages and with prolonged use [10]. Triamterene, used as a diuretic, has a low affinity for DHFR. However some cases of megaloblastic anemia have been reported in patients with preexisting depletion of folate stores, such as in alcoholics [11]. Trimethoprim is a potent inhibitor of bacterial DHFR, and thus exerts an efficient antibacterial effect but it can weakly inhibit mammalian DHFR. The occurrence of megaloblastic anemia is rare except in patients receiving high doses and with depleted folate stores [12]. Concomitant administration of folinic acid, which does not interfere with the antibacterial effect of trimethoprim, should be considered in patients at high risk for folate deficiency and with lowered folate stores.

Drugs Interfering with Folate Absorption. Sulphasalazine, currently utilized in patients with inflammatory bowel diseases such as ulcerative colitis, may induce a depletion of folate stores by two mechanisms: inhibition of intestinal absorption of polyglutamates and monoglutamates [13] and inhibition of some hepatic enzymes which catalyze reactions requiring different folate coenzymes, probably by interfering with a folate recognition site common to these enzymes [14]. Sulphasalazine reduces red cell folate levels in patients on long-term maintenance with the drug [15]. Prophylaxis with folic acid should therefore be considered in patients receiving long-term treatment with sulphasalazine, especially if associated with another cause of folate deficiency (e.g., malnutrition, hemolysis, pregnancy). Cholestyramine, a nonabsorbable anion-exchange resin, may interfere with folate absorption and inhibit the enterohepatic cycle [16]; however, no case of megaloblastic anemia has been reported after treatment with cholestyramine.

Drugs Inducing Folate Deficiency by Unknown Mechanisms. Long-term treatment with anticonvulsivants usually induces a decrease of folate in serum, red cells, and cerebrospinal fluid ([9] and Chap. 11, this volume). The mechanisms involved in this deficiency are complex. Malabsorption of polyglutamates by inhibition of intestinal conjugase has been reported [17] but not confirmed [18]. Modification of intraluminal pH changing the microclimate has been proposed to explain malabsorption related to phenytoin [19], but this hypothesis has not been confirmed by other studies [20]. Increased catabolism and increased folate losses can also contribute to folate deficiency [21]. The desirability of administering folic acid supplements during long-term treatment by anticonvulsivants has been raised. In some epileptic patients, folic acid therapy has seemed to precipitate seizures. Whereas some well-controlled trials have shown no effect of folic acid supplements on the frequency of seizures ([22] and Chap. 11, this volume). Folate deficiency, and even in some cases megaloblastic anemia, have been reported in some women taking oral contraceptives [23]. The incidence of megaloblastic anemia in such women is very low except when another cause of folate deficiency is present [16]. However, about 20% of women taking oral contraceptives have been found to have megaloblastic changes on smears of cervical epithelium which disappeared after folic acid therapy [24].

Drugs Impairing Cobalamin Metabolism

Several drugs have been well shown to inhibit the absorption of cobalamins in humans, mainly biguanides, neomycin, para-aminosalicylic acid, colchicine, [7]. Histamine H_2-receptor antagonists may also induce malabsorption of protein-bound cobalamins (Chap. 2, 4, 5). However, megaloblastic anemia or neurological disturbances secondary to this cobalamin malabsorption is very uncommon. In contrast, the effect of nitrous oxide on the occurrence of megaloblastic anemia by impairing cobalamin metabolism and inactivating methionine synthase is well established [25, 26]. The mechanism of action of nitrous oxide is extensively described in Chap. 3.

Hematologic Effects of Alcoholism

Macrocytosis is the most common feature in alcoholic patients [27]. It is sometimes associated with megaloblastic anemia, the incidence of which varies according to the series. Besides megaloblastic changes, other abnormalities may be present, including ring sideroblasts and vacuoles in the cytoplasm of erythroid or white cell precursors [28]. The latter two findings are commonly seen after acute ethanol intoxication [29]. In many patients, bone marrow cells may be megaloblastic without anemia [28]. Marrow abnormalities revert rapidly when alcohol ingestion is stopped or when a normal diet is reinstituted.

Megaloblastic anemia in alcoholics is usually associated with folate deficiency. In blood and in bone marrow, cellular and biochemical abnormalities are those found in folate deficiency (see Chap. 2). Folate in serum and red cells are decreased. In some patients, levels of serum folate may be normal with anemia, which is responsive to folic acid [30]. Serum folate is often decreased despite the absence of cellular abnormalities found in folate deficiency [31]. Indeed, serum folate levels can undergo considerable short-term fluctuations due to diet variations or alcohol ingestion and are not reliable in evaluating folate stores. Red cell folate, however, has a slow turnover and is an important tool for defining folate stores. In some alcoholic patients, anemia can be shown to correlate with low red cell folate without megaloblastic changes. In these last mentioned conditions, anemia is usually not responsive to folic acid [31].

Levels of serum cobalamin are normal or more often increased [27]. This increase in alcoholic does not reflect tissue cobalamin stores, which are in fact often decreased.

Indeed, in addition to its action on folate and cobalamin metabolism, alcohol exerts direct toxicity on bone marrow [32] by suppressing hematopoietic progenitor cell proliferation [33]. Alcohol may also induce thrombopenia and neutropenia, not always related to folate deficiency, sepsis, or hypersplenism [16].

Pathogenesis of Alcohol-Induced Folate and Cobalamin Deficiency

Two main factors are involved in the genesis of folate deficiency in alcoholic patients: malnutrition and the antifolate action of alcohol per se [34, 35]. Megaloblastic anemia is commonly present in alcoholics who eat poorly and drink wine or whisky, which contain little folate; beer, on the other hand, is rich in this vitamin [36, 37]. A normal diet, however, can prevent the toxic effect of alcohol on hematopoiesis.

Mild malabsorption of folic acid has been reported in alcoholics, but in these cases, the relative contributions of alcohol intoxication and folate deficiency are difficult to define. Both can induce impairment of small-bowel functions [35], and folic acid absorption reverts to normal after folic acid administration even if there is no reduction of alcohol intake [27]. Inhibition of some intestinal enzyme activities by alcohol is reversible after folic acid supplements [36, 38, 39].

Toxicity of alcohol on folate metabolism also seems related to the interruption of the enterohepatic cycle of this vitamin [35]. Alcohol stops the delivery of storage folate from the liver into the circulation. Alcohol administration induces a rapid drop in serum folate and consequently a depletion of available folate. These metabolic abnormalities revert to normal after withdrawal of alcohol.

Many factors secondary to alcohol ingestion may induce disturbances of cobalamin metabolism. An increased incidence of atrophic gastritis has been reported in chronic alcoholics, usually in association with malnutrition [40]; however, this has not been demonstrated to cause lack of intrinsic factor. Chronic pancreatitis, a well-known complication of chronic alcoholism may cause an abnormal Schilling test in about half of patients (see Chap. 5, 7), but this malabsorption of cobalamin is rarely associated with megaloblastic anemia. Serum cobalamin levels are usually increased in alcoholics because of defective cobalamin storage by liver and hepatic stores are often decreased [41].

Despite these various abnormalities, hematological and neurological features related to cobalamin deficiency are rare in alcoholics except when pernicious anemia is associated.

Macrocytosis of Alcoholism

Increased mean corpuscular volume (MCV) is a very common finding in alcoholics even when anemia and/or folate deficiency are absent. Macrocytosis has been found in more than 80% of chronic alcoholics [42–44]; it may be used as one of the most sensitive screening test for alcoholism associated with assay of serum γ glutamyl-transpeptidase. Macrocytosis is usually mild and not correlated with the severity of liver dysfunction. It may be seen in alcoholics with normal biopsies.

The mechanism of macrocytosis is complex and not well defined [27]. Possible causes of this macrocytosis include folate deficiency, reticulocytosis in response to hemolysis or to gastrointestinal bleeding, and abnormalities of red cell membranes due to increased deposit of cholesterol and phospholipides secondary to disturbances in serum lipoproteins [45]. Other mecanisms, not yet identified, may also be involved.

Folate deficiency is variable and not of sufficient severity to explain macrocytosis, which often persists after folic acid administration. Normal levels of serum and red cell folate in alcoholics exclude folate deficiency as the cause of macrocytosis. In contrast, decreased folate levels are not always associated with abnormalities of DNA synthesis, as shown by the deoxyuridine suppression test, which often is normal. Macrocytosis may be partially corrected with folic acid [46].

Indeed, macrocytosis diminished after withdrawal of alcohol but the decrease is slow and MCV does not become completely normal until after several weeks or months [42, 43].

References

1. Rosenberg IH (1976) Absorption and malabsorption of folates. Clin Haematol 5:589–618
2. Leslie GI, Rowe PB (1972) Folate binding by the brush broder membrane proteins of small intestinal epithelial cells. Biochem J 11:1696–1703
3. Steinberg SE, Campbell C, Hillman RS (1979) Kinetics of the normal folate entero-hepatic cycle. J Clin Invest 64:83–88
4. Foo SK, Shane B (1982) Regulation of folylpolyglutamate synthesis in mammalian cells. In vivo and in vitro synthesis of pteroylpolyglutamates by chinese hamster ovary cells. J Biol Chem 257:1358–1392
5. McGuire JJ, Bertino JR (1981) Enzymatic synthesis and function of folylpolyglutamates. Mol Cell Biochem 38:19–48
6. Steinberg SE (1984) Mechanisms of folate homeostasis. Am J Physiol 246:G319–324
7. Scott JM, Weir DG (1980) Drug-induced megaloblastic change. Clin Haematol 9:587–606
8. Krakoff IH, Brown NC, Reichard P (1968) Inhibition of ribonucleoside diphosphate reductase by hydroxyurea. Cancer Res 28:1559–1565
9. Stebbins R, Bertino JR (1976) Megaloblastic anaemia produced by drugs. Clin Haematol 5:619–630
10. Hughes A, Gatus B (1979) Severe megaloblastic anemia induced by maloprim. J Trop Med Hyg 82:120–121
11. Remacha A, Gimferrer E, Baiget M et al. (1983) Triamterene-induced megaloblastosis. Report of two cases and review of the literature. Biol Clin Hematol 5:127–134
12. Kobrinsky NL, Ramsay NKC (1981) Acute megaloblastic anemia induced by high dose trimethoprim – sulfamethoxazole. Ann Intern Med 94:780–781
13. Halsted CH, Gandhi G, Tamura T (1981) Sulfasalazine inhibits the absorption of folates in ulcerative colitis. N Engl J Med 305:1513–1517
14. Selhub J, Dhar GJ, Rosenberg IH (1978) Inhibition of folate enzymes by sulfasalazine. J Clin Invest 61:221–224
15. Longstreth GF, Green R (1983) Folate status in patients receiving maintenance doses of sulfasalazine. Arch Intern Med 143:902–904
16. Lindenbaum J (1983) Drug-induced folate deficiency and the hematologic effects of alcohol. In: Lindenbaum J (ed) Nutrition in hematology. Churchill Livingstone, Edinburgh pp 33–58
17. Hoffbrand AV, Necheles TF (1968) Mechanism of folate deficiency in patients receiving phenytoin. Lancet ii:528–530
18. Houlihan CM, Scott JM, Boyle PH, Weir DG (1972) The effect of phenytoin on the absorption of synthetic folic acid polyglutamate. Gut 13:189–190
19. Benn A, Swan CHS, Cook WT, Blair JA, Matty AJ, Smith MG (1971) Effect of intraluminal pH on the absorption of pteroylmonoglutamic acid. Br Med J i:148–150

20. Doe WF, Hoffbrand AV, Reed PI, Scott JM (1971) Jejunal pH and folic acid. Br Med J ii:669–670
21. Kelly D, Weir D, Reed B, Scott J (1979) Effect of anticonvulsivant drugs on the rate of folate catabolism in mice. J Clin Invest 64:1089–1096
22. Norris JW, Pratt RF (1974) Folic acid deficiency and epilepsy. Drugs 8:366–385
23. Shojania AM (1982) Oral contraceptives: effects on folate and vitamin B_{12} metabolism. Can Med Assoc J 126:244–247
24. Whitehead N, Reyner F, Lindenbaum J (1973) Megaloblastic changes in the cervical epithelium: association with oral contraceptive therapy and reversal with folic acid. JAMA 226:1421–1424
25. Amess JAL, Burman JF, Rees GM, Nancekievill DG, Mollin DL (1978) Megaloblastic haemopoiesis in patients receiving nitrous oxide. Lancet ii:349–352
26. Nunn JF, Sharer NM, Gorchein A, Jones JA, Wickramasinghe SN (1982) Megaloblastic haemopoiesis after multiple short-term exposure to nitrous oxide. Lancet i:1379–1381
27. Lindenbaum J (1980) Folate and vitamin B_{12} deficiencies in alcoholism. Semin Hematol 17:119–129
28. Hines JD, Cowan DH (1974) Anemia in alcoholism. In: Dimitrov NV, Nodine J (eds) Drugs and hematologic reactions. Grune and Stratton. New York, pp 141–153
29. Eichner ER, Hillman RS (1971) The evolution of anemia in alcoholic patients. Am J Med 50:218–232
30. Eichner ER, Buchanan Smith JW (1972) Variations in the hematologic and medical status of alcoholics. Am J Med Sci 263:35–42
31. Klipstein FA, Lindenbaum J (1965) Folate deficiency in chronic liver disease. Blood 25:443–456
32. Sullivan LW, Herbert V (1964) Suppression of hematopoiesis by ethanol. J Clin Invest 43:2048–2062
33. Meagher RC, Sieber F, Spivak JL (1982) Suppression of hematopoietic-progenitor-cell proliferation by ethanol and acetaldehyde. N Engl J Med 307:845–849
34. Halsted CH (1980) Folate deficiency in alcoholism. Am J Clin Nutr 33:2736–2740
35. Hillman RS, Steinberg SE (1982) The effects of alcohol on folate metabolism. Ann Rev Med 33:345–354
36. Herbert V (1963) A palatable diet for producting experimental folate deficiency in man. Am J Clin Nutr 12:17–20
37. Wu A, Chanarin I, Slavin G (1975) Folate deficiency in the alcoholic its relationship to clinical and haematological abnormalities liver disease and folate stores. Br J Haematol 29:469–478
38. Baraona E, Lindenbaum J (1977) Metabolic effects of alcohol on the intestine. In: Lieber CS (ed) Metabolic aspects of alcoholism. MTP Lancaster pp 81–115
39. Greene HL, Stifel FB, Herman RH, Herman YF, Rosensweig NS (1974) Ethanol-induced inhibition of human intestinal enzyme activities: reversal by folic acid. Gastroenterology 67:434–440
40. Dinoso VP, Chey WY, Braverman SP (1972) Gastric secretion and gastric mucosal morphology in chronic alcoholics. Arch Intern Med 130:715–719
41. Halsted JA, Carroll J, Rubert S (1959) Serum and tissue concentrations of vitamin B_{12} in certain pathologic states. N Engl J Med 260:575–580
42. Wu A, Chanarin I, Levi AJ (1974) Macrocytosis of chronic alcoholism. Lancet i:829–831
43. Unger KW, Johnson D (1974) Red blood cell mean corpuscular volume: a potential indicator of alcohol usage in a working population. Am J Med Sci 267:281–290
44. Buffet C, Chaput JC, Albuisson F (1975) La macrocytose dans l'hépatite alcoolique chronique, histologiquement prouvée. Arch Fr Mal App Dig 64:309–315
45. Cooper BA, Arner EC, Wiley JS, Shattir SJ (1975) Modification of red cell membrane structure by cholesterol – rich lipid dispersions. A model for the primary spur cell defect. J Clin Invest 55:115–126
46. Belaiche J, Zittoun J, Marquet J (1978) La macrocytose de l'alcoolisme chronique est-elle due à un trouble de synthèse de l'ADN lié à une carence en folates. Gastroenterol Clin Biol 2:597–602

Chapter 11

Neuropsychiatric Illness and Deficiency of Vitamin B_{12} and Folate

M. I. BOTEZ

Deficiency of either folate or vitamin B_{12} (cobalamin) causes megaloblastic anemia. Reports that folate supplements might aggravate the neurological defects in deficiency of vitamin B_{12} were interpreted during the 1970s to suggest that folate deficiency did not induce neurological abnormality [47]. This is consistent with the observation that the various neurological and psychiatric abnormalities reported in patients deficient in folate could not be readily demonstraded as secondary to the folate deficiency [43].

In this chapter, the neuropsychiatric defects associated with deficiency of these two vitamins are considered under three headings: (a) illness caused (or probably caused) by the vitamin deficiency, (b) illness associated statistically with the vitamin deficiency, and (c) illness coincident with or consequent to the vitamin deficiency.

Neuropsychiatric Disorders Related to Cobalamin Deficiency

The causes of deficiency of vitamin B_{12} are described elsewhere in this volume. Neuropsychiatric effects of such deficiency include subacute combined degeneration of the spinal cord (SCDC), peripheral neuropathy, optic neuritis, and dementia [27]. Any combination of these may be present in a patient to different degrees and may not correlate with the severity of megaloblastic anemia.

Pathologic Anatomy. Lesions in the spinal cord (predominantly in the dorsocervical region) are found with descending frequency in the posterior columns and in the pyramidal and spinocerebellar tracts. These are characterized by demyelination, with loss of definition of myelin borders, swelling of white matter and axones, degenerative changes, and phagocytosis (status spongiosus). This myelin degeneration is followed by gliosis with giant astrocytes, with involvement of perivascular mesenchyme by scarring [77]. In peripheral nerves, myelin sheaths become smaller [57], with myelin degeneration and increased numbers of Schwann cells. Spinal root ganglia show secondary degenerative changes. The large fibers are more affected than are the small, with both demyelination and axonal degeneration visible by electron microscopy [8, 34]. These changes may also be observed in cranial nerve roots. Demyelination and spongiosus degeneration had been observed in the optical nerve [39]. Spongiosis changes may also be observed in the deep white matter of the cerebral hemisphere and in the brain stem. Lesions are symmetrical, and similar lesions have been observed in some experimental animals under selected conditions [3, 40] with the addition of anterior horn cell degeneration.

Clinical Aspects

Subacute Combined Degeneration of the Spinal Cord. Clinical manifestations of SCDC include progressive loss of deep sensation, with pyramidal tract involvement, sometimes with cerebellar signs and loss of sphincter control. The initial symptoms are usually those of paresthesias, described as numbness, pins and needles, burning, or sensations of hot or cold running water. These are sometimes associated with myalgias and lightening pains. Physical examination at this stage reveals loss of vibration sense in the toes and ankles, with intact position sense. Quantitative measurement of vibration sense has been described [77].

Two clinical syndroms may be distinguished: one with predominance of posterior column signs, and the other with more pyramidal tract involvement. In the posterior-column syndrome, vibration sense is lost up to the iliac crest (a finding characteristic of this abnormality), with loss of position sense, usually more severe in the legs [79]. Ataxia (with loss of balance on closing the eyes) is present, with the sensation of walking on a carpet. Lhermitte's sign (a sensation of electric discharge down the spine on bending foreward the neck) has been reported in about one-four of cases of SCDC and is not pathognomonic of multiple sclerosis. In addition, muscular tone may decrease, and deep tendon reflexes may be decreased. The pyramidal-tract syndrome is usually mild and produces up turning toes (Babinski response) without much disability. Decreased tendon reflexes in the legs and the absence of clonus are usually due to associated posterior-column disease. Thigh and calf weakness is usually demonstrable, however, and patients may complain of abrupt falls as their legs give way. In a few patients, spastic paraplegia may appear. In a few cases, mixtures of these abnormalities may produce rapidly developing spasticity, severe paresthesias without motor involvement, a cerebellar syndrome, or other combinations of symptoms.

SCDC has been reported to occur in 16% of cases of pernicious anemia [82] and some authors have suggested that up to 50% of patients with vitamin B_{12} deficiency caused by malabsorption may be affected [77].

Polyneuropathy. The peripheral neuropathy-polyneuropathy, is predominantly sensory, developing insidiously, with burning sensations and paresthesias. Signs include stocking-glove hypoesthesia and decreased vibration sense and deep tendon reflexes in the lower extremities. Plantar responses characteristically remain flexor. Muscular weakness may develop late in the course. Decreased vibration sense in the legs is observed in both polyneuropathy and SCDC, although its extension to the iliac crests is more characteristic of the latter. Its association with stocking-glove hypoesthesia characterizes the polyneuropathy syndrome whereas the Babinski response characterizes SCDC. It has been reported that polyneuropathy may appear very early in deficiency of vitamin B_{12} [29, 41], that it may precede or coexist with SCDC, and that it may occur when the deficiency has induced no detectable hematological abnormalities [36]. One report described electrophysiological evidence of neuropathy in one-third of patients with pernicious anemia despite absence of subjective symptoms in all but 7% of cases [29].

Dementia and Psychosis. Both organic dementia and organic psychosis may be considered to be caused by deficiency of vitamin B$_{12}$ if deficiency is present, if the psychiatric abnormality was not present before deficiency appeared, and if the illness is corrected or improved by treatment of the deficiency. Mental illness has been reported in 37%–72% of patients with SCDC or polyneuropathy [27], and 13% of patients developing deficiency of vitamin B$_{12}$ after gastrectomy have been reported to have transient decrease in mental capacity [77]. In most patients these syndromes are characterized by decreased initiative and concentration, depression, and problems with abstract thought, but confusion is sometimes also observed [42, 83].

Affective psychoses associated with deficiency of vitamin B$_{12}$ have ranged from mild to severe and from depression to paranoid behavior. Delirium and hallucinations have been reported. Rarely have these not been associated with either hematological or neurological evidence of deficiency of vitamin B$_{12}$ [62, 97]. Psychotic depression with organic brain syndrome thus requires investigation for deficiency of vitamin B$_{12}$ [97].

Optic Neuritis. This manifestation of deficiency of vitamin B$_{12}$ has been reported in 2% of deficient patients [27, 77]. The complex of anemia, optic atrophy, and spastic paraplegia was first described in 1897 [27]. The severity of optic neuropathy does not correlate with that of anemia and may occasionally occur without hematological abnormalities [77]. The clinical syndrome is one of a chronic retrobulbar neuritis, which may be associated with mesencephalic defects, including ophthalmoplegia and symmetrical cranial nerve palsies [2, 39, 79]. Tobacco amblyopia has been considered by some to be related to the optic neuritis of vitamin B$_{12}$ deficiency, but the former neuropathy develops more slowly, with an enlarging central scotoma and is probably not related to the neuropathy of vitamin B$_{12}$ deficiency, although hydroxocobalamin therapy should be tried [58].

Differential diagnosis includes spinocerebellar heredoataxias, vascular, paraneoplastic, and other myelopathies, and multiple sclerosis. Investigations includes those for deficiency of vitamin B$_{12}$ and may sometimes be aided by analysis of vitamin B$_{12}$ concentrations in spinal fluid [93]. The biological importance of the dissociation of spinal fluid and serum levels of vitamin B$_{12}$ in chronic epileptics [10, 59] remains unclear.

Neurophysiological Investigations

Evoked Potentials. In patients with normal nerve conduction and SCDC, somato-sensory evoked potentials may be abnormal due to demyelination in central pathways. Visual evoked potentials may be abnormal, but auditory evoked potentials are normal [56, 66].

Electromyography (EMG) Nerve Conduction, Nerve Biopsies. In B_{12}-deficient polyneuropathy, sensory and motor nerve conduction is usually normal in the presence of diminished sensory action potentials [56]; this phenomenon reflects the dying back process [77]. Sensory nerve conduction may be decreased, with a prolonged H wave [26]. Polyphasic potentials may be observed in EMG studies, and the characteristic primary axonal degeneration followed by demyelination [34, 36, 56] may be observed in biopsies of the sural nerve [34].

Electroencephalography (EEG). Nonspecific abnormalities such as slow dysrhythmias have been reported in almost half of deficient patients studied [27]. In patients with the severe mental syndromes described above, EEG abnormalities are usually present.

Tomodensitometry (Computerized Axial Tomography, CAT Scan). Other than cerebral atrophy, which does not correlate with mental illness in deficient patients, this procedure provides no positive diagnostic information at this time.

Physiopathology

The mechanism by which deficiency of vitamin B_{12} affects the brain and peripheral nerves is unknown [34]. It is presumed that the vitamin acts as a coenzyme for a process necessary for myelination of axonal survival.

It has been suggested that several mechanisms may mediate brain damage in deficiency of vitamin B_{12} [6]: accumulation of methylmalonyl-CoA causing interference with arachidonate synthesis; deranged relationship between phosphatidylcholine and phosphatidylethanolamine caused by decreased methionine-dependent methylation. The neurological disease appears to be aggravated by folate or folic acid administration [95], and methionine supplements have prevented the progressive neurological lesions in animals treated chronically with nitrous oxide (which blocks vitamin B_{12}-dependent methionine synthesis). It also has been suggested [6] that the neurological disease is due to accumulation of fatty acids (C15–C17) due to interference with methylmalonyl-CoA metabolism. Such neurological defects have not been reported, however, in children with congenital defects in the adenosylcobalamin or methylmalonyl-CoA mutase pathways (B. Cooper 1986, personal communication). Some severe congenital hydrocephalus could never causes major neuropsychiatric disease.

Treatment and Prognosis

Prompt diagnosis of vitamin B_{12} deficiency is necessary because irreversible paraplegia and dementia are to be prevented. Neurological symptoms seem to appear when the body's reserves decrease below about 10% of the normal content of 2 mg vitamin B_{12} [53]. Treatment requires vitamin B_{12} replacement. A proportion of injected vitamin B_{12} is lost in the urine [67], and in the absence of gastriinstrinsic factor, less than 1% of oral vitamin B_{12} is absorbed from the gut. In the opinion of the present author, the modest doses of vitamin B_{12} required to correct hematological abnormalities are insufficient for reversal of the neurological disease. It is our practice, therefore, to treat such patients with injections of 1 mg hydroxocobamin every 2–3 days for 3 weeks, twice weekly for 2 months, weekly for 6 months, and then monthly thereafter [79]. We recommend a supplement of 300 mg ferrous sulfate daily in addition.

Spinal cord lesions and others that are present for many weeks usually do not regress. Sensation in SCDC often improves strikingly, but pyramidal tract manifestations and deep tendon reflexes do not improve [2, 77]. In polyneuropathy, about half of affected patients do not improve [36], and in many with subjective improvement, neurophysiological parameters do not change [7, 77]. Mental symptoms and psychoses appear to respond regularly to vitamin B_{12} therapy (75%–90%) [2, 7, 77], and this improvement parallels improvements in EEG.

Neurological and Psychiatric Illness Associated or Coexisting with Cobalamin Deficiency

The clinical forms of tropical myeloneuropathies include tropical ataxic neuropathy (TAN; Strachan's syndrome) and tropical spastic paraplegia (TSP). TAN is characterized by a feeling of burning in the legs, with loss of deep superficial sensation and diminished deep tendon reflexes in the legs. It was frequent in prisoners of war in the Far East during World War II (5% of previous prisoners have sequelae) and is associated with malabsorption syndrome in India, Africa, and the Himalayan area [76]. Many of these patients respond to treatment with vitamin B complex. The role of vitamin B_{12} in this syndrome is probably minimal. TSP is caused by lathyrism. In some patients with a similar syndrome in Jamaica, serum levels of vitamin B_{12} have been below normal [76], but the syndrome did not respond to therapy with vitamin B_{12}.

In addition to patients with known deficiency of vitamin B_{12}, decreased serum levels of vitamin B_{12} have been reported in a proportion of patients institutionalized for psychiatric illness (1% [27]; 5.8% and 10%). In some cases, clinical improvement has been reported following therapy with vitamin B_{12}. Various reports have suggested widespread deficiency of vitamin B_{12} in the aged, based on measurements of serum vitamin B_{12} concentration [1]. Most studies note such deficiency with other nutritional deficiencies, primarily in freely-living aged patients living alone and becoming inable to cope. Some patients with Alzheimer's disease have been observed to have deficient levels of vitamin B_{12} in serum without other evidence for deficiency. The relationship of this to the disease is unclear, but deficiency of vitamin B_{12} must be excluded in such patients.

Neurologic and Psychiatric Illness Caused by Folate Deficiency

Unlike deficiency of vitamin B_{12}, which usually appears without associated deficiency of other factors, folate deficiency is rarely pure [78]. Although malnutrition, intestinal malabsorption and alcoholism usually produce mixed deficiencies, in the cases described by Botez et al. [20, 15] no evidence of deficiency of factors other then folate was detected. As with vitamin B_{12}, neuropsychiatric illness observed in patients deficient in folate may be either caused by the folate deficiency or fortuitously associated with the deficiency (e.g., folate deficiency in the elderly). A further possibility is that the deficiency is caused by the neuropsychiatric illness (e.g., folate deficiency in depressed patients) or is secondary to the effect on folate metabolism of treatment regimens or drugs. In depression, the symptoms may be aggravated by the secondary folate deficiency [46], producing a vicious circle of symptomatology: depression, poor diet, folate deficiency, and aggravation of depression.

Pathologic Anatomy. Most studies have dealt with cases of axonal polyneuropathies (see below). A number have focused on the pathology of the central nervous system. Hansen et al. [65] described an epileptic with folate-deficiency megaloblastic anemia, clinical signs of myelopathy, with demyelination of the posterior spinal colums (more severe in the cervical region) and cerebellar lesions. Robertson et al. [74] described autopsy studies of a patient with classical SCDC in whom they reported exclusion of deficiency of vitamin B_{12}. Folate levels were not, unfortunately, determined. Henry et al. [68] described a young woman with folate deficiency, a consequence of chronic alcoholism, who had ascending polyneuronitis (Landry's paralysis) and Korsakoff's psychosis. Autopsy revealed lesions in the mammillary bodies of the type reported in deficiency states and segmental axonal degeneration and demyelination of peripheral nerves, not typical of alcoholic neuropathy [49, 68] but resembling lesions observed in patients deficient in vitamin B_{12}. Anterior horn neuronal degeneration was also observed secondary to the latter. Marcus and Freedman [31] described cerebral atrophy which may have been secondary to severe folate deficiency. Cerebral atrophy, determined by radionuclide cisternograms with indium 111 was described by Botez et al. [15] in folate-deficient patients in whom the abnormality improved during treatment with folate. We have not found histological studies of such cases reported.

Folate deficiency as the cause of neuropsychiatric illness. In our opinion, the association of classical hematological features of folate deficiency with certain neuropsychiatric syndromes effectively links the deficiency and the pathogenesis of the syndrome. In the absence of hematological features of folate deficiency, such linkage depends on demonstration of folate deficiency by measurement of serum and erythrocyte folate levels, exclusion of cobalamin deficiency using serum levels, and other criteria. It is recommended [43, 49] that one must (a) exclude conditions other than folate deficiency which may have caused the neuropsychiatric syndrome; (b) exclude other deficiency states, especially those known to induce the neuropsychiatric syndrome; and (c) observe improvement in the syndrome during therapy with folate (e.g., 10–30 mg per day [15, 79]. This is especially useful in patients in whom supplements of other vitamins have been shown not to improve the condition of their illness [33, 49].

Clinical Studies

Subacute Combined Degeneration of the Spinal Cord. A syndrome identical to SCDC has been reported in about 30 cases associated with folate deficiency caused by alcoholism, malnutrition, intestinal malabsorption (including celiac disease, Crohn's disease), and gastrectomy [43, 49, 52]. In some, polyneuropathy has been associated. These cases have been reported to respond to folate therapy [43], and their response has appeared to be faster and more complete (including some pyramidal-tract lesions) [13, 33, 49] than observed in the vitamin B_{12}-deficient cases treated with vitamin B_{12}. As with deficiency of vitamin B_{12}, the neurological lesion has been reported in the absence of megaloblastic anemia [13, 49].

Polyneuropathy. Characteristics of polyneuropathies associated with folate deficiency include the following [17, 54]: (a) They are mixed motor-sensory neuropathies which, although more commonly affecting the lower limbs, may affect the upper limbs in some cases; (b) electrophysiological data suggest than the neuropathy is indistinguishable from that observed in deficiency of vitamin B_{12}; (c) hematological manifestation of folate deficiency may be absent or minimal, but both serum folate and erythrocyte folate levels are in the deficient range (dissociation is the rule betwen severity of hematological effects of folate deficiency and the neuropathy); (d) polyneuropathy is more common in folate deficiency caused by intestinal malabsorption than by other factors; and (e) clinical improvement, when it occurs, occurs 2–8 weeks after beginning folate therapy. Neurophysiological improvement usually lags behind subjective improvement.

Cerebellar Syndromes. Several reports have described cerebellar syndromes associated with SCDC or with polyneuropathy [33, 68]. It is possible that cerebellar pathology is related to the observation of Haltia [64], who reported decreased RNA synthesis in folate-deficient Purkinje's cells.

Depression with Minimal Neurological Signs. A group of 50 patients have been described with the following syndrome [15]: (a) fatigue, lassitude, irritability, insomnia, and restless legs, associated with depressive symptoms, loss of recent memory, and poor concentration; (b) abnormal levels of serum folate (50/50), with low levels of erythrocyte folate in 14; (c) long-standing gastrointestinal illness (colitis, irritable colon, gastritis, small-intestinal surgery), sometimes associated with the taking of sulfasalazine; and (d) decreased vibration sense in the lower extremities (measured quantitatively), with decreased ankle jerks. Neuropsychological studies showed characteristic organic brain syndrome; 68% had some cerebral atrophy by CAT scan and radionuclide cisternograms with indium 111, and in 20% small-bowel atrophy was documented. These patients have been considered to have prolonged folate deficiency, with depression and some evidence of peripheral neuropathy. Treatment for 6 months with folic acid resulted in improved neuropsychologic performance, with less cerebral atrophy by radionuclide cisternograms in ³/₇ patients 2 years later and less jejunal atrophy in ²/₁₀ patients after 2–3 years of treatment. The syndrome is consistent with the observations of Herbert [69], who during experimentally induced folate deficiency noted somnolence, loss of recent memory, and irritability. We suggested [15] that the jejunal atrophy apparently induced by the folate deficiency induces a vicious circle (Fig. 1), with folate malabsorption unassociated with steatorrhea and not responding to gluten free diet [50].

Restless Legs Syndrome. The syndrome of restless legs may be the earliest sign of peripheral neuropathy. It has been reported to be associated with malabsorption syndrome, celiac disease, and malnutrition both in the tropics [61] and during pregnancy [12]. The familial syndrome (autosomal dominant transmission) is not related to folate or vitamin B_{12} deficiencies.

Dementia. Folate deficiency in serum is common among demented patients [46]. Reynolds has suggested [77] that such deficiency induces progressive apathy, followed by depression and leading to dementia. Such progression has not been observed in epileptics (see below). The folate-induced dementia is not correlated with the degree of megaloblastic anemia. The clinical syndrome of the small number of demented patients whose dementia improved with folate therapy has included rapid mental deterioration (as with deficiency of vitamin B_{12}) in contrast to the usual course in Alzheimer's disease. During treatment, abnormal EEGs have been reported to improve [15, 17, 35, 52, 89].

Folate Deficiency Associated with Neurologic Illness; Psychiatric Disorders of the Elderly

Folate deficiency appears to be secondary to the primary neuropsychiatric illness in those patients who have such disorders.

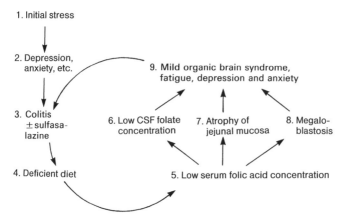

Fig. 1. Hypothetical mechanism showing the pathophysiological vicious circle underlying occult and previously undiagnosed folate deficiencies in nonalcoholic patients [15].

Clinical Studies

Anticonvulsant Therapy. Megaloblastic anemia has been associated with treatment with antiepileptic medication since 1952 [70]. The frequency of low serum folate levels varies from 27% to 91% of patients receiving different antiepileptic drugs, whereas megaloblastic anemia is observed in only about 1% of cases [46, 70]. It has been reported that the folate levels in spinal fluid of patients receiving antiepileptic medication is only about 50% of the level in untreated epileptics and normal subjects [48]. In another study, whereas the frequency of low serum folates was greater in treated epileptics than in new epileptics, the folate level in the spinal fluid was not different between the two groups. The mechanism for this is unknown [28]. Folate deficiency is significantly more common among treated, institutionalized epileptics and those with mental illness than among freely living epileptics [70, 75]. No clear correlation between mental illness and folate deficiency in this population has been found [70]. Folate treatment has been reported in uncontrolled trials to improve the well-being and performance in psychological tests of such patients [38, 71], whereas prospective double-blind studies giving 15 mg/day for 3-6 months have revealed no such effect [32, 44, 63]. It would appear that the epilepsy or abnormal mental state produce the folate deficiency, and not the reverse. The peripheral neuropathy reported in some treated epileptics [55] appears to be secondary to the effect of the drugs and, likewise, not due to folate deficiency [91].

Psychiatric Illness. Depressive illness and affective psychoses are frequently associated with low levels of serum folate [82]. Up to 86% of hospitalized psychiatric patients and 20%-30% of patients with depression, organic psychosis, or schizophrenia were reported to have subnormal levels of serum folate [21]. The suggestion that endogenous depression can be caused by folate deficency [60] has not been proved. Depression, by interfering with nutrition, can induce folate deficiency [31]. It is probable that this deficiency may aggravate preexisting symptoms [42, 45]. Reports that postpartum depression is caused by folate deficiency and responds to folate [92] have not been substantiated by a careful study of these patients [84]. The mechanism by which oral contraceptives cause folate deficiency is unknown [37, 84]; this is discussed elsewhere in this volume.

Geriatrics and Geriatric Psychiatry. Many factors may produce folate deficiency in the aged, some physiological and some psychological [31]. Low levels of serum folate have been reported among groups of aged in Great Britain [78] and among the economically deprived elderly in the United States [31]. Studies suggest that such low levels of folate may be found in 20%-80% of elderly persons [1, 78]; especially among inmates of old-age institutions. The elderly who are psychologically ill are frequently found to have decreased levels of folate in serum [1] and spinal fluid [21].

Other Neurological Illnesses. Two reports have indicated an association between folate deficiency and both pyramidal-tract lesions and organic brain syndrome

[35, 72]. In animals, neural tube lesions may be induced in the fetus by folate deficiency in the mother [86]. Smithells et al. [86] had reported decrease of neural tube defects in a population at risk by prepregnancy folate supplementation. This is presently being evaluated in a prospective study. The predominance of neural tube defects in certain regions and in lower socioeconomic groups [81] strongly suggests a multifactorial cause for this syndrome.

Kearns-Sayre Syndrome. Patients with Kearns-Sayre syndrome (external ophthalmoplegia, retinitis pigmentosa, cardiac conduction abnormalities, defective muscle mitochondria, and spongiform encephalopathy), have been reported frequently to be folate deficient without folate malabsorption [4], and there are reports of improvement after folate therapy [51]. It has been suggested that in these patients, folate transport into the CNS may be defective, based on three patients with smaller than normal gradient between spinal fluid and serum folate [30]. This awaits further study, as does the case of hypotonia reported, to be associated with folate deficiency and apparently responding to folate therapy [80].

Folates in Cerebrospinal Fluid

The level of folate in spinal fluid is greater than that in plasma or serum [96]. It has not been demonstrated whether spinal fluid folate reflects folate available to the brain more precisely than does plasma folate [48]. Spector and Lorenzo described the function of the choroid plexus in maintaining this folate gradient [88] although it has been reported that folate don't cross blood-brain barrier [87]. At normal levels of serum of plasma folate, spinal fluid folate is three to four times higher [48]. The relationship between folate concentration in plasma and that in spinal fluid is sometimes disturbed, with several cases reported in whom spinal fluid folate was within normals limits despite decreased levels of folate in serum [10]. For this reason, it is our practice to measure spinal fluid folate levels in patients suspected of folate-dependent neuropsychiatric disease [11]. This appears not to be necessary in patients with folate deficiency megaloblastic anemia.

Pathophysiology

Neuropsychiatric illness caused by folate deficiency is not due to an associated deficiency of vitamin B_{12}, since associated deficiency often is not present, and folate deficiency does not interfere with absorption of vitamin B_{12} [22].

Folates are required in the biosynthesis of purines and contribute to choline synthesis. Deficiency might thus be expected to affect synthesis of RNA and myelin [94]. The demonstrated association of folate deficiency with depression [82] and the role of serotonin in depressive states [9] suggest that metabolism of serotonin is in some way implicated. Serotonin metabolism has been reported affected in children with inherited defects of folate metabolism [85], in patients with acquired deficiencies [19], and in deficient experimental animals [14]. Synthesis of tetrahydrobiopterin (BH4), which is a cofactor in serotonin synthesis, is stimulated in extracts of rat brain by addition of 5-methyltetrahydrofolate [9] by an unknown mechanism. It has been reported that BH4 synthesis may be decreased in depressed subjects [9].

Folate and cobalamin metabolism are interrelated through methionine resynthesis. S-adenosylmethionine (SAM), which is both a cofactor and an eventual product in this reaction, is an important source of methyl groups in brain metabolism. Folate deficiency decreases brain SAM [73], whereas SAM or folate administration increases the level of 5-hydroxyindolacetic acid in spinal fluid [9, 73]. Possible relationships between folate and acetylcholine have not been confirmed by experimental data [18].

Treatment

If neurological or psychiatric symptoms are caused by folate deficiency, both are readily reversed by folate supplements [31, 46]. In patients in whom the association is secondary to the neuropsychiatric disease, folate therapy will correct only the secondary symptoms caused by the deficiency.

It is our practice to treat such patients orally with 15–20 mg folic acid daily, adding weekly injections of the same dose in patients with malabsorption [13, 17, 79]. We continue this treatment for 1–3 months. Treatment is begun as soon as the association is established [35]. If after 3 months no improvement is noted, treatment is discontinued. If the patient improves, we continue the same treatment for an additional 3 months, followed by a prolonged period of treatment with 5 mg folic acid daily. The precautions taken include the following: (a) vitamin B$_{12}$ supplements are given to prevent secondary decrease of cobalamin level [25]; (b) care is taken with epileptics despite failure of controlled studies to support the previous impression that seizure frequency may increase with folate therapy [24]; (c) we do not give such large doses of folate in patients with neoplastic disease; and (d) the above regimen is restricted to patients in whom the association between folate deficiency and the neuropsychiatric illness is considered highly probable. Usually, patients with psychiatric illness caused by folate deficiency improve after 2–3 weeks of therapy and those with SCDC after 1–2 months.

Comparison of Neuropsychiatric Illness Caused by Deficiency of Folate and of Vitamin B$_{12}$

In many patients with either folate or vitamin B$_{12}$ deficiency, no neuropsychiatric illness appears. Comparison of the effect of such deficiency on neuropsychiatric health reveals that [82]: (a) neuropathy is three times as common in deficiency of vitamin B$_{12}$ than of folate; (b) SCDC is *very* uncommon in folate deficiency; (c) the frequency of organic brain syndrome is equal in both deficiencies; (d) depression is more common in deficiency of folate than that of vitamin B_{12}; (e) optic atrophy has not been reported as a consequence of folate deficiency; and (f) the hematological manifestations of deficiency are independent of the neuropsychiatric in both deficiencies.

Acknowledgements. The author's works cited in this chapter were supported by the Fondations de l'Hôtel-Dieu de Montréal and by the Fonds de la Recherche en Santé du Québec. The author thanks Mrs. Thérèse Botez-Marquard for her professional and moral support during the last 16 years of work on nutritional and metabolic neuropsychology.

References

1. Abalan F, Subra G, Picard M, Boueih P (1984) Fréquences des déficiences en vitamine B_{12} ou en acide folique chez les patients admises en gérontopsychiatrie. Encephale X:9–12
2. Abramsky O (1972) Common and uncommon neurological manifestations as presenting symptoms of vitamin B_{12} deficiency. J Am Geriatr Soc 20:93–96
3. Agamanolis DP, Chester EM, Victor M, Kark JA, Hines JD, Harris JM (1976) Neuropathology of experimental vitamin B_{12} deficiency in monkeys. Neurology 26:905–914
4. Allen RJ, Dimauros S, Coulter DL, Papadimitriou A, Rothenberg SP (1983) Kearns-Sayre syndrome with reduced plasma and cerebrospinal fluid folate. Ann Neurol 13:679–682
5. Areekul S, Cheeramakara C (1985) Cerebrospinal fluid folate activity in patients with *Plasmodium Falciparum* cerebral malaria. Rrop Geogr Med 37:227–230
6. Bender DA (1984) B-vitamins in the nervous system. Neurochem Int 6:297–321
7. Berguignan FX, Peyronnard JM, Charron LF, Botez MI (to be published) Polyneuropathy in vitamin B_{12} deficiency (manuscript in preparation)
8. Bischoff A, Lütschig G, Meier C (1975) Polyneuropathie bei vitamin B_{12}. Muench Med Wochenschr 117:1593–1598
9. Blair JA, Morar C, Hamon CGB et al. (1984) Tetrahydrobiopterin metabolism in depression. Lancet 2:163–164
10. Botez MI (1986) Etudes neurochimiques et comportementals sur les effets secondaires de la phénytoïne chez les épileptiques chroniques. 6 mars 1986, présentation à la Société française de Neurologie, Paris
11. Botez MI, Bachevalier J (1981) The blood-brain barrier and folate deficiency. Am J Clin Nutr 34:1725–1730
12. Botez MI, Lambert B (1978) A possible correlations between restless legs syndrome and folate deficiency in pregnancy. Nutr Rep Int 18:143–146
13. Botez MI, Cadotte M, Beauzlieu R, Pichette LP, Pison C (1976) Neurologic disorders responsive to folic acid therapy. Can Med Assoc J 115:217–222
14. Botez MI, Young SN, Bachevalier J, Gauthier S (1979) Folate deficiency and decreased brain 5-hydroxytryptamine synthesis in man and rat. Nature 278:182–183
15. Botez MI, Botez T, Léveillé J, Bielmann P, Cadotte M (1979) Neuropsychological correlates of folic acid deficiency: facts and hypotheses. In: Botez MI, Reynolds HE (ed) Folic acid in neurology, psychiatry and internal medicine. Raven, New York pp 435–461
16. Botez MI, Peyronnard JM, Charron L (1979) Polyneuropathies responsive to folic acid therapy. In: Botez MI, Reynolds EH (eds) Folic acid in neurology, psychiatry and internal medicine. Raven, New York pp 401–412
17. Botez MI, Peyronnard JM, Berube L, Labrecque R (1979) Relapsing neuropathy, cerebral atrophy and folate deficiency: a close association. Appl Neurophysiol 42:171–183
18. Botez MI, Bachevalier J, Tunnicliff G (1980) Dietary folic acid and the acticity of brain cholinergic and gamma aminobutiric acid (GABA) enzymes. Can J Neurol Sci 7:133–134
19. Botez MI, Young SN, Bachevalier J, Gauthier S (1982) Effect of folic acid and vitamin B_{12} deficiencies on 5-hydrixyindoleacetic acid in human cerebrospinal fluid. Ann Neurol 12:479–484

20. Botez MI, Botez T, Maag U (1984) The Wechsler subtests in mild organic brain damage associated with folate deficiency. Psychol Med 14:431–437
21. Carney MWP (1979) Psychiatric aspects of folate deficiency. In: Botez MI, Reynolds EH (eds) Folic acid in neurology psychiatry and internal medicine. Raven, New York pp 475–482
22. Cattan D, Belaiche J, Zittoun J, Yvart J (1982) Effect of folate deficiency on vitamin B$_{12}$ absorption. Ann Nutr Metab 26:367–373
23. Cole MG, Prchal JF (1984) Low serum vitamin B$_{12}$ in Alzheimer-types dementia. Age Ageing 13:101–105
24. Hommes OR, Hollinger JL, Jansen MJT, Schoofs M, Weil T, Kok JCN (1979) Convulsant properties of folate compounds: some considerations ans speculations. In: Botez MI, Reynolds EH (eds) Folic acid in neurology, psychiatry and internal medicine. Raven, New York; pp 285–316
25. Hunter R, Barnes J, Matthews DM (1969) Effect of folic acid supplement on serum vitamin B$_{12}$ levels in patients on anticonvulsants. Lancet 2:666–667
26. Kayser-Gatchalian MC, Neundorfer B (1977) Peripheral neuropathy with vitamin B$_{12}$ deficiency. J Neurol 214:183–193
27. Kunze K, Leitenmayer K (1976) Vitamin B$_{12}$ deficiency and subacute combined generation of the spinal cord. In: Vinken PJ, Bruyn WG (eds) Handbook of clinical neurology vol 28, no 6. Elesvier, Amsterdam, pp 141–198
28. Lambie DG, Johnson RH (1985) Drugs and folate metabolism. Drugs 30:145–155
29. Lockner D, Reizenstein P, Wenngerg P et al. (1969) Peripheral nerve function in pernicious anaemia before and after treatment. Acta Haematol (Basel) 41:257–263
30. Macron J Л, Mizon JP, Rosa A (1983) Troubles du métabolisme des folates au cours d'un syndrome de Kearns et Sayre. Rev Neurol 139:673–678
31. Marcus DL, Freedman ML (1985) Folic acid deficiency in the elderly. J Am Geriatr Soc 33:552–558
32. Mattson RH, Gallagher BB, Reynolds EH, Glass D (1973) Folate therapy in epilepsy: a controlled study. Arch Neurol 29:78–81
33. Mauffroy B, Zittoun J, Barrois A, Poisson M, Zittoun R (1975) Carences en folates et troubles neurologiques. Sem Hop Paris 51:237–241
34. McCombe PA, McLeod JG (1984) The peripheral neuropathy of vitamin B$_{12}$ deficiency. J Neurol Sci 66:117–126
35. Melamed E (1979) Neurological disorders related to folate deficiency. In: Botez MI, Reynolds EH (eds) Folic acid in neurology, psychiatry and internal medicine. Raven, New York; pp 427–434
36. Melson H, Kornstad S, Abildgaard U (1977) Reversible neuropathy of vitamin B$_{12}$ deficiency with normal hemoglobin serum vitamin B$_{12}$. Lancet 1:803
37. Mountifield JA (1985) Effects of oral contraceptive usage on B$_{12}$ and folate levels. Can Fam Physicans 31:1523–1526
38. Neubauer C (1970) Mental deterioration in epilepsy due to folate deficiency. Br Med J 2:759–761
39. Olivarius B, Jensen L (1961) Retrobulbar neuritis and optic atrophy in pernicious anemia. Acta Ophtalmol (Copenh) 39:190–197
40. Oxnard CE, Smith WT (1966) Neurological degeneration and reduced serum vitamin B$_{12}$ levels in captive monkeys. Nature 210:507–509
41. Pallis CA, Lewis PD (1974) The neurology of gastrointestinal disease. Saunders, London
42. Petrie WM, Ban TA (1985) Vitamins in psychiatry – do they have a role? Drugs 30:58–65
43. Pincus JH (1979) Polic acid deficiency: a cause of subacute combined system degeneration. In: Botez MI, Reynolds EH (eds) Folic acid in neurology, psychiatry and internal medicine. Raven, New York pp 427–434
44. Ralston AJ, Snaith RP, Hinley JB (1970) Effects of folic acid on fit-frequency and behavious in epileptics on anticonvulsants. Lancet 1:867–868
45. Raphael JC, Choutet P, Barois A, Nouailhat F, Goulon M (1975) Myelopathie et anémie macrocytaire associées à une carence en folates. Guérison par l'acide folique. Ann Med Intern (Paris) 126:339–348

46. Reynolds EH (1976) Neurological aspects of folate and vitamin B_{12} metabolism. Clin Haematol 5:661–696

47. Reynolds EH (1979) Folic acid, vitamin B_{12} and the nervous system: historical aspects. In: Botez MI, Reynolds EH (eds) Folic acid in neurology, psychiatry and internal medicine. Raven, New York, pp 1–6

48. Reynolds EG (1979) Cerebrospinal fluid folate: clinical studies: In: Botez MI, Reynolds EH (eds) Folic acid in neurology, psychiatry and internal medicine. Raven, New York, pp 195–204

49. Contamin F, Ollat H, Levy VG, Thiermann Duffaud D (1983) Involvement of the pyramidal and peripheral nervous system associated with Crohn disease – decisive role of a folic acid deficiency. Sem Hop Paris 1983; 59:1381–5

50. Davidson GP, Townley RRW (1977) Structural and functional abnormalities of the small intestine due to nutritional folic acid deficiency in infancy. J Pediatr 90:590–594

51. Dougados M, Zittoun J, Laplane D, Castaigne P (1983) Folate metabolism disorder in Kearns-Sayre syndrome. Ann Neurol 13:687

52. Enk C, Hougard K, Hippe E (1980) Reversible dementia associated with folate deficiency 16 years after partial gastrectomy. Scand J Haematol 25:63–66

53. Fao/Who Expert Group (1970) Requirements of ascorbic acid, vitamin D, vitamin B_{12}, folate and iron. WHO Technical Report Series no 452 WHO, Geneva

54. Fehling C, Jagerstad M, Lindstrand K, Elmquvist D (1974) Folate deficiency and neurological disease. Arch Neurol 30:263–265

55. Figueroa AM, Johnson RH, Lambie DG, Shakir RA (1980) The role of folate deficiency in the development of peripheral neuropathy caused by anticonvulsants. J Neurol Sci 48:315–323

56. Fine EJ, Hallett M (1980) Neurophysiological study of subacute combined degeneration. J Neurol Sci 45:331–336

57. Foster DB (1945) Degeneration of peripheral nerves in pernicious anemia. Arch Neurol Psychiatry 42:53–75

58. Freeman AG (1984) The visual evoked potential in tobacco-alcohol and nutritional amblyopia. Am J Ophtamol 97:110–111

59. Frenkel EP, McCall MS, Sheehan RG (1983) Cerebrospinal fluid folate, and anticonvulsant-induced megaloblastosis. J Lab Clin Med 81:105–115

60. Ghadirian AM, Ananth J, Engelsmann F (1980) Folic acid deficiency and depression. Psychosomatics 21:926–929

61. Girard PL, Pele J, Bidaut J, Dumas M, Heraut L (1974) Acide folique et neuropathies dites "nutritionnelles". Bull Soc Med Afr Noire Lang Fr 19:384–391

62. Goggans FC (1980) A case of mania secondary to vitamin B_{12} deficiency. Am J Psychiatr 141:300–301

63. Grant RHE, Stores OPR (1970) Folic acid in folate-deficient patients with epilepsy. Br Med J 4:644–648

64. Haltia M (1970) The effect of folate deficiency on neuronal RNA content. Br J Exp Pathol 51:191

65. Hansen AA, Nordquvist P, Sourander P (1964) Megaloblastic anemia and neurologic distrubances combined with folic acid deficiency. Observation on an epileptic patient treated with anticonvulsants. Acta Med Scand 176:243–251

66. Havelius U, Hindfelt B, Rosen I (1982) Reversibility of neurological deficits in vitamin B_{12} deficiency. Arch Psychiatr Nervenkr 232:473–478

67. Heinrich HC (1967) Die experimentellen Grundlagen einer hochdosierten oralen vitamin B_{12}-Therapie beim Menschen. Ergeb Inn Med Kinderheilkd 25:1–24

68. Henry P, Bonnaud E, Laporte A, Orgogozo JM, Vital C, Loiseau P (1977) Place du déficit en acide folique dans les affections carentielles du système nerveux. A propos de 12 cas dont un anatomo-clinique. Sem Hop Paris 53:1530–1537

69. Herbert V (1962) Experimental folate deficiency in man. Trans Assoc Am Physicans 75:307–320

70. Reynolds EH, Trimble MR (1985) Adverse neuropsychiatric effects of anticonvulsant drugs. Drugs 29:570–581

71. Reynolds EH, Wales HD (1967) Effects of folic acid on the mental state and fit frèquency of drug treated epileptic patients. Lancet 2:1086–1088

72. Reynolds EH, Rothfeld P, Pincus JH (1973) Neurological disease associated with folate deficiency. Br Med J 2:398–400

73. Reynolds EH, Carney MWP, Toone BK (1984) Methylation and mood. Lancet 2:196–198

74. Robertson DM, Dinsdale HB, Campbell RJ (1971) Subacute degeneration of the spinal cord not associated with vitamin B$_{12}$ deficiency. Arch Neurol 27:203–209

75. Rodin E, Schmaltz S (1983) Folate levels in epileptic patients. In: Parsonage M, Grant RHE, Craig A, Ward AA (eds) The XIVth epilepsy international symposium. Raven, New York pp 143–149

76. Roman GC, Spencer PS, Schoenberg BS (1985) Tropical myeloneuropathies – the hidden endemias. Neurology 35:1158–1170

77. Roos D (1978) Neurological complications in patients with impaired vitamin B$_{12}$ absorption following partial gastrectomy. Acta Neurol Scand (Suppl– 58:1–77

78. Runcie J (1979) Folate deficiency in the elderly. In: Botez MI, Reynolds EH (eds) Folic acid in neurology, psychiatry and internal medicine. Raven, New York, pp 493–500

79. Schmitt J, Jacquier A (1982) Scléroses combinées de la moelle. Encycl Med Chir Paris 17068A[10]:6

80. Shapira Y, Zvi AB, Statter M (1978) Folic acid deficiency: a reversible cause of infantile hypotonia. J Pediatr 93:984–985

81. Shojania AM (1984) Folic acid and vitamin B$_{12}$ deficiency in pregnancy and in the neonatal period. Clin Perinatol 11:433–459

82. Shorvon SD, Carney MWP, Chanarin I, Reynolds EH (1980) The neuropsychiatry of megaloblastic anemia. Br Med J 281:1036–1038

83. Shulman R (1972) The present status of vitamin B$_{12}$ and folic acid deficiency in psychiatric illness. Can Psychiatr Assoc J 17:205–216

84. Shulman R (1979) An overview of folic acid deficiency and psychiatric illness. In: Botez MI, Reynolds EH (eds) Folic acid in neurology, psychiatry and internal medicine. Raven, New York, pp 463–474

85. Singer HS, Butler I, Rothenberg S, Valle O, Freeman J (1980) Interrelationship among serum folate, CSF folate neurotransmitters, and neuropsychiatric symptoms. Neurology 30:419

86. Smithells RW, Sheppard D, Schorah CJ et al. (1981) Apparent prevention of neural rube defects by periconceptional vitamins supplementation. Arch Dis Chid 56:911–918

87. Spaans F (1970) No effects of folic acid supplement on CSF folate and serum vitamin B$_{12}$ in patients on anticonvulsants. Spilepsia 11:403–411

88. Spector R, Lorenzo AV (1975) Folate transport in the central nervous system. Am J Physiol 229:777–782

89. Strachan RW, Henderson JG (1967) Dementia and folate deficiency. Q J Med 36:189–204

90. Swift TE, Gross LA, Ward LD, Crout BO (1981) Peripheral neuropathy in epileptic patients. Neurology 31:826–831

91. Taylor JW, Murphy MJ, Rivay MP (1985) Clinical and electrophysiologic evaluation of peripheral nerve function in chronic phenytoin therapy. Epilepsia 26:416–420

92. Thornton WE (1977) Folate deficiency in puerperal psychosis. Am J Obstet Gynecol 129:222–223

93. Tiggelen CJM, Peperkam JPC, Tertoolen JEW (1984) Assessment of vitamin B$_{12}$ status in CSF. Am J Psychiatry 141:136–137

94. Turner AJ (1983) The fluctuating fortunes of folates. Trends Pharmacol Sci 4:191–192

95. Van Der Westhuyzen J, Fernandez Costa F, Metz J (1982) Cobalamin inactivation by nitrous oxide produces severe neurological impairment in fruit bats: protection by methionine and aggravation by folates. Life Sci 31:2001–2010

96. Wells DG, Casey HJ (1967) *Lactobacillus casei* CSF folate activity. Br Med J 3:834–836

97. Zucker DK, Livingston RL, Nakra R, Clayton PL (1981) B$_{12}$ deficiency and psychiatric disorders: case report and literature review. Biol Psychiatry 16:197–205

Chapter 12

Folate Deficiency During Pregnancy and Lactation

J. METZ

Introduction

Possibly the commonest vitamin deficiency worldwide is that of folate. Although the incidence is highest in the lower socio-economic groups, no group is immune. Pregnancy is probably the severest physiological stress that can be imposed on body folate stores, and pregnant and lactating women therefore comprise the major group at risk for folate deficiency. It has been suggested that up to one-third of women in the world have evidence of folate deficiency during pregnancy. More cases of folate deficiency have been reported in pregnant or puerpural women than in any other population group, and this deficiency is thus a serious problem during pregnancy. When the increased demand for folate during pregnancy and lactation cannot be met because of inadequate diet, negative folate balance ensues, and depending on the size of folate stores, deficiency of the vitamin may result. Thus the folate deficiency is conditioned by poor diet and precipitated by pregnancy or lactation, or both.

Folate deficiency associated with pregnancy is a problem particularly of, but not restricted to, developing populations where poor socio-economic conditions, adverse diet, high parity and prolonged lactation are contributory factors. Deficiency is especially common in populations subsisting on cereal-based diets (rice, maize) where the intake of fresh unboiled vegetables rich in folate is often low. Such diets do not provide adequate amounts of folate for women during pregnancy. This is especially so where refined cereal is favoured. For example, with maize meal such refining may result in the loss of up to two-thirds of the folate content of the crude meal [31].

Folate Economy in Pregnancy

The maternal requirement for folate increases in pregnancy to provide for foetal and placental growth, the increase in red cell volume (up to 20%–30%) and the blood loss at delivery. There is evidence that the need of the foetus for folate takes precedence over that of the mother. The plasma and red cell folate concentrations in the foetus at all stages of pregnancy, and in cord blood, are substantially higher than in the mother [13]. Folate-deficient mothers give birth to babies with normal folate levels. To achieve this favoured status for the foetus vis-à-vis maternal folate needs, it is necessary to invoke an active transport mechanism whereby folate is transferred from mother to foetus [26]. A high-affinity folate binding protein is present in cord serum [23], and high-affinity binding receptor sites for folate have been detected in human placenta [1, 19]. The placenta may concentrate folates (and so provide high concentrations to the foetus) as suggested by the higher concentrations of serum folate found in the intervillous space of the placenta compared with those in the newborn or their mothers [18]. Both total and unsaturated folate binding capacity are increased in the serum of pregnant women [8, 16], but these changes are the result of hormonal influences in pregnancy and are not related to folate nutritional status [16]. Their role, if any, in the delivery of folate to the foetus is unclear.

The amount of folate which the mother is required to obtain from her diet to provide for the increased needs of pregnancy is thought to be about double that of the non-pregnant state. Thus the recommended daily allowance (RDA) for folate in pregnant women has been set at 800 µg.

Serum and Red Cell Folate Levels in Normal Pregnancy

Fall in serum folate concentration as pregnancy progresses occurs in most women irrespective of socio-economic class. In the first trimester the fall is small, followed by a gradual decrease in the second trimester and an accelerated decline after 36 weeks, reaching the lowest level at delivery. Following delivery, the level usually rises to that before pregnancy in well-nourished populations. The decline in serum folate level in pregnancy does not necessarily reflect the development of folate depletion. When the substantial increase in plasma volume (up to 50%) that occurs in pregnancy is taken into consideration, it can be calculated that there is usually no significant fall in *total* folate in the plasma during pregnancy [20], and the fall in *concentration* may be due mainly to the increase in plasma volume. Furthermore, in many pregnant subjects with low serum folate levels, red cell folate is normal [21]. There is no evidence that low levels of serum folate are related to a fall in haemoglobin or any complication of pregnancy, and a low value early in pregnancy does not have predictive value for the subsequent development of megaloblastic anaemia in the pregnancy. Thus, low folate levels in pregnancy in themselves do not necessarily suggest folate deficiency and are not an indication for folic acid supplementation.

Whereas the pattern of fall in serum folate levels in pregnancy is almost universal, less uniformity in the behaviour of red cell folate has been reported. Some studies showed a fall only during the last 2 months of pregnancy [14] or only after delivery in women who breast feed [44], or no significant fall at all. Most workers who have studied red cell folate levels in pregnancy have reported a steady fall in mean values, but not as steep as that of serum folate.

As red cell folate is more closely related to the state of folate stores than serum folate, different behaviour patterns in red cell folate would be anticipated during pregnancy in populations subsisting on diets of varying folate content. It seems that red cell folate values may be a useful guide to the state of folate stores in pregnancy [6].

Folate Deficiency and Megaloblastic Anaemia of Pregnancy and the Puerperium

The incidence of folate deficiency associated with pregnancy varies not only with the socio-economic status of the population under study, but particularly with the diagnostic criteria used. As stated, low serum folate values are not necessarily indicative of deficiency and reflect, at best, suboptimal dietary folate intake. Low red cell values are a better index of folate nutrition but do not necessarily reflect deficiency of clinical significance, and there is still substantial overlap in values between patients with normoblastic and megaloblastic marrows. To some extent red cell folate is a retrospective index reflecting the state of nutrition at the time the red cells were produced, i.e. some 2–3 months before drawing a sample. Nonetheless, red cell folate is probably the most practical way of assessing the incidence of folate deficiency in pregnancy in a given population. Megaloblastic anaemia is a late manifestation of folate deficiency and is an insensitive index of deficiency. However, in the individual pregnant patient, there is probably no substitute for marrow morphology or the diagnostic deoxyuridine (dU) suppression test in determining folate status [29].

The most serious complication of folate deficiency in pregnancy is the development of megaloblastic anaemia. Depending on the degree of dietary folate deficiency and the size of maternal folate stores, megaloblastic anaemia may develop towards the end of pregnancy. The reported incidence varies from a negligible number in populations with high nutritional standards to up to 50% in undernourished people. As with folate deficiency *per se,* incidence of megaloblastic anaemia varies not only in relation to the nutritional status of the population, but also with the different diagnostic criteria that have been used. Diagnosis should rest upon the presence of anaemia associated with overt megaloblastic changes in the bone marrow, and when such changes are marked in both developing erythroid and myeloid cells, the diagnosis is unequivocal. However, the significance of more subtle morphological changes or the presence of myeloid changes only, is open to individual interpretation. Furthermore, when anaemia accompanies megaloblastic change in the marrow, the association is not necessarily causal, in that iron deficiency is the commonest cause of anaemia in pregnancy in any population.

Megaloblastic anaemia in pregnancy is almost always the result of folate deficiency, even in vegetarians whose intake of vitamin B_{12}-rich foods is low. It is probably only in pregnant women, who have rigidly excluded all sources of B_{12} from the diet (vegans) for a number of years, or in rare malabsorption syndromes affecting predominantly the terminal ileum that B_{12}-deficiency megaloblastic anaemia associated with pregnancy is likely.

The cause of folate-deficiency megaloblastic anaemia in pregnancy and in the puerperium is basically inadequate amounts of dietary folate to meet the demands of pregnancy. Haematological remission can be induced by dietary means alone, e.g. by the oral administration of natural folate in the form of lettuce [2]. It has been postulated that impaired folate absorption may be a significant factor in the development of folate deficiency in pregnancy. However, a consistent defect in absorption during pregnancy has not been demonstrated [24], and the absorption of folic acid in patients with megaloblastic anaemia following pregnancy is normal [40].

There is a seasonal variation in the incidence, for the disease tends to occur commonly in spring following the winter period when green vegetables are scarce and expensive. Clinically the anaemia presents usually during the 6-week period to either side of delivery. In the Western world, it is said to occur during pregnancy and to recover spontaneously following delivery. Yet in some such groups, about half the cases are diagnosed post-partum, mostly in the first few weeks after delivery [5]. In populations where prolonged lactation is common, as in parts of rural Africa, the disease is most commonly seen after delivery, even up to a year, in nursing mothers. When megaloblastic anaemia manifests itself shortly after delivery, pregnancy alone has probably been the precipitating factor, and the condition is termed megaloblastic anaemia of pregnancy or the puerperium. However, if it presents many months after delivery, an association with lactation should be recognized.

The symptoms of the disease are related to anaemia in general. The haematological features are similar to those of megaloblastic anaemia from other causes, except that concomitant iron deficiency is relatively frequent.

Treatment of megaloblastic anaemia associated with pregnancy with folic acid (about 1 mg/day) is simple and effective, but the emphasis should be on prophylaxis in susceptible populations.

The Diagnosis of Folate Deficiency in Pregnancy

Folate deficiency may be manifest by subnormal serum folate (less than 3.0 ng/ml); subnormal red cell folate (less than 140 ng/ml), anaemia, neutropenia and thrombocytopenia; macrocytosis (particularly macro-ovalocytosis); a 'right shift' with the presence of hypersegmented neutrophils in the peripheral blood; and megaloblastic change in the bone marrow and a specific abnormality in the diagnostic dU suppression test. There are stages and degrees of deficiency. Low serum folate *per se* is indicative only of negative folate balance; subnormal red cell folate and an abnormal dU suppression test indicate tissue depletion. Morphological changes in the blood and bone marrow and anaemia indicate that haemopoiesis has been affected.

In pregnant patients, the diagnostic value of some of these features may be less than in non-pregnant subjects. Low serum folate value in pregnancy may be a reflection of the increase in plasma volume that occurs in pregnancy and could be unrelated to folate nutrition. Macrocytosis (but not macro-ovalocytosis) may be physiological in pregnancy and also unrelated to folate nutrition. The average neutrophil lobe count is of little diagnostic value, but the presence of a significant number of hypersegmented neutrophils (5 or more lobes) probably is diagnostically important.

Some of the manifestations of folate deficiency may be masked by the concomitant iron deficiency which is common in pregnancy, especially in women who are not taking supplemental iron. When iron deficiency coexists, macrocytosis and megaloblasts may be masked, but the myeloid changes of folate deficiency are not usually affected.

The Need for Folic Acid Supplementation in Pregnancy

Prevention of Folate Deficiency and Megaloblastic Anaemia Associated with Pregnancy and Lactation. Particularly in poorly nourished populations where evidence of folate deficiency during pregnancy is common, folic acid supplementation during pregnancy results in an improvement in tests of folate nutrition and a reduction in the incidence of megaloblastic anaemia during and after pregnancy [12, 17]. However, supplementation has no significant effect on mean haemoglobin values, even in populations subsisting on a poor diet.

Reduction in the Incidence of Prematurity (Low Birth Weight Infants). Folic acid administered to pregnant women whose diet is low in folate may be associated with a significant reduction in the incidence of prematurity [3, 25]. Such an effect is often not demonstrable in patients subsisting on an average Western diet. However, it is of interest that a positive correlation between the folate nutrition of mothers and/or their infants with the birth weight of the infants'/length of gestation has also been reported in populations in England and the United States and more recently in Denmark [35], Norway [13] and France [46]. In the Danish study [35], an effect of maternal supplementation with folic acid on infant birth weight was also demonstrated.

It is undeniable that there is a need for folic acid supplementation of pregnant women subsisting on a poor diet to reduce the consequent infant morbidity and mortality from low birth weight babies. The need may not be restricted to such groups, however, but may also apply to some apparently well-nourished populations.

In populations where undernutrition is common, supplementation with folic acid during pregnancy increases the concentration of folate in the breast milk during subsequent lactation. Thus the beneficial effect on the infant of such supplementation is not confined to the time spent *in utero* but may continue during the period of breast feeding.

Reduction of the Incidence of Neural Tube Defects in Newborns. There is evidence that maternal folic acid deficiency may play a role in the pathogenesis of neural tube defects (NTD) in the infant. Such defects are commoner in the infants of women of lower socio-economic status. Serum and red cell folate values in the first trimester of pregnancy are lower in women who give birth to infants with NTD, and folic acid supplementation before conception and in the early weeks of pregnancy may reduce the incidence of infants with NTD in women with an increased risk of having an infant suffering from this condition [27, 38]. However, other studies have failed to note any relationship of serum folate concentrations in early pregnancy to NTD in infants [32].

Critical examination of these studies indicates that the evidence that folic acid supplementation around the time of conception reduces the risk of having an

NTD infant is suggestive, but not sufficient to warrant supplementation of all women at risk [45]. In view of the uncertainly of the role, if any, of maternal folate deficiency in the causation of NTD, the Medical Research Council (MRC) in Great Britain is currently conducting a large randomised, multicentre, international trial in an attempt to resolve the issue.

In the interim, it would seem justified to administer folic acid tablets around the time of conception to women at higher risk of producing an infant with NTD, expecially as such supplementation is harmless.

Prevention of Obstetrical Complications. In spite of numerous attempts to demonstrate a relationship between folate deficiency and obstetrical complications, there is little convincing evidence that the routine administration of folic acid to pregnant women reduces complications such as abortion and abruptio placenta.

Prevention of Gingivitis. An interesting role for folic acid has been postulated in the exaggerated gingivitis which may be associated with pregnancy. Improvement in such gingivitis can be achieved by use of a mouth wash containing folic acid but not by the administration of folic acid tablets, suggesting that localised end-organ deficiency of folate may be involved [34].

Should All Pregnant Women Receive Folic Acid Supplements, and How Much?

Few would dispute that severe folate deficiency in pregnancy, with morbidity for both mother and infant, is relatively common in undernourished women in developing countries. Routine supplementation with folic acid should be provided for all pregnant women in such populations.

But how applicable are the results of studies on folate nutrition in poor populations to women in the comparatively affluent Western societies? The question of routine supplementation in such populations is controversial. The issue hinges on whether the average Western diet contains sufficient amounts of folate to provide for the increased need in pregnancy. There is evidence that this may not always be so. Although the availability of dietary folate cannot readily be measured, an analysis of the diet in Denmark has revealed that the folate intake in pregnant women is less than a third of that of the RDA [15]. The average diet in the United States contains sufficient vitamins, with the exception of folate, the intake of which is only marginal. In that country, females of reproductive age are at greatest risk for developing deficiency, with an incidence of low serum and red cell folates of 15% and 13% respectively [36]. A significant incidence of low red cell folate levels early in pregnancy has been reported in a New York community in women from low income families [21].

Thus it could be argued that even in affluent populations, most diets fail to provide the increased amounts of folate required by pregnant women, and all should receive folic acid supplementation. The contrary view favours supplementing only in those whose diet is clearly inadequate, in individuals with a multiple pregnancy or haemolytic anaemia (such as sickle cell disease), or in those taking anticonvulsant drugs.

There is no universally determined amount of folic acid that should be given during pregnancy, for the amount required to prevent folate deficiency developing in pregnancy is dependent on body stores and dietary intake. The diet of various populations differs significantly in regard to folate content, and the supplement needed varies according to the amount of folate present in the basal diet. The amount of daily supplement recommended has varied from 100 to 1000 µg. In well-nourished pregnant women, a supplement of about 100 µg a day is probably adequate, while in populations subsisting on a diet poor in folate 200–300 µg is required [11]. The supplement is usually given as a single tablet which contains iron as well. In the pregnant women receiving anti-epileptic drugs, doses of this order appeared sufficient to prevent folate deficiency without increasing the frequency of seizures [22].

Fortification of Food with Folic Acid

Despite the effectiveness of folic acid supplementation in the prevention of folate deficiency in pregnancy, in some populations the issue of folic acid tablets at pre-natal clinics may fail to prevent folate deficiency in a significant number of pregnant women. This is partly due to lack of adequate pre-natal facilities, to erratic attendance at clinics, or to non-compliance in the taking of tablets issued. In Nigeria, for example, a third of pregnant women who received folic acid tablets during pregnancy still showed megaloblastic changes in the marrow later in the pregnancy as a result of poor patient compliance [33]. Even in urban areas in the United States and Great Britain compliance in the taking of supplemental tablets may be poor [7]. This suggests that an alternative to the pre-natal administration of folic acid tablets should be sought in areas where there is a high incidence of folate deficiency in pregnancy.

Food supplements can provide much of the RDA for folate in pregnant women, and this approach provides a convenient alternative to the pre-natal administration of folic acid tablets, which is often not feasible in large rural populations with inadequate pre-natal services. Absorption studies indicate that the amount of folate available from fortified staple items of diet such as maize meal, rice and bread, make them appropriate vehicles for folic acid fortification [10]. Giving maize meal fortified with folic acid to pregnant subjects prevents folate deficiency in pregnancy [9]. Fortified meal also induces haematological remission when fed to patients with folate-deficiency megaloblastic anaemia. Bread fortified with folic acid has a similar effect in preventing deficiency in pregnancy [28]. Additional folate can also be provided in the form of folic acid fortified breakfast cereals [42].

Folate Deficiency Associated with Lactation

The daily folate requirement in lactating women is some 100–300 µg, which is about two to three times greater than that of non-pregnant, non-lactating women. There is much evidence that the parasitic effect of the foetus on maternal folate nutrition during pregnancy continues into the period of lactation, and that the supply of folate to the infant through breast milk takes precedence over the mother's own need for folate, even when maternal deficiency is extreme. The plasma and red cell folate levels are higher in breast-fed infants than in their mothers [41]. Folate is secreted in breast milk, and there is apparently a mechanism to maintain folate levels in the milk, despite maternal folate insufficiency, possibly by the firm binder for folic acid which is present in milk but not in serum [30].

Thus, lactation may act as a conditioning factor for the development of folate deficiency and megaloblastic anaemia. If the deficiency is not severe enough to result in megaloblastic anaemia during pregnancy, these patients nonetheless enter lactation with diminished folate stores. The added folate requirement during lactation in a mother whose folate intake is low, and whose folate stores have been depleted by pregnancy, may then cause megaloblastic anaemia. Prolonged lactation is a rarity in developed populations, and even with prolonged lactation these women are unlikely to become folate deficient because of good diet and the widespread practice of folic acid supplementation during pregnancy.

In developing populations, such as in Africa, the situation is often quite different – dietary folate intake is suboptimal, attendance at pre-natal clinics irregular, and prolonged lactation common. In such populations, there is but a transient rise in serum folates following pregnancy, and deterioration in folate nutrition occurs as lactation progresses [37], such that the incidence of folate deficient megaloblastic anaemia is highest in the 12-week period following delivery.

There is evidence of folate deficiency during lactation in populations other than in Africa. In Navajo women in the United States, serum folate decreases as lactation progresses, and a third of such women have red cell folate levels in the low-deficient range [4].

The deterioration in folate nutrition with prolonged lactation can be prevented by folic acid supplementation of the diet of lactating women. Such supplementation raises the milk folate concentration in women of low socio-economic status [39], but not in well-nourished women [43], yet another example of how socio-economic status determines the response to supplementation.

In view of the increased maternal need for folate during lactation and the need to make up any folate deficiency that occurs during pregnancy, it could be argued that folic acid supplementation should be continued during lactation in subjects whose diet is low in folate. Adequate supplementation of the diet with folic acid during the pregnancy is probably sufficient to render maternal stores sufficient to provide for the needs during lactation of average duration. However, with prolonged lactation, as practised in some developing populations, supplementation of the diet with folic acid may well be indicated, particularly where there is evidence of continued deficiency after delivery, possibly related also to poor patient compliance with the routine administration of tablets during the pregnancy.

References

1. Antony AC, Utley C, Van Horne KC, Kolhouse JF (1981) Isolation and characterization of a folate receptor from human placenta. J Biol Chem 256:9684–9692
2. Baumslag N, Metz J (1964) Response to lettuce in a patient with megaloblastic anaemia associated with pregnancy. S Afr Med J 38:611–614
3. Baumslag N, Edelstein T, Metz J (1970) Reduction of incidence of prematurity by folic supplementation in pregnancy. Br Med J 1:16–17
4. Butte NF, Calloway DH, Van Duzen JL (1981) Nutritional assessment of pregnant and lactating Navajo women. Am J Clin Nutr 34:2216–2228
5. Chanarin I, MacGibbon BM, O'Sullivan WJ, Mollin DL (1959) Folic-acid deficiency in pregnancy. The pathogenesis of megaloblastic anaemia of pregnancy. Lancet 2:634–639
6. Chanarin I, Rothman D, Ward A, Perry J (1968) Folate status and requirement in pregnancy. Br Med J 2:390–394
7. Colman N (1982) Addition of folic acid to staple foods as a selective nutrition intervention strategy. Nutr Rev 40:225–233
8. Colman N, Herbert V (1976) Total folate binding capacity of normal human plasma, and variations in uremia, cirrhosis and pregnancy. Blood 48:911–921
9. Colman N, Barker M, Green R, Metz J (1974) Prevention of folate deficiency in pregnancy by food fortification. Am J Clin Nutr 27:339–344
10. Colman N, Green R, Metz J (1975) Prevention of folate deficiency by food fortification. Part II. Absorption of folic acid from fortified staple foods. Am J Clin Nutr 28:459–464
11. Cooper BA, Cantlie GSD, Brunton L (1971) The case for folic acid supplements during pregnancy. Am J Clin Nutr 23:848–854
12. Edelstein T, Stevens K, Baumslag N, Metz J (1968) Folic acid and vitamin B_{12} supplementation during pregnancy in a population subsisting on a suboptimal diet. J Obstet Gynaec Br Cmmwth 75:133–137
13. Ek J (1982) Plasma and red cell folate in mothers and infants in normal pregnancies: relation to birth weight. Acta Obstet Gynecol Scand 61:17–20
14. Ek J, Magnus EM (1981) Plasma and red blood cell folate during normal pregnancies. Acta Obstet Gynecol Scand 60:247–251
15. Elsborg L, Rosenquist A (1979) Folate intake by teenage girls and by pregnant women. Int J Vitam Nutr Res 49:70–76
16. Fernandes-Costa F, Metz J (1981) The specific folate binding capacity of serum. Evidence that levels are not directly related to folate nutrition but influenced by hormonal status. J Lab Clin Med 98:119–126
17. Fleming AF, Martin JD, Hahnel R, Westlake AJ (1974) Effects of iron and folic acid antenatal supplements on maternal haematology and fetal wellbeing. Med J Aust 2:429–436
18. Giugliani ERJ, Jorge SM, Goncalves AL (1985) Serum and red blood cell folate levels in parturients, in the intervillous space of the placenta and in full-term newborns. J Perinat Med 13:55–59
19. Green T, Ford HC (1984) Human placental microvilli contain high-affinity binding sites for folate. Biochem J 218:75–80
20. Hall MH, Pirani BBK, Campbell D (1976) The cause of the fall in serum folate in normal pregnancy. Br J Obstet Gynaecol 83:132–136
21. Herbert V, Colman N, Spivack M et al. (1975) Folic acid deficiency in the United States: folate assays in a prenatal clinic. Am J Obstet Gynecol 123:175–179
22. Hillesmaa VK, Teramo K, Granstrom M-L, Bardy AH (1983) Serum folate concentrations during pregnancy in women with epilepsy: relation to antiepileptic drug concentrations, number of seizures, and fetal outcome. Br Med J 287:577–579
23. Holm J, Hansen SI, Lyngbye J (1980) A high affinity folate binding protein in umbilical cord serum. Scand J Clin Lab Invest 40:523–527

24. Iyengar L, Babu S Folic acid absorption in pregnancy. Br J Obstet Gynaecol 82:20–23
25. Iyengar L, Rajalakshmi K (1975) Effect of folic acid supplements on birth weights of infants. Am J Obstet Gynecol 122:332–336
26. Landon MJ, Eyre DH, Hytten FE (1975) Transfer of folate to the fetus. Br J Obstet Gynaecol 82:12–19
27. Laurence KM, James N, Miller MH, Tennant GB, Campbell H (1981) Double-blind randomised controlled trial of folate treatment before conception to prevent recurrence of neural-tube defects. Br Med J 282:1509–1511
28. Margo G, Barker M, Fernandes-Costa F, Colman N, Green R, Metz J (1975) Prevention of folate deficiency by food fortification. Part VII. The use of bread as a vehicle for folate supplementation. Am J Clin Nutr 28:761–763
29. Metz J (1984) The deoxyuridine suppression test. Crit Rev Clin Lab Sci 20 (3):205–241
30. Metz J, Zalusky R, Herbert V (1968) Folic acid binding by serum and milk. Am J Clin Nutr 21:289–297
31. Metz J, Lurie A, Konidaris M (1970) A note on the folate content of uncooked maize. S Afr Med J 44:529–541
32. Molloy AM, Kirke P, Hillary I, Weir DG, Scott JM (1985) Maternal serum folate and vitamin B_{12} concentrations in pregnancies associated with neural tube defects. Arch Dis Child 61:660–665
33. Okafor LA, Diejomaoh FME, Oronsaye AU (1985) Bone marrow status of anaemic pregnant women on supplemental iron and folic acid in a Nigerian community. Angiology 36:500–503
34. Pack ARC, Thomson ME (1980) Effects of topical and systemic folic acid supplementation on gingivitis in pregnancy. J Clin Periodontol 7:402–414
35. Rolschau J, Date J, Kristoffersen K (1979) Folic acid supplement and intrauterine growth. Acta Obstet Gynecol Scand 58:343–345
36. Senti FR, Pilch SM (1985) Synopsis. Analysis of folate data from the second national health and nutrition examination survey (Nhanes II). J Nutr 115(ii):1398–1402
37. Shapiro J, Alberts HW, Welch P, Metz J (1965) Folate and vitamin B_{12} deficiency associated with lactation. Br J Haematol 11:498–504
38. Smithells RW, Seller MJ, Harris R et al (1983) Further experience of vitamin supplementation for prevention of neural tube defect recurrences. Lancet I:1027–1031
39. Sneed MS, Zane C, Thomas MR (1981) The effects of ascorbic acid, vitamin B_6, vitamin B_{12} and folic acid supplementation on the breast milk and maternal nutritional status of low socioeconomic lactating women. Am J Clin Nutr 34:1338–1346
40. Stevens K, Metz J (1964) The absorption of folic acid in megaloblastic anaemia associated with pregnancy. Trans R Soc Trop Med Hyg 58:510–516
41. Tamura T, Yoshumura Y, Arakawa T (1980) Human milk folate status in lactating mothers and their infants. Am J Clin Nutr 33:193–197
42. Thenen SW (1982) Folacin content of supplemental foods for pregnancy. J Am Diet Assoc 80:237–241
43. Thomas RT, Sneed SM, Wei C, Nail AP, Wilson M, Sprinkle EE (1980) The effects of vitamin C, vitamin B_6, vitamin B_{12}, folic acid, riboflavin and thiamin on the breast milk and maternal status of well-nourished women at 6 months postpartum. Am J Clin Nutr 33:2151–2156
44. Tso SC, Wong RC (1980) Folate status in pregnant Chinese women in Hong Kong. Int J Gynaecol Obstet 18:290–294
45. Wald NJ, Polani PE (1984) Neural-tube defects and vitamins: the need for a randomized clinical trial. Br J Obstet Gynaecol 91:516–523
46. Zittoun J, Blot I, Hill C, Zittoun R, Papiernik E, Tchernia G (1983) Iron supplements versus placebo during pregnancy: its effects on iron and folate status on mothers and newborns. Ann Nutr Metab 27:320–327

Chapter 13

Folic Acid Deficiency in Developing Nations

J. D. RAIN, I. BLOT, and G. TCHERNIA

Although folate deficiencies are much less prevalent than iron deficiencies in the Third World, they nonetheless represent a major public health problem among two high-risk groups: pregnant women and young children. In most instances, the folate deficiency is not isolated but is part of a multiple deficiency syndrome. Anemia is the one visible symptom, appearing long after a slight or even unequivocal deficiency; it is only the "top of the iceberg."

Frequency of Folate Deficiency

Apart from a few large surveys performed with Western methods, the incidence of folate deficiencies in developing countries has not been well investigated and is not well known. Results concerning the prevalence of the deficiency differ according to the diagnostic criteria used, even in the same risk group, depending on whether one considers red cell and serum folate levels in the anemias that result in a visit to a health care center. Table 1 gives an indication of the frequencies of folate deficiencies and anemias in various Third World countries. Taking into consideration large variations among these results, one can still note a high frequency of folate deficiencies and anemias in pregnant women in late pregnancy. The folic acid deficiency never seems to be the only reponsible factor for the anemias discovered during or aggravated by pregnancy. It is almost always coupled with iron deficiency, the latter being usually more prominent. Occasionally, the folate deficiency is unmasked after an iron supplementation that causes increased erythropoiesis [30]. Folate folic acid deficiency, however, must be considered as responsible for megaloblastic bone marrow. B_{12} deficiency is much less frequent [3], (except, possibly, in India).

The deficiency appears more widespread and more severe in Asia and India than in Africa. In Asia depending on the diagnostic criteria and the regions, it involves from 22% to 66% of all women during pregnancy [3]. The highest prevalence is noted in Southern India [35]. In Africa, on the other hand, some populations seem completely deficiency-free during pregnancy [36].

In children, the incidence of folate deficiency varies greatly from country to country but appears to be lower than in adults [19]. Folate level seems influenced by socioeconomic status and by age: the lowest levels are in those aged 13–18 months [1].

Epidemiology

Knowledge of the epidemiology of folate deficiency in developing nations enables us to understand why they are so frequent, and how they could be prevented. As emphasized by Herbert et al. [18] three factors often coexist: decreased intake caused by poor nutrition, impaired absorption, and increased requirements.

The diet in developing countries is based essentially on cereals. Foods with high folate content such as meat, especially liver, green vegetables, yeast, represent a very small part of the diet. Furthermore, the green vegetables that could, by their low cost, contribute to an important part of folic acid intake, are not available during the dry season, which sometimes lasts 7–8 months. It is at the end of this dry season that one observes a peak in the incidence of folate deficiency. Moreover, the measurements of folic acid contents of foodstuffs cannot be a valid indication of the true folate intake, in as much as folates are heat labile, and because 50%–95% of them may be destroyed by excessive cooking, in particular by boiling in large amounts of water [16].

Unfortunately, the green-leafed plants that grow in the tropics and represent an important source of folic acid often have a high fiber content and are barely edible unless they are softened by prolonged cooking [28]. This source of intake varies considerably from one population to another [25], reflecting the diverse culinary and nutritional customs. The same vegetable may or may not bring a non negligible amount of active folates [14], depending on its method of preparation or on the composition of the soil. Survey results differ greatly from dry to rainy seasons and from urban to rural zones, where picked fruit and vegetables can, at little cost, prevent or correct deficiencies. The great disparity in levels of folic acid intake explains the observations of different authors that iron replace-

Table 1. Frequency of folic acid deficiencies and anemias in developing countries

Country (Reference)	Population	Serum folate < 3 ng	Anemia Hb < 11 g
Senegal (O.R.A.N.A. [26])	Pregnant women[a]	51%	23%
Nigeria (Flemming [15])	Pregnant women[b]	Megaloblastosis	85%
Ivory Coast (Rain et al. [29])	Women of childbearing age[b] Children[b]	66% –	– 14%
Burma (Aunz-Than-Batu et al. [2])	Pregnant women[a]	57%	–
India (Baker and De Maeyer [4])	Pregnant women[a]	73% (< 6 ng)	66%

[a] Mass survey
[b] Inpatients

ment therapy alone can correct the anemia of pregnancy [33], that the additions of folic acid increases hemoglobin levels [5], and that the association of both nutrients is necessary [22].

The absorbability of the various forms of polyglutamates is not perfectly known. A small percentage (approximatively 10% [17]) of these dietary polyglutamates is absorbed. Polyglutamate folates require a transformation into monoglutamates by an intestinal conjugase in order to be absorbed. This intestinal conjugase can be inhibited by nutrients such as orange juice and by inhibitors found in beans and other vegetables [18]. In pregnant women, however, polyglutamate folates may be as efficiently absorbed as monoglutamates [24]. Folate intake in developing nations is thus often quantitatively inadequate or is destroyed by cooking and night not be readily absorbable because of possible the intestinal conjugase inhibitors.

The absorption processes can also be disturbed, in particular in tropical sprue, which is absent in Africa but is present in the Caribbean area and in Asia. In tropical sprue, there is frequently a folic acid deficiency. The malabsorption can be detected when glucose or galactose are added to the diet [11]. In sprue, conjugase activity in the jejunum is decreased. In many poverty-sticken tropical countries, it is difficult to establish the respective roles played by undernutrition and malabsorption [10].

The daily allowances recommended by the World Health Organization WHO [34]; (Table 2) vary according to age, gestational status, and lactation. Folic acid requirements are higher in developing countries than in the developed countries. Pregnancies there are more frequent and occur at shorter intervals to one another, and maternal lactation is customary and usually more prolonged.

Furthermore, as soon as intense erythropoietic activity is present, the requirements are increased. Such is the case in malaria and hemoglobinopathies. Malaria prophylaxis increases the serum folate levels and gives partial protection from the severe anemias of pregnancy [15]. With hemoglobinopathies, particularly sickle cell anemia, a folate deficiency frequently coexists; its correction allows a partial correction of the anemia. Megaloblastic anemia may sometimes reveal a hemoglobinopathy in the child [12].

Table 2. Daily folate requirements as recommended by the World Health Organization [34]

Age	Folate requirement (μg/day)
0–6 months	40–50
6–12 months	120
1–12 years	100
Over 13 years	200
Gestation	400
Lactation	300

Clinical Diagnostic Problems in Developing Nations

In public health surveys on the incidence of folate deficiencies, diagnostic problems arise concerning the presence of anemia, which is a late manifestation of deficiency. Peripheral macrocytosis is not constant. The anemia can be normocytic. On blood smears microcytic and hypochromic red blood cell may coexist with macrocytic and normochromic cells. These dimorphic anemias are not infrequent in tropical areas because of the high frequency of combined (iron and folates) deficiencies. In a survey performed in Abidjan on anemias in women of childbearing age, less than half of the folate-depleted women had a mean corpuscular volume (MCV) over 95 μm^3 even though this sample included patients with severe anemias and serum folate levels of less than 3ng/ml [13].

Bone marrow examination is of great diagnostic help if it shows megaloblastosis or macroblastosis. This anomaly however, is not always present, and bone marrow examination is not performed routinely in health centers of developing countries. In some cases, erythroblastopenia is found or maturational arrest which predominates in white cell lineage [9]. Often, stigmata of folate deficiency (anomalies of maturation, megaloblastosis) and of iron deficiency (poorly hemoglobinized and spiculated cytoplasm) are both present. One or the other aspect may predominate on the erythroblasts, leading to very polymorphic images.

Slight hematological signs can help in the diagnostic procedure, such as hypersegmentation of polymorphonuclear leukocytes and giantism and abnormal morphology of the medullary white cell precursors. The substantiation of folate deficiency by measuring the red cell and serum folate levels and by the deoxyuridine (dU) suppression test is seldom feasible in developing nations apart from university hospitals in the capitals or in cooperation Western countries where the samples may be sent. The diagnosis is often only hypothetical and can be further confirmed by folate supplementation, which serves as a diagnostic test.

In developing nations, it is important not to accept that a single etiology explains anemia. The diversity of causes of anemias in developing countries is well established. Nutritional anemia are rarely due to a single deficiency. The deficiency is often multiple e.g., folic acid and iron in the pregnant woman or folic acid, iron, and protein in the small child. In undernourished children, folate deficiency participates in the genesis of the nutritional anemia; its role is not as prominent as that of protein and iron deficiencies and may be unmasked only when the child receives food supplements [3, 23]. These deficiencies often arise on a special background, which itself causes anemia (e.g., hemoglobinopathy, G6PD deficiency) that the additional deficiency enhance.

Bacterial and especially parasitic infections, especially malaria and ancylostomiasis, can further add to the deficiencies and abnormal backgrounds. In an Ivory Coast survey [29] on severe anemias of the African infant and child, over half of the children showed at least two causes responsible of anemia.

This etiologic diversity implies that a range of potential factors must be taken into account.

Particular Consequences of the Deficiency

Besides megaloblastic anemia, curable by folate treatment, profound folic acid deficiency can be responsible in the postpartum period for systemic and digestive symptoms, associated in a severe picture, with anemia, fever, and profuse diarrhea. Serum folate levels are extremely low. The symptoms resist to antibiotic treatment and rapidly regress during oral folate replacement therapy [7].

The consequences of this deficiency in the course of pregnancy have been discussed by several authors. In South Africa. Hibbard [20] found that folate-deficient women had more premature infants, abortions, and retroplacental hematomas. Some of these factors remain difficult to assess in populations with multiple nutritional deficiencies and strenuous living conditions during pregnancies. A relation between birth weight or gestational age and folate status does however appear in several studies. Folic acid supplementation has no effect on the birth-weight of white newborns in South Africa but leads to a significant decrease in the incidence of hypotrophic infants among Bantus [6]. In Asia, folic acid supplementation during pregnancy, although having no effect on the maternal hemoglobin level, does significantly increase the mean birth-weights of infants as well as the placental weight [21].

These results, which seem of particular significance in lower socioeconomic groups, are not, however, relevant exclusively to Third World countries; in our experience, a child's birth-weight increases with longer gestations and not with improved nutritional status [32]. Folate supplementation has an indirect role. The increase in birth weights is explained by longer gestation.

Treatment

The treatment of folate deficiency in developing nations should be primarily prophylactic. Better nutritional status is often difficult to obtain due to socioeconomic or climatic barriers or to long-standing culinary habits. The requirements for folate supplements decrease progressively with development in the country. The best preventive method is supplementation with easily absorbable folate in high-risk groups: pregnant women starting at the 3rd month of gestation and young children. For these groups replacement therapy should combined with iron supplementation.

Curative therapy requires a daily intake of folic acid in the form of pteroyl(mono)glutamate at the dose of 5 mg per day (or one tablet) for 3 months. This dosage of 5 mg for each tablet is more than actually required [31]. To this treatment one must add that of the associated deficiencies, in particular iron and protein, and of the often associated parasitic infections, ankylostomiasis and malaria.

In the undernourished child, bone marrow megaloblastosis recedes with folic acid replacement therapy but can coexist with normal serum folate levels, probably due to impaired cellular folate utilization linked to other deficiencies [8, 27].

References

1. Akinsete FI, Boyo AE (1977) Studies on folic acid in Nigerian infants and pre-school children. Environ Child Health 23:202–205
2. Aung-Than-Batu, Thane-Toe, Hla-Pe, Khin-Kyi-Nyunt (1976) A prophylactic trial on iron and folic acid supplements in pregnant Burmese women. Isr J Med Sci 12:1410–1417
3. Baker SJ (1981) Nutritional anemia: tropical asia. Clin Haematol 10:843–871
4. Baker SJ, De Maeyer EM (1979) Nutritional anaemia: its understanding and control with special reference to the work of the World Health Organization. Am J Clin Nutr 32:368–417
5. Basu RN, Sood SK, Ramachandran K, Mathur M, Ramalingaswami V (1973) Etiopathogenesis of nutritional anemia in pregnancy: a therapeutic approach. Am. J Clin. Nutr 36:591–594
6. Baumslag N, Edelstein T, Metz J (1970) Reduction of incidence of prematurity by folic acid supplementation in pregnancy. Br Med J 1:1–6
7. Blot I, Diop-Mar I (1974) Anémie fébrile par carence folique du post-partum. Bull Soc Med Afr Noire Lang Fr XIX:148–156
8. Blot I, Tchernia G, Becart-Michel R, Zucker JM (1972) L'anémie au cours du Kwashiorkor. Nouv. Rev. Fr. Hematol 12:424–441
9. Chanarin I, Davey A (1964) Acute megaloblastic arrest of hemopoiesis in pregnancy. Br. J Haematol 10:314–319
10. Chatterjea JB (1966) Nutritional megaloblastic anaemia in tropical zones. In: R. J. Walsh, W. R Pitney (Eds) plenary sessions, XIth congress of international society of haematology. Blight, Sydney, p 120
11. Corcino JJ, Coll G, Klipstein FA (1975) Pteroylglutamic acid malabsorption in tropical sprue. Blood 45:577–580
12. Danel P, Girot R, Tchernia G (1983) Thalassémie majeure révélée par une anémie mégaloblastique par déficit en folates. Arch. Fr. Pediatr 40:799–801
13. Dexemple P, Monnier A, Tea D, Rain JD et al. (1985) Anémie par carence en folates et vitamine B_{12} en Côte d'Ivoire. Sem. Hop. Paris 61:2515
14. Dutta SK, Russel RM, Chowdhury B (1980) Folate content of north Indian breads. Nutr Rep Int 21:251–255
15. Fleming AF (1978) Nutritional anemia fue to folate deficiency. In: XVIIème Congrès de la Société Internationale d'Hématologie, Paris. Abstracts II, 636
16. Herbert V (1963) A palatable diet for producing experimental folate deficiency in man. Am J. Clin. Nutr 12:17–20
17. Herbert V (1968) Nutritional requirements for vitamin B_{12} and folic acid. Am J. Clin. Nutr 21:743–752
18. Herbert V, Colman Neuville, Jacob E (1977) Nutritional anemias overview. Megaloblastic anemias. In: Gordon AS (ed) The year of heamtology 1977. Plenum, New York p 549–581
19. Hercberg S, Chauliac M, Galan P et al. (1986) Relationship between anemia iron and folacin deficiency, hemoglobinopathics and parasitic infection. Hum. Nutr. Clin. Nutr 40c:371–379
20. Hibbard BM (1975) Folates and the fetus. S. Afr. Med. J 49:1223–1226
21. Iyengar L, Rajalakshmi K (1975) Effect of folic acid supplement on birthweights of infants. Am. J. Obstet Gynecol 122:332–336
22. Izak G, Levy SH, Rachmilewitz M, Grossowicz N (1973) The effect of iron, folic acid therapy on combined iron and folates deficiency anemia: the results of a clinical trial. Scand. J. Haematol 11:236–240
23. Kaimel K, Waslien CI, El-Ramly Z et al. (1972) Folate requirements of children. II. Response of children recovering from protein calorie malnutrition to graded doses of parenterally adminstrated folic acid. Am. J. Clin. Nutr 25:152–165

24. MacLean FW, Heine NW, Held D, Streiff RR (1970) Folic acid absorption in pregnancy: comparison of the pteroyl polyglutamate and pteroyl monoglutamate. Blood 5:628–631
25. Masawe AEJ (1981) Nutritional anemia: tropical Africa. Clin Haematol 10(3):815–843
26. Organisme de Recherches sur l'Alimentation et la Nutrition Africaines (1975) Enquête sur les anémies de la femme enceinte à Nouakchott. Dakar TL Fournie, AMN-Diaye Eds
27. Pereira S, Baker SJ (1966) Haematologic studies in kwashiorkor. Am. J. Clin. Nutr 18:413–420
28. Pitney WR (1973) Anaemia in the tropics. Clin. Haematol 2:337–356
29. Rain JD, Plo Kouie J, Tea Daignekpo N, Cagnard V, Aholi P (1982) Les anémies sévères du nourrisson et de l'enfant en milieu tropical africain. Ann Pediatr 29:289–295
30. Srisupandit S, Pootrakul P, Areekul S (1983) A prophylactic supplementation of iron and folate in pregnancy. Southeast Asian J Trop Med Public Health 14:317–323
31. Streiff RR (1970) Folic acid deficiency anemia. Semin Haematol 7:23–39
32. Tchernia G, Blot I, Rey A, Kaltwasser JP, Zittoun J, Papiernik E (1982) Maternal folate status birth weight and gestational age. Dev. Pharmacol. Ther 4 (Suppl 1):58–65
33. Valyasevi A, Tanphaichite V, Veriyepanich T, Sucharitakul S, Dhanamitta S (1980) Communication to the annual meeting of international nutritional anaemia consultative group, Bangkok 1980
34. World Health Organization (1972) Série de rapports techniques no 503. Anémies nutritionnelles. WHO, Geneva
35. Yusufji D, Mathan VI, Baker SJ (1973) Iron, folate and vitamin B_{12} nutrition in pregnancy. A study of 100 women from southern India. Bull WHO 48:15–22
36. Zittoun J (1985) Folate and nutrition. Chemioterapia 4:388–392

Chapter 14

Folate Deficiency in Geriatric Patients

P. BROCKER and J.C. LODS

The elderly are extremely susceptible to nutritional deficiencies, especially vitamin deficiencies. Among these, folate deficiency is associated with the highest rate of consequences, both because of the considerable number of patients concerned, and because of the various nonspecific clinical and humoral disorders that it can cause. Studies conducted in the 1970s established the significance of this deficiency and its role in neuropsychiatric disorders that had erroneously been ascribed to the process of ageing. While folate deficiency is found in 6%–46% of those over 60 years of age, living at home, and in good health [17, 19, 35, 42], the prevalence increases in the institutionalized, among whom it has been estimated at 16%–80% (Table 1) [3, 19, 26, 36, 43].

While the correlation between hypofolatemia and associated disorders is sometimes difficult to evaluate, it is reasonable to assume that it has an effect on

Table 1. Folate deficiency frequency in the elderly

	n	Mean age	Setting	SF	EF	Limit values (ng/ml)	Method
Read et al. [36]	50	60	G	80%	–	<6	MB
Hurdle and Picton-Williams [26]	72	70	H	39%	–	<5	MB
Girwood et al. [19]	72	75	A	6.9%	–	<3	MB
	39	78	H	17.9%	–	<3	MB
Batata et al. [3]	99	60	H	20%	–	<2	MB
Elwood et al. [17]	533	65	A	30%	–	<5	MB
Sneath et al. [43]	113	–	H	16%	–	–	–
Chapuy et al. [14]	144	79	H	63%	41%	<4 <200[a]	MB
Runcie [42]	250	–	A	46%	–	–	–
Munasinghe and Pritchard [35]	250	65	A	21.6%	–	<3	MB
Boles et al. [5]	480	59	H	44%	–	<3	RD
Abalan et al. [1]	27	79	H	92.6%	–	<3	RD
Roudier et al. [41]	124	81	H	58%	75%	<2 <200[a]	RD
Brocker et al. [11]	1000	81	H	75%	–	<5	MB

SF, Serum folate; EF, erythrocyte folate; G, geriatric home; H, hospital; A, ambulatory; MB, microbiological; RD, radiodilution assay.
[a] Figures refer to EF.

the health of elderly patients presenting with such pathological conditions as healing problems, repeated infections, and behavioral troubles.

A serum-folate (SF) assay helps detect a potential deficiency, and such a finding is confirmed both by a decrease in the folate tissue levels and by the reversal of clinical symptoms after corrective treatment.

Frequency of Folate Deficiency

While a survey on the vitamin status of the French population showed 20% of those aged 18–44 years and in good health to suffer from folate deficiency [31], the figures on this are considerably higher in geriatric patients. Out of 1000 hospitalized elderly patients (mean age, 81 ± 7 years) 75% exhibited a folate serum level under 5 ng/ml [11]. Among these deficient patients, five groups were found to be at particular risk: patients with malnutrition (86% showing hypofolatemia), cancer patients (82%), surgical patients (82%) patients with multiple infections (81%), and patients with varicose ulcers (77%). These data are consistent with those of Read et al. [36], although the exact frequencies may vary (Table 1).

Apparent discrepancies among various studies are due to the heterogeneity of the population studied and to variation in the methods of assessment used. Specific factors here include the following:

- The age of patients vary, as does the number studied. Their socioeconomical status is often overlooked and, more importantly, their daily vitamin consumption from food is not evaluated. (Such a survey is particularly difficult to carry out; the published reports cover a period of about 20 years and have been conducted in various parts of the world.)
- The immediate setting of those studied varies from study to study, subjects being either in a geriatric institution [36], in a hospital [3, 5, 11, 19, 26, 42, 43], in a prolonged catering department [14, 41], in a geriatric psychiatric department [1], or at home [17, 19, 35, 42].
- The assay method may use either the microbiological method with *Lactobacillus casei* or the radiodilution assay.
- Minimal values for folatemia associated with suspected deficiency vary between 2 and 6 ng/ml. Some authors have suggested that deficiency is certain when values are less than 3 ng/ml and probable between 3 and 6 ng/ml. (In our study, we selected the value of 5 ng/ml as a limit, based on clinical improvement data in patients with 4.5 ng/ml; these patients presented with symptoms presumably associated with folate deficiency and which were improved after corrective therapy.)
- Two groups have used the erythrocyte-folate (EF) assay, but these have resulted in inconsistent data. The prevalence of deficiency for Chapuy et al. [14] was estimated at 63% by SF and at 41% by EF; Roudier et al. [41], on the other hand, reported levels, respectively, of 58% and 75%. The latter figures may perhaps have been influenced by a low reference value for SF, since tissue folate levels are usually decreased at a later stage. According to Runcie [42],

the prevalence of folate deficiency is 30% in a pooled population of elderly people in the United Kingdom.

Early Symptoms of Deficiency

No early biological or clinical symptoms of folate deficiency has been described. The classical sign of hypofolatemia is macrocytic anemia. However, this develops only at a very late stage, and is not significant in the elderly [7, 50] since macrocytosis is often concealed by associated iron deficiency. Of 750 elderly patients with folate deficiency, we found only 29 (3.9%) with macrocytic anemia (Hb ≤ 11 g/l in females, Hb ≤ 12 g/l in males; MCV ≥ 100 fl). Macrocytosis was not correlated with the deficiency; in two groups, one with and one without deficiency, similar percentages of macrocytosis were found (18% and 21.6; Table 2).

Iron levels in the blood were found to be low in 113 patients (15.1%) in our deficient population. In 47 cases this was due to true iron deficiency; in 66 cases it may have been related to an inflammatory syndrome. Of 138 patients with macrocytosis, 44 had low iron blood levels. There was a relationship between macrocytosis and low iron blood levels in folate-deficient patients (Table 3). The prevalence of iron deficiency in geriatric patients may have been overestimated in the past as suggested by Roudier et al. [41], who found only 10% of elderly patients with low levels of iron in the blood.

The polymorphism of clinical disorders generated by folate deficiency is such so that no symptoms should be considered as specific in elderly patients.

Table 2. Frequency of macrocytosis or macrocytic anemia among 1000 elderly patients

	SF < 5 ng/ml		SF $\geqslant$ 5 ng/ml	
	n	%	n	%
MCV $\geqslant 100$	138	18	54	21.6
MCV < 100	612	82	196	78.4

$\chi^2 = 1.24$ (ns).
Macrocytic anemia: 29/750 deficient patients (3.9%).

Table 3. Effect of low iron level on macrocytosis in folate-deficient patients

	Low iron level		Normal iron level	
	n	%	n	%
MCV $\geqslant 100$	44	39	94	15
MCV < 100	69	61	543	85

$\chi^2 = 37.4$ ($p < 0.001$).

Causes of Deficiency

A number of mechanisms are involved in the genesis of folate deficiency; some are present throughout the life span while others develop with in ageing. Regarding, first, the former, a number of common etiologies for folate deficiency [2, 21] have been described, with well-established mechanisms. These include the following:

Insufficient food intake
Alcoholism
Digestive diseases
   Proximal small-intestine disease (e.g., Whipple's disease, sprue)
   Inflammatory diseases (e.g., Crohn, ulcerative colitis)
   Small-bowel surgery
   Intraluminal microbiological pullulation
Iatrogenic disorders
   Malabsorption syndrome (e.g., with cholestyramine)
   Enzymatic inhibition (e.g., with trimethoprime, triamterene)
Inflammatory and neoplastic diseases
Hemodialysis

Among the causes that are particularly relevant in geriatrics, intake deficiencies are common; these are more frequent and more specific in the elderly whose food intake is quantitatively reduced. This may be due to several reasons. Anorexia, taste disturbances, monotony of diet, absence of conviviality, solitude, dental problems, fatigue, loss of interest, and, to a lesser extent, socio economical conditions often result in a diminished consumption of fresh folate-rich foods. In addition, a loss of thermolabile food polyglutamates is caused by prolonged cooking, which is customary in institutions, and individuals who live at home have a tendency to prefer deep-frozen or ready-made foods that they warm up a number of times.

Regarding malabsorption, ageing itself is an aggravating factor. The elderly have a reduced volume of gastric juice, with a decrease in acidity and in gastric and jejunal pH [29], which in turn decreases the activity of the pH-dependent folic conjugase. Another factor associated with ageing, and sometimes with malnutrition, is hypozincemia; we found this in over 50% of our elderly patients. A zinc decrease could reduce the activity of pteroylglutamyl hydrolase (folic conjugase), a zinc-dependent enzyme [2]. Furthermore, under normal conditions, competition between folates and zinc has been found to occur in intestinal transport sites [18]. Finally, it was found in the rat that ageing was associated with a decreased secretion of folic conjugase, which plays a key role in the hydrolysis of food polyglutamates into absorbable monoglutamates [27].

The extent of increasing folate needs among the elderly is not well understood but is important in diseased and postsurgical patients. The consequences of increasing needs is more severe in the presence of a low-grade deficiency developing with the ageing process. Besides this purely nutritional factor, certain acute pathological situations are known to determine increased needs for folic acid, especially in patients with cancer [10] and in those after surgery [4], during intensive care, or with acute infections [6].

We have compared the mean folatemia of 16 elderly deficient patients with superinfected bedsores to that of 100 other deficient individuals as controls. After corrective therapy with 150 mg folinic acid, in divided doses of 50 mg intramuscularly at days 1, 2, 11, and 21, plasma folates were assayed at day 30 and then every 2nd week until day 135. The two curves show a parallel decrease in the two groups, with mean values for infected patients significantly lower at all points in time ($p < 0.001$) and a deficiency status reappearing more rapidly, thus suggesting folate overconsumption (Fig. 1).

Consequences

Given the physiological role of folates, it can be assumed that folate deficiency is associated with a number of clinical disturbances, all the more severe as they occur in particularly fragile subjects whose defense mechanisms are physiologically diminished.

Hematopoiesis. Hematopoietic distrubances were the first to be described. Macrocytic anemia is the most suggestive condition; this follows a prolonged and severe deficiency and is not commonly observed. Furthermore, elderly subjects also exhibit latent iron deficiency that conceals macrocytosis. These biological problems have little significance in the elderly because neurological and psycho-

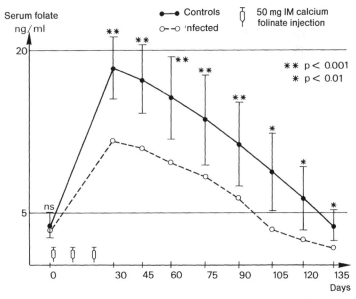

Fig. 1. Mean levels of folatemia in elderly folate-deficient patients with superinfected bedsores ($n = 16$) and in folate-deficient but noninfected controls ($n = 100$)

logical symptoms take precedence over them [13]. The earliest hematological changes are hypersegmentation of polymorphonuclear neutrophils and reticulocytopenia [24].

Digestive Troubles. Although classical complaints, digestive troubles are not common. These include glossitis and malabsorption through a modified jejunal mucosa with megalobastosis. This creates a vicious circle favoring the deficiency but one that is reversible after folic acid therapy, as shown by Lederer et al. [30] in a 77-year-old patient. Parenteral administration appears to be more effective.

Decreased Resistance to Infections. Folate deficiency was found to be responsible for a reduction of activity and phagocytosis in polymorphonuclear cells [49] and for a depression of cell-mediated immunity [20], thus rendering the elderly more vulnerable to infections.

Delayed Healing. The fundamental role played by folates in cell multiplication accounts for the fact that certain deficient elderly individuals present otherwise unexplained healing abnormalities, occurring after surgery or in the presence of varicose ulcers. Folinic acid treatment improves the healing process.

Neurologic Disorders. Problems of a neurologic nature are analyzed in detail in the chapter by Botez (this volume). Several nonspecific pictures have been reported: sensory-motor polyneuropathies, restless leg syndrome, spinal cord and cerebellar syndromes, and a very few cases of subacute combined cord degeneration [7, 22, 32].

Psychiatric Repercussions. The role of folates in the genesis of psychiatric disorders was long unknown or even denied. Only with recent reports of case studies [23, 45] and a better understanding of neurochemistry have folic acid deficiency come to be considered responsible for some forms of severe mental illness. In a recent study [11], we examined the mental evolution of 50 deficient elderly patients with mental disorders such as depression and dementia (DSM III criteria). A geriatric scale (NOSIE) as well as psychiatric one (Hamilton) were used for the follow-up of these patients. After a 3-week wash-out period for all medication with cerebral action, the deficiency was corrected by use of 50 mg parenteral calcium folinate administration at days 1, 11, and 21. By day 30 significant de-

Table 4. Evolution on geriatric and psychiatric scales among 50 deficient elderly with behavior abnormalities before and after folate administration

Period	Improved	Dissociated	Stable	Impaired
Day −21 to day 0	0	5	45	0
Day 0 to day 30	7	32	10	1

On days 1, 11, 21 administration of 50 mg parenteral calcium folinate.
$\chi^2 = 49.97$ ($p < 0.001$).

crease in depressive symptoms was observed in four patients; three patients with dementia recovered total autonomy ($X^2 = 49.97$; $p < 0.001$). This occurred only following treatment for folate deficiency (Table 4). Reynolds et al. has also observed this type of mental improvement in depressed epileptic patients [37], as well as Botez et al. [9] and Carney [13] in depressive younger patients. Coppen and Abou-Saleh [16] recommend a systematic folate supplement in all usual antidepressive treatments because of the better results they obtained than with antidepressive treatment alone. The frequency of folate deficiency is particularly high in geriatric-psychiatric homes [1], where one should systematically search for and correct it.

The basic clinical phenomenon is one of folate-related "dementia," corrected by vitamin supplementation even if there is no specific clinical picture. It is now opinion of some, that 15%–20% of dementia cases are related to nutritional deficiencies. Thornton and Thornton [46] have also suggested that folate deficiency may alter the mental status of patients, and Reynolds et al. [40] observed a close relationship between hypofolatemia and dementia in the elderly. Our own cases are not isolated; Strachan and Henderson [44] published the cases of two elderly malnourished patients whose dementia improved after folate treatment alone. Similarly, Melamed et al. [33] reported the recovery from senile dementia of an epileptic folate-deficient 75-year-old woman with electroencephalographic normalization after treatment with folic acid for several weeks. Botez et al. [8], in young deficient patients presenting cerebral atrophy, obtained more or less complete normalization of comprehension and memory scales without any change in computed tomography scan of the brain. They assumes that a very long history of folic acid deficiency may be responsible for cerebral atrophy, which as in alcoholics, could improve with therapy [12].

Such an effect, if present, could be explained by N-5-methyltetrahydrofolate (N5-methyl-THF), or abnormal neurotransmitter synthesis (e.g., dopamine, noradrenalin, serotonin), decreased levels of which are thought to be responsible for dementia and/or depressive symptoms [47, 48]. Other folate derivatives may be involved in different metabolisms, such as glutamic acid, serine, and choline. The role of these components on the central nervous system has still to be elucidated [28]. Laborit [28] also assumes that in vitro, incubation products of catecholamines, indolamines, and N5-methyl-THF could be alkaloids (e.g., tetrahydroisoquinoline or tetrahydro-B-carnoline) but not biogenic amines. These alkaloids seem to have a neuromodulating action in limiting the synthesis of biogenic amines even if the exact cerebral physiopathologic role remains unknown. Another (complementary) hypothesis is as follows [25, 39]: folate deficiency causes S-adenosylmethionine (SAM), decrease which could produce a depressive state [47] by reducing noradrenaline and serotonin synthesis as well as a modification of membrane phospholipid and thereby reducing neurotransmitter receptor sensitivity. Thus, folates may be substantially involved in neuropsychiatric pathology in the elderly by altering neurotransmitter metabolism.

Relationship Between Folates and Vitamin B_{12}

Vitamin B_{12} deficiency is far less frequent in geriatric patients than is folate deficiency: 22% according to Elsborg [16b] 21% in Chapuy et al. [14]; 5% in Chulmann [1]; and 9% in our own series.

The role of vitamin B_{12} deficiency is well-known in hematologic and neurologic consequences. However, dementia has also been reported in cases of severe deficiency, as have been mood alterations, but less than in folic acid deficiency [38]. When both cobalamin and folic deficiencies are involved, it is difficult to determine the exact role of each vitamin because of the close relationship. Neurologic complications in pernicious anemia may be due to folate metabolism blocking [15]. Excessive folic acid treatment may create or impair neuropathy in patients with vitamin B_{12} deficiency, whether recognized or not [38]. Combined vitamin B_{12} and folic acid deficiencies can alter cerebral neurotransmitters and SAM metabolism, thus explaining a slow clinical improvement, especially with long-standing deficiency.

Conclusion

Folic acid deficiency is one of the most frequent deficiencies encountered in geriatrics. A folate assay is the only way to confirm the deficiency, but it is unfortunately difficult to handle in clinical practice. It is of major importance to suspect such a vitamin deficiency in elderly persons, all of whom are at high risk, especially if there is either acute pathology (infection, cancer, surgery, bedsore, varicose ulcers), chronic malnutrition or behavioral disorders (apathy, anorexia, depression, dementia) which may wrongly be attributed to ageing. Folate defi-

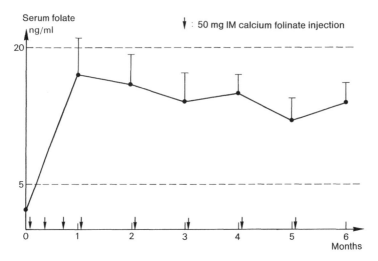

Fig. 2. Folinate dosage needed to establish constant level of serum folate (patients n = 25)

ciency with or without cobalamin deficiency must be considered as a possible etiology of senile dementia.

In our opinion, systematic supplementation should be undertaken. One can use either improvement of food intake, when possible and sufficient, or medical treatment. The latter consists of 5 mg folic acid/day, per os for 20 days (increasing to 10 mg in high-risk patients), then 5 mg/day, 10 days per month. In patients with clinical symptoms of deficiency, however, this should involve 50 mg folic acid by parenteral route, three times during the 1st month, then a monthly injection in order to obtain constant plasma levels (Fig. 2). This type of treatment has two advantages: it is easy to control in elderly persons with multiple medication, and it avoids zinc chelation due to folic acid [18, 34].

It seems preferable to check vitamin B_{12} levels or to add 1000 µg monthly cobalamin therapy to prevent any neuroanemic syndrome in predisposed patients.

References

1. Abalan F, Subra G, Picard M, Boueilh P (1984) Fréquence des déficiences en vitamine B_{12} ou en acide folique chez les patients admis en géronto-psychiatrie. Encephale 10:9–12
2. Audebert M, Le Parco JC, Gendre JP (1978) Absorption des folates: physiologie et pathologie. Gastroenterol Clin Biol 2:309–320
3. Batata M, Spray GH, Bolton F, Higgins G, Wolner L (1967) Blood and bone marrow changes in elderly patients with special reference to folic acid, vit B_{12}, iron and ascorbic acid. Br Med J 2:667–669
4. Benhamou G, Casadei M, Puis P, Berta JL, Lemoan P (1980) Variations du taux des folates sériques dans les suites précoces de la chirurgie digestive. Nouv Presse Med 9:185–186
5. Boles JM, Youinou P, Garre M, Jouquan J, Legendre JM, Morin JF, Vaurette D, Le Menn G (1982) La carence en acid folique: fréquence chez 480 malades. Conséquences sur la poagocytose des polynucléaires neutrophiles. Rev Med Interne 3:51–56
6. Boles JM, Morin JF, Garo B, Youinou P, Garre M (1984) Relation entre acide folique sanguin, acide folique érythrocytaire et infection en réanimation. Gastroenterol. Clin Biol 8:70
7. Botez MI, Peyronnard JM, Charron L (1979) Polyneuropathies responsives to folic acid therapy. In: Botez MI, Reynolds EH (eds) Folic acid in neurology, psychiatry, and internal medicine. Raven, New York, pp 401–412
8. Botez MI, Botez T, Leveille J, Bielmann P, Cadotte M (1979) Neurophysiological correlates of folic acid deficiency: facts and hypotheses. In: Botez MI, Reynolds EH (eds) Folic acid in neurology, psychiatry, and internal medicine. Raven, New York, pp 435–461
9. Botez MI, Botez T, Maag V (1984) The Wechsler subtest in mild organic brain damage associated with folate deficiency. Psychol Med 14:431–437
10. Brocker P, Hogu N, Lods JC (1985) Carences en folates chez les sujets âgés hospitalisés. Fréquence, clinique, traitement, résultats thérapeutiques. Rev Geriatr 10:167–172
11. Brocker P, Lebel C, Maurin H, Lods JC (1986) Carences en folates chez les sujets âgés: intérêt de leur correction dans le traitement des troubles du comportement. Sem Hop Paris 62:2135–2139
12. Carlen PL, Wortzman G, Holgate RC, Wilkinson DA, Rankin JG (1978) Reversible cerebral atrophy in recently abstinent chronic alcoholics measured by computes tomography scans. Science 200:1076–1078
13. Carney MWP (1979) Psychiatric aspects of folate deficiency. In: Botez MI, Reynolds EH (eds) Folic acid in neurology, psychiatry, and internal medicine. Raven, New York, pp 475–482

14. Chapuy P, Ramet JC, Karlin R (1977) Intérêt du dosage de la vitamine B_{12} et des folates. Rev Geriatr 2:151–156

15. Cooper BA (1979) Vitamin B_{12} – folate interrelationship in bone marrow cells. In: Botez MI, Reynolds EH (eds) Folic acid in neurology, psychiatry, and internal medicine. Raven, New York, pp 81–87

16. a) Coppen A, Abou-Saleh MT (1982) Plasma folate and affective morbidity during long term lithium therapy. Br J Psychiatry 141:87–89

16. b) Elsborg L, Lund V, Basturd-Madsen P: Serum vitamin B_{12} levels in the aged. Acta Med Scand 1976, 200, 309–314

17. Elwood PC, Shinton NK, Wilson CID, Sweetman P, Frazer AC (1971) Haemoglobin, vit B_{12} and folate levels in elderly. Br H Haematol 21:557–563

18. Ghishan FK, Said HM, Wilson PC, Murrell JE, Greene HL (1986) Intestinal transport of zinc and folic acid: a mutual inhibitory effect. Am J Clin Nutr 43:258–262

19. Girwood RH, Thomson AD, Williamson J (1967) Folate status in the elderly. Br Med J 1:670–671

20. Gross RL, Reid JV, Newverne PM, Burgess B, Maston R, Hiet W (1975) Depressed cell-mediated immunity in megaloblastic anemia due to folic acid deficiency. Am J Clin Nutr 28:225–232

21. Halsted CH (1980) Folate deficiency in alcoholism. Am J Clin Nutr 34:2736–2740

22. Henry P, Bonnaud E, Laporte A, Orgogozo JM, Vital C, Loiseau P (1977) Place du déficit en acide folique dans les affections carentielles du système nerveux. Sem Hop Paris 53:1530–1537

23. Herbert V (1962) Experimental folate deficiency in man. Trans Assoc Am Physicians 75:307–320

24. Herbert V, Colman N (1979) Hematological aspects of folate deficiency. In: Botez MI, Reynolds EH (eds) Folic acid in neurology, psychiatry, and internal medicine. Raven, New York, pp 63–74

25. Hirata F, Axelrod J (1980) Phospholipid methylation and biological signal transmission. Science 209:1082–1090

26. Hurdle ADF, Picton-Williams TC (1966) Folic acid deficiency in elderly patients admitted to hospital. Br Med J 2:202–205

27. Kesavan V, Noronha JM (1983) Folate malabsorption in aged rats related to low level of pancreatic folyl conjugase. Am J Clin Nutr 37:262–267

28. Laborit HM (1979) The role of folic acid in central nervous system physiology. In: Botez MI, Reynolds EH (eds) Folic acid in neurology, psychiatry, and internal medicine. Raven, New York, pp 249–266

29. Lamy PP, Kitler MB (1985) Gastric erosions in the elderly: increased nedd for cytoprotection. Gastroenterol Clin Biol 9:102–105

30. Lederer J, Mathieu F, Bataille JP, Kumps J, Willcox R, Lammens P (1982) Malabsorption réversible par carence nutritionnelle en acid folique chez un vieillard. Rev Fr Gastroenterol 181:5–10

31. Lemoine A, Le Devehat C, Herbeth B, Bourgeay-Causse M, Delacoux H, Mareschi JP, Martin J, Miravet L, Potier De Courcy G, Zittoun J (1986) Esvitaf: Enquête sur le statut vitaminique de trois groupes d'addultes francais (témoins, obèses, buveurs excessifs). Ann Nutr Metab 30 (Suppl 1):15–94

32. Mauffroy B, Zittoun J, Barrois A, Poisson M, Zittoun R (1975) Carences en folates et troubles neurologiques. Sem Hop Paris 51:237–241

33. Melamed E, Reches A, Hershko C (1975) Reversible central nervous system dysfunction in folate deficiency. J Neurol Sci 25:93–98

34. Milne DB, Canfield WK, Mahalko JR, Sanstead HH (1984) Effect of oral folic acid supplements on zinc, copper and iron absorption and excretion. Am J Clin Nutr 39:535–539

35. Munasinghe DR, Pritchard JG (1978) The relation between mean corpuscular volume, serum B_{12} and serum folate status in aged persons admitted to a geriatric unit. Br J Clin Pract 32 (1):16–18

36. Read AE, Gough AR, Pardone JL, Nicholas A (1965) Nutritional studies on the entrants to and old people's home with particular reference to folic acid deficiency. Br Med J 2:843–848

37. Reynolds EH (1967) Effects of folic acid on the mental state and fit-frequency of drug treated epileptic patients. Lancet 1:1086–1088
38. Reynolds EH (1979) Interrelationships between the neurology of folate and vitamin B_{12} deficiency. In: Botez MI, Reynolds EH (eds) Folic acid in neurology, psychiatry, and internal medicine. Raven, New York, pp 501–515
39. Reynolds EH, Stramentinoli G (1983) Folic acid, S-adenosylmethionine and affective disorder. Psychol Med 13:705–710
40. Reynolds EH, Rothfield P, Pincus JH (1973) Neurological disease associated with folate deficiency. Br Med J 2:398–400
41. Roudier M, Hercberg S, Soustre V, Galan P, Kasouche J, Abramovitz C (1984) Statut en fer et en folates d'une population de sujets âgés vivant en hôpital gériatrique de long séjour. Cah Nutr Diet 19:27–29
42. Runcie J (1979) Folate deficiency in the elderly. In: Botez MI, Reynolds EH (eds) Folic acid in neurology, psychiatry, and internal medicine. Raven, New York, pp 493–499
43. Sneath P, Chanarin I, Hodkinson HM, Mc Pherson CK, Reynolds EH (1973) Folate status in a geriatric population and its relation to dementia. Age Ageing 2:177–182
44. Strachan RW, Henderson JG (1967) Dementia and folate deficiency. Q J Med 142:189–204
45. Thornton WE (1977) Folate deficiency in puerperal psychosis. Am J Obstet Gynecol 129:222–223
46. Thornton WE, Thornton BP (1978) Folic acid, mental function and dietary habits. J Clin Psychiatry 39:315–320
47. Van Praag HM (1982) Neurotransmitters and CNS disease, depression. Lancet 2:1259–1264
48. Yotis A (1979) Catécholamines cérébrales, maladies mentales, vieillissement et démences séniles; considérations biochimiques et pharmacologiques. Encephale 5:339–358
49. Youinou PY, Garre MA, Menez JF, Boles JM, Morin JF, Pennec Y, Miossec PJ, Morin PP, Le Menn G (1982) Folic acid deficiency and neutrophil dysfunction. Am J Med 73:652–657
50. Zittoun J, Zittoun R, Marquet J, Bilski-Pasquier G, Sultan C (1978) Anémies par carence en vitamine B_{12} ou en folates: valeur diagnostique du test de suppression par deoxyuridine. Nouv Presse Med 7:1617–1620

Chapter 15

Folate Deficiency in Intensive Care Patients

E. DE GIALLULY, B. CAMPILLO, and J. ZITTOUN

Folate deficiency is now considered to be the most common vitamin deficiency in the Western world. Because of the physiological role of the vitamin, particularly in intermediate metabolisms and in process of replication and cell growth, investigation of folate status in patients who have suffered from acute trauma is necessary in order to identify in these subjects any feature indicative of vitamin deficiency. Patients hospitalized in an intensive care unit often present with several conditions that predispose to the occurrence of this deficiency. Several recent studies note the adverse effects, haematological in particular, which are associated with folate deficiency.

The Frequency of Folate Deficiency in Intensive Care Patients

Folate deficiency is well-known in some high-risk populations (pregnant women, elderly and alcoholic subjects). Patients admitted to intensive care unit constitute a further group such exposed to vitamin deficiency.

Boles et al. analysed serum levels of folate in a total of 480 hospitalized patients and found that 40.35% of the 342 nonselected patients admitted to an internal medicine department had abnormally low levels of serum folate; the frequency was higher still (52%) in 138 patients who were in intensive care [8]. Audebert et al. observed low serum folate levels in 42% of 62 patients, none of whom were alcoholics, and who were to undergo surgery of the gastrointestinal tract (correction of an inguinal hernia, cholecystectomy for gallstones, resection and anastomosis for colon cancer). Six of these patients had extremely low serum levels of folate [3]. In a study conducted under similar conditions by Benhamou et al. on 30 patients, 50% had hypofolatemia prior to surgery [7]. The folate levels were very depressed (below 3 μg/l) in 23% of the patients.

It could be objected, however, that none of these studies included determinations of levels of folate in erythrocytes, a parameter that is considered to provide an accurate index of folate stores present in body tissues. During a survey designed to examine the vitamin status of patients receiving parenteral nutrition during the post-operative period because of complications, Stromberg et al. observed a low serum folate level in 12 out of the 15 patients who were analysed [20]. Two of the patients also had low erythrocyte levels of folate. Barker et al. found hypofolatemia in 10 out of 20 patients examined before they were put on total parenteral nutrition [5].

In our own study on 49 young patients (average age 23), none of whom were alcoholics, and who had been admitted to the intensive care unit of trauma department within hours following trauma, 75.5% had a serum folate level below 5 µg/l, and 40% had extremely low levels (below 3 µg/l). The average serum folate level in this group was 3.83 ± 1.91 µg/l compared to 8.42 ± 2.93 µg/l in 20 control subjects in the same age group who were examined at the same time. All measurements of serum levels were conducted within the first 36 h after admission, and none of the patients had received folic acid supplements. On the other hand, the folate levels in erythrocytes did not differ from those found in the control group. The serum levels did not correlate with the nature of the injuries (head injuries alone, multiple injuries), the time elapsed from admission to collection of blood sample, or with general anesthesia or blood transfusion. We observed a significant and marked fall in serum and erythrocyte levels of folate in 10 of these patients who we were able to reexamine 1 week later, and who had not received folic acid supplementation. In the particular case of the young patients who had suffered multiple injuries, we were unable to elucidate the mechanism which would explain the frequency of abnormally low serum levels in the first few hours following injury and the rapid fall in erythrocyte levels of folate in the following week.

In another survey, among 105 patients who were studied when they were admitted to a general intensive care unit (this group did not include patients with multiple injuries), 53.3% had serum levels of folate below 5 µg/l (normal value: 7.70 ± 2.50 µg/l). In, addition, 19% had low erythrocyte levels, below 170 µg/l (normal value: 230 ± 60 µg/l). A significant difference emerged in both serum and erythrocyte folate levels between febrile (51 cases) and non-febrile cases. Hypofolatemia was more marked and more common in febrile patients.

Factors Responsible for Folate Deficiency in Intensive Care Patients

Patients admitted to intensive care units most often present with several factors involved in the pathogenesis of folate deficiencies. Reduced food intake is common, associated with anorexia or withdrawal of all kinds of food, because of the patient's underlying disease. In some cases, patients have a disorder which has been progressing for several weeks or several months before a state of malnutrition is produced. Increased frequency of folate deficiency in some populations will naturally be aggravated in the group of patients hospitalized in intensive care departments: old people, patients with malignancies, alcoholic subjects. Malabsorption is also found to be associated with inflammatory disease of the gastrointestinal tract, extensive resections of the small intestine, or when some drugs are being taken, e.g. cholestyramine, salazopyrine, diphenylhydantoin. Inhibition of dihydrofolate reductase can impair folate utilization; this effect is exerted by antifolic agents such as pyrimethamine, triamterene, pentamidine, trimethoprim and particularly by methotrexate.

In addition to these classical mechanisms, other factors which predispose to folic acid deficiencies are also encountered in intensive care patients, such as losses which result from a fistula in the digestive tract, gastric aspiration, diarrhoea, renal dialysis, as well as increased requirements which are particularly involved in the course of infections, stimulation of bone marrow activity as compensatory response to anemia, inflammatory diseases, or tissue repair. Such increased demands are probably involved during all "hypermetabolic" states. Thus, in our own study related to folate status in young patients who had suffered multiple injuries, none of the causes classically known to produce folate deficiencies could explain the observed results. Other factors must be sought, such as stress in the widest sense, as a cause of changes in folate metabolism, perhaps by increasing methylation reactions.

Parenteral nutrition may be by itself the sole factor responsible for the fall in serum folate levels. A number of studies have shown intravenous administration of amino acids to produce a significant fall in serum folate levels [21-23]. This finding could be relevant to the intake of glycine and/or methionine, amino acids which are metabolized together with folic acid [9].

Barbiturates, which are often used in intensive care and in neurological surgery especially to provide a cerebral protection effect, interfere with folate metabolism by some unknown mechanism and may thus predispose to folate deficiency.

Finally, surgery by itself may produce a fall in serum folate within 24 h, an effect which is more serious when the patient's presurgical condition predisposes to folate deficiency [2, 3, 7]. The mechanism responsible for this hypofolatemia is unknown, and recovery to normal levels is always very protracted [7]. Disorders of folate metabolism have also been described in patients who receive N_2O anaesthesia [1, 10, 19]. A large number of cases have shown megaloblastic bone marrow changes and abnormal deoxyuridine (dU) suppression tests. These abnormalities can be prevented by routine prior administration of folic acid.

Consequences of Folate Deficiencies in Intensive Care Patients

Many studies have testified to the adverse effects associated with folate deficiencies in patients hospitalized in intensive care units. The most dramatic feature is the sudden onset of thrombocytopenia. Beard et al. reported four such cases in patients who had undergone surgery for aneurisms of the abdominal aorta, and who developed acute platelet depression between 3 and 10 days after surgery, with a bone marrow pattern suggesting vitamin deficiency [6]. A dU suppression test showed the disorder to be due to folate deficiency in two of the patients. Folate administration corrected these haematological abnormalities. Easton presented the case of two patients in whom the onset of acute thrombocytopenia was associated with a severe haemorrhagic syndrome and was related to an acute folate deficiency [12]. The same findings were obtained by Ibbotson et al. in two patients whose thrombocytopenia disappeared when folic acid was given [17]. Mant et al. observed 13 patients with thrombocytopenia associated with a megaloblastic bone marrow, probably related to folate deficiency [18]. In our own study, conducted in nonspecialized intensive care unit, we saw two cases of acute thrombocytopenia that were directly related to a folate deficiency. In one of these two cases, neutropenia was also present. These haematological disorders were corrected by administration of folinic acid [8 bis].

In all these observations, the rapid onset of severe acute thrombocytopenia in intensive care patients was seen to be corrected by folinic acid. The laboratory test data excluded other possible causes of thrombocytopenias.

Observations of thrombocytopenias associated with folic acid deficiency have also been reported in patients who had been receiving total parenteral nutrition over several weeks [4, 11, 14]. Genin et al. reported two cases of pancytopenia and a confusional syndrome, associated with extremely low serum folate levels in patients in an intensive care unit. All symptoms regressed promptly when folate supplement was given [13].

Folate deficiency also interferes with white cell functions. Youinou et al. showed abnormalities in white cell functions (phagocytosis and bactericidal activity) in patients with low serum folate levels [24]. There was a correlation between the decrease in serum folic acid levels and the severity of observed white cell abnormalities. Correction of folic acid deficiency was accompanied by normalization of phagocytosis but not of bactericidal activity.

Special Aspects of Folate Deficiencies Occurring Under Intensive Care Conditions

Some features which are more specific to these deficiencies emerge from the published observations. The manifestations which are associated with folate deficiencies often appear at a very early stage, within 1–2 weeks after the onset of the disorder originally responsible for the patient's admission to hospital, even though folate stores should in theory have been adequate to cover physiological requirements for several weeks. The usual biological abnormalities (macrocytosis in particular) are rarely observed; this is because the onset of folate deficiency is sudden, and the time which elapses is too short for changes in erythrocyte population to be overt. The first haematological manifestation is often thrombocytopenia, which may or may not be associated with neutropenia, the bone marrow often being megaloblastic. Serum folate level is usually very low, but folate level in erythrocytes may still be normal. These haematological features quickly appear in pathological situations in which folate requirements increased, but where the body seems unable to mobilize its reserves, particularly liver stores. The earliest effects involve tissues in which turnover is rapid; this would explain the high frequency of megaloblastic bone marrow in the absence of peripheral macrocytosis. Platelet counts are also decreased. Thus, in severe infections, for example, increased metabolic demands will sometimes impose an additional load, beyond the high peripheral consumption of platelets. Folate deficiency will induce insufficient compensation by the bone marrow, with thrombocytopenia as a result. The low platelet count can be in this condition corrected by folate supplementation. It is probable that some of the unexplained cases of thrombocytopenia which are observed under intensive care conditions are related to such a mechanism.

Folic Acid Supplementation

The prevention and treatment of folic acid deficiency is simple, in contrast with the severity of the complications which are sometimes seen, and is based on regular and adequate folate supplements to high-risk patients. The doses cited by various authors for the prevention and treatment of these deficiencies vary widely. This lack of agreement illustrates the difficulty of assessing real a requirements in these acutely-ill patients.

Herbert determined minimum requirements in healthy subjects during experimentally induced folate deficiency [14]. In fact, these requirements correspond to the minimal amount of folates necessary to prevent a "biochemical" megaloblastosis and are estimated at 50 µg pteroylglutamic acid (or equivalent amount of other forms of folate). The requirements recommended by the United States Food and Drug Administration (FDA) or National Institutes of Health (NIH) are higher, ranging from 300 to 400 µg daily in adults. Barker et al. have examined the changes in serum levels of folates in 20 patients under total parenteral nutrition [5]. Half the patients received daily replacements of 0.2 mg folinic acid by intravenous route for 1 week; the other half received 1.2 mg. Serum folate levels at the begining of the study did not differ significantly between the two groups. By day 8, however, a significant fall appeared in folate levels in the blood of all patients in the first group, whereas those receiving 1.2 mg/24 h showed a rise in serum folate levels. This daily dose of 1.2 mg was enough to maintain or improve serum folate levels but not enough to correct a pre-existing deficiency. It should be noted that no detail of the clinical conditions of the patients in this study were given.

In the study of Stromberg et al., conducted in 15 patients receiving total parenteral nutrition and given 0.2 mg of folic acid every 24 h, 12 were already hypofolatemic at the beginning [20]. Despite significant increase in the mean levels of serum folate in the meantime, seven still had abnormally low serum folate levels at the end of the trial (the duration of which was not stated). This study showed that although the doses which are usually recommended are adequate to cover daily requirements of folate, they are still too low in cases of pre-existing deficiency.

Howard examined vitamin requirements in six patients under total parenteral nutrition at home [16]. The doses of folinic acid used (5 mg intravenously twice a week = 1.4 mg/24 h) appeared to be adequate for maintaining serum levels at normal values. The study was, however, very short (limited to 3 days after injection).

In a study of nutritional supply in 48 elderly patients who had been hospitalized for orthopedic surgery, Hessov and Elsborg showed that their actual folate intake was less than 200 µg per 24 h in all cases. The average daily intake was about 20% of the amounts generally recommended [15].

In our own work in 105 patients admitted to a general intensive care unit, we assessed on folic acid status the effect of the administration of folinic acid at various doses and with two dosage schemes: 5mg/24 h intravenously for 7 days in sequence, or 50 mg given in single intravenous dose at the start of the trial. We found that a daily injection of 5 mg of folinic acid always normalized serum and erythrocyte levels of folate, whereas a single dose of 50 mg did not regularly correct folate levels when measured on the 8th day, at least in patients who were initially deficient. Like the majority of authors, we found that pre-existing hypo-folatemia is not always corrected in patients who receive only 200–330 µg/24 h folic acid given orally.

Conclusions

Folate deficiency is of frequent occurrence in intensive care patients. This vita-min deficiency is serious because of the possibility of hematological complica-tions which can prejudice vital prognosis. Although such acute manifestations are uncommon, they are not an exceptional event (two patients out of 105 in our own survey) and may appear within the first few days of hospitalization, parti-cularly if metabolic requirements are increased.

An intake corresponding to physiological requirements was found to be inad-equate to correct low serum folate levels.

Considering the frequency of latent folate deficiency in intensive care popula-tions (around 50% of low serum folate and around 20% of low erythrocyte folate) it seems reasonable to make a large supplementation. The dose and schedule of administration remain a subject of discussion. We showed daily supplements of 5 mg folinic acid to be sufficient in all cases in preventing or correcting a latent folate deficiency, and this appears slightly more effective than a weekly dose of 50 mg. These patients, whose folate requirements seem to be increased, were in any case receiving inadequate doses of this vitamin.

In fact, neither the solutions used for total parenteral alimentation nor indeed the multivitamin preparations contain folic or folinic acid. The dietary prepara-tions intended for oral administration are generally at too low dosages. Thus, it would advisable to adopt routine daily supplements of 2–5 mg folic or folinic acid, given orally or parenterally to all these patients, since the optimum dose is still difficult to define.

References

1. Amos RJ, Amess JAL, Hinds CJ, Mollin DL (1982) Incidence and pathogenesis of acute megaloblastic bone-marrow change in patients receiving intensive care. Lancet II:835–839
2. Audebert M, Gendre JP, Zittoun J Le Quintrec Y (1976) Carence en folates et chirurgie digestive. Etude de 50 cas. Arch Fr Mal App Dig 65:455–462
3. Audebert M, Huguier M, Zittoun J, Gendre PM, Le Quintrec Y (1980) Carence en folates après intervention chirurgicale. Gastroenterol Clin Biol 4:911–914
4. Ballard HS, Lindenbaum J (1974) Megaloblastic anemia complicating hyperalimentation therapy. Am J Med 56:740–742

5. Barker A, Hebron BS, Beck PR, Ellis B (1984) Folic acid and total parenteral nutrition. J. Parent Ent Nutr 8:3-7
6. Beard MEJ, Hatipov CS, Hamer JW (1980) Acute onset of folate deficiency in patients under intensive care. Crit Care Med 8:500-503
7. Benhamou G, Casadei M, Puis P, Berta P, Lemoan P (1980) Variations du taux de folates sériques dans les suites précoces de la chirurgie digestive. Nouv Presse Med 9:185-186
8. Boles JM, Youinou P, Garre M et al. (1982) La carence en acide folique: fréquence chez 480 malades Conséquence sur la phagocytose des polynucleaires neutrophiles. Rev Med Interne 3:51-56
8 bis Campillo B, Zittoun J, de Gialluly E (1988) Prophylaxis of folate deficiency in acutely ill patients: results of a randonized clinical trial. Intensive Care Med. 14:640-645
9. Connor H, Newton QJ, Prestin FE, Woods HF (1978) Oral methionine loading as a cause of acute serum folate deficiency: its relevance to parenteral nutrition. Postgrad Med 54:318-320
10. Deacon R, Lumb MJ, Perry J (1982) Vitamin B_{12} folate and nitrous oxide. Med Lab Sci 39:171-178
11. Denburg J, Bensen W, Ali MAN, McBeide J, Ciok J (1977) Megaloblastic anemia in patients receiving total parenteral nutrition without folic acid or vitamin B_{12} supplementation. JAMA J 117:144-147
12. Easton DJ (1984) Severe thrombocytopenia associated with acute folic acid deficiency and severe hemorrhage in two patients. Can Med Assoc J 130:419-423
13. Genin R, Tanter Y, Freysz M, Guy H, Rifle G (1979) Carences aigues en folates survenant au cours de la nutrition parentéraie exclusive. Nouv Presse Med 8:137
14. Herbert V (1962) Experimental nutritional folate deficiency in man. Trans Assoc Am Physicans 75:307-320
15. Hessov I, Elsborg L (1976) Nutritional studies on long-term surgical patients with special reference to the intake of vitamin B_{12} and folic acid. Int J Vitam Nutr Res 46:427-432
16. Howard L, Bigquette J, Chu R, Krenzer BE, Smith D, Tenny C (1983) Water soluble vitamin requirements in home parenteral nutrition patients. Am J Clin Nutr 37:421-428
17. Ibbotson RM, Colvin BT, Colvin MP (1975) Folic acid deficiency during intensive therapy. Br Med J 4:145
18. Mant MJ, Connolly T, Gordon PA, King EG (1979) Severe thrombocytopenia probably due to acute folic acid deficiency. Crit Care Med 7:297-300
19. O'Sullivan H, Jennings F, Ward K, McCann S, Scott JM, Weir DG (1981) Human bone marrow biochemical function and megaloblastic hematopoiesis after nitrous oxide anesthesia. Anesthesiology 55:645-649
20. Stromberg P, Shenkin A, Campbell RA et al. (1981) Vitamin status during total parenteral nutrition. J Parent Ent Nutr 5:295-299
21. Tennant GB, Smith RC, Leinster SJ, O'Donnell JE, Wardrof CAJ (1981) Aminoacid infusion induced depression of serum folate after cholecystectomy. Scand J Haematol 27:333-338
22. Tennant GB, Smith RC, Leinster SJ, O'Donnell CAJ, Wardrop CAJ (1981) Acute depression of serum folate in surgical patients during preoperative infusion of ethanol-free parenteral nutrition. Scand J Haematol 27:327-332
23. Wardrof CJ, Lewis MH, Tennant GB, Williams RHP, Hughes LE (1977) Acute folate deficiency associated with intravenous nutrition with aminoacid-sorbitol-ethanol: prophylaxis with intravenous folic acid. Br J Haematol 37:521-526
24. Youinou PY, Garre MA, Menez JF (1982) Folic acid deficiency and neutrophil dysfunction. Am J Med 73:652-657

Chapter 16

Inherited Defects of Folate Metabolism

B. A. COOPER

Disorders of folate metabolism affecting the newborn include inherited abnormalities in the utilization of folate and abnormalities relating to the concentration of folates within the infant. The latter include disorders caused by levels of folate in the body below those considered normal but may also include abnormalities which are not properly defects in folate metabolism or accumulation, but which may be affected by treatment with large doses of folates. Examples of the latter include the effect of large supplements of folic acid in pregnant women in producing an apparent increase of mean birth weight of infants [1] and in possibly reducing the frequency of congenital defects of the neural tube among susceptible infants [2], as well as the absence of the fragile site on the X chromosome in cultured cells from children with mental deficiency associated with the 'fragile-X' syndrome when the cells were cultured in excess folate [3].

This review concentrates on inherited defects of the metabolism of folate without detailed examination of the other conditions referred to above. Other reviews dealing with this subject should also be consulted [4, 5, 6].

Folate Defects Producing Folate Deficiency

Folate deficiency produces megaloblastic anemia, mouth ulcers, diarrhea; it interferes with immune function by preventing the clonal proliferation of stimulated lymphocytes; and in young animals it retards growth in general and of the nervous system in particular. Its effect on neurotransmitter function and brain function is postulated but not proved.

Maternal Folate Deficiency. Although a number of children born by mothers deficient in vitamin B_{12}, have suffered from deficiency of this vitamin, folate deficiency in the infant secondary to deficiency in the mother is unusual, probably because fetuses unable to extract adequate folate from the mother may not survive. Folate levels in umbilical-cord blood exceed maternal levels, although folate concentration in the newborn liver is low, and thus total body folate in the newborn is probably not greater than in the adult. Although a number of reports have implicated folate deficiency as a cause of premature terminations of pregnancy [7], a causal relationship between these has not been proved.

Inadequate Transport. Folate transport across cell membranes appears mediated by a single major transport system [8] which utilizes a folate binder to transfer all types of monoglutamate folate. In intestine, all monoglutamate folates are transported with the same efficiency and velocity. In some other cells, the concentration of folic acid required to generate adequate concentrations of intracellular folates is 100–200 times that of reduced folates such as 5-methyl tetrahydrofolate (THF) and 5-formyl THF (folinic acid) [9]. 5-methyl THF and folinic acid are commercially available as racemic mixtures, only part of which are biologically active. In some cells, the inactive isomer [10] is transported into the cell but not utilized; at least one folate-dependent enzyme reaction is inhibited by the presence of the inactive isomer. Radiodilution assays measure both biologically active and inactive isomers; microbiological assays measure only active isomers. 5-Methyl THF is labile to oxidation and is difficult to store.

Most of the folate accumulated by the fetus accumulates during the last 4 weeks of pregnancy [11]. Infants born prematurely and those nourished by defective placentae (e.g., sickle cell disease, erythroblastosis fetalis, small-for-dates infants) may have inadequate folate stores at birth. Growth rates of such infants during the first months of life have been shown to be increased significantly when they receive supplements of folic acid [12]. Such folate supplementation, however, has no effect on their weight at birth or at age 1 year or on their frequency of illness during the 1st year of life.

Seven cases have been described in six families [13–18] of children with inherited defects of absorption of folates from the intestine. All children thus far described have been girls. This female predominance would be improbable by chance alone ($p < .01$), suggesting that the defect may be fatal in utero in boys. These children were found with megaloblastic anemia at age 8–14 weeks. In all the children tested, accumulation of folates in the CSF was defective, suggesting the presence of a transport system shared between choroid plexus and small intestine. Absorption of all folate forms tested has been defective, except in one case in which folinic acid was absorbed better than folic acid and another in which 5-methyl THF was absorbed better than either folic acid or folinic acid.

Because folate deficiency was present during early life, mental deficiency is frequent in these children, and calcifications have been observed in basal ganglia of some of them. It is unknown whether this cerebral defect is due to failure of CNS growth as a result of inadequate folate or to toxic effects of a metabolite which accumulates in such children.

Regarding treatment, it is vital to maintain folate levels in serum, erythrocytes, and CSF above levels associated with 4, 150, and 15 ng/ml, respectively. This may be achieved in some children with large daily doses (5 mg or more) of folic or folinic acid or 5-methyl THF. Monitoring of serum and of erythrocyte and CSF folate to determine whether normal levels are achieved is vital. Oral doses may be increased to 100 mg or more per day if necessary. If oral treatment does not produce normal folate levels in all three fluids, treatment must be instituted with daily injections (subcutaneous or intramuscular) of folinic acid. Weekly intrathecal injections may be required if CSF folate cannot be increased to normal by other means. It is most important to individualize therapy for each patient and to maintain normal levels of folates in all three fluids if mental development, resistance to infection, and growth are to be made optimal.

One family has been described [19] in which 14 of 105 members had severe hematological disease, including acute leukemia and life-threatening leukopenias in early life, with an additional 20 members whose history suggested the presence of severe hematological disease. Of the 16 children produced by four affected members, severe hematological disease was observed in 7 compared with only 4 cases among the 74 children produced by the 17 unaffected member who had children. Accumulation of 5-[^{14}C]methyl THF was below normal in lymphocytes and bone marrow cells from the patient and from 4 of 8 unaffected family members. Folate transport in erythrocytes and intestine were normal in this patient, as was transport and accumulation in his fibroblasts. This suggests that the defect in this family was restricted to certain hemopoietic cells (erythroblasts, T-lymphocytes) and may have been secondary to a clonal disease of hemopoiesis related in some way to Fanconi's anemia. Although the patient's hemoglobin improved when he was treated with large doses of folate, he eventually died of his disease.

The phenomenon appears to be unique to this family, but treatment with large doses of folinic acid to achieve normal levels of intracellular folate may be attempted in any patient with hematological abnormality and deficient levels of folate. Because of the more efficient accumulation by bone marrow derived cells of reduced (tetrahydro-) folates than of folic acid, treatment with the former would seem appropriate in such patients. It is unknown whether such treatment may accelerate the appearance of acute leukemia in patients prone to this. Although folate has been reported to accelerate growth of some leukemias, correction of folate deficiency appears to benefit patients with neoplastic diseases.

Increased Destruction or Loss of Folate. No documented cases of folate deficiency due to increased destruction of folate have been reported. In pregnant women, increased urinary loss of folate has been reported, but this represents only a small fraction of total folate requirements and appears not to be quantitatively significant. The marked lability to oxygen by THF and 10-formyl THF probably explains the decreased plasma folate in hyperoxygenated states and may be the normal mechanism of folate destruction. Smoking has been reported to increase folate clearance, and serum folate levels of smokers have been reported to be lower than those of nonsmokers [21]. The relationship between this effect and the effect of maternal smoking on the fetus has not been examined.

Inappropriate Storage. Although 12–16 weeks of deficient folate intake are required to produce nutritional megaloblastic anemia in the well-nourished subject, this can be achieved within 15–20 days if alcohol is administered. Based on studies of rats and on double-isotope studies in man, it has been suggested that alcohol acts in this situation to prevent the mobilization of folate stored in the liver. Normally, the enterohepatic circulation of hepatic folate maintains serum folate during periods when folate is not being ingested [24]. Whether a similar mechanism is responsible for some of the effects of alcohol or other drugs on the fetus is unknown.

Enzyme Pathway Defects Fig. 1

Methylene THF Reductase Deficiency. The cytoplasmic enzyme, methylene THF reductase mediates the conversion of 5-10-methylene THF to 5-methyl THF, and within the cell probably exclusively utilizes polyglutamyl folate. It binds flavine adenine dinucleotide (FAD) and utilizes nicotinamide adenine dinucleotide phosphate reduced, (NADPH) as electron donor [25]. In vitro the reaction is bidirectional, but due to intracellular conditions it is essentially unidirectional in the direction of 5-methyl THF within the cell. In vitro, it is usually assayed in the reverse direction using menadione as electron acceptor but can be assayed in the "forward" direction [26]. Under the latter conditions, the concentration of *S*-adenosylmethionine (SAM) required for inhibition of the reaction is considerably smaller than required for inhibition of the reverse reaction [27]. In vitro, the enzyme reaction is inhibited by dihydrofolate and by dihydrobiopterin [28], pro-

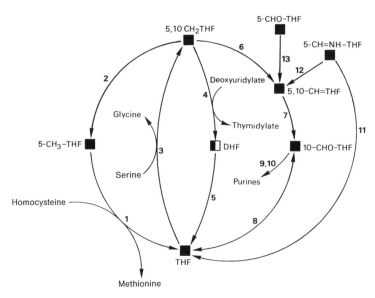

Fig. 1. Simplified scheme of folate metabolism. Symbols represent the pteridine structure: *shaded locations*, in tetrahydro or reduced form; *unshaded* locations, in oxidized forms. The active side chain is indicated for each form. Under physiological conditions, folate enters the cell in the form of 5-methyl (CH3-) tetrahydrofolate (THF) and is converted to THF by reaction *1* which generates methionine from homocysteine. This reaction utilizes vitamin B$_{12}$. THF can receive a methylene group (-CH2) (reaction *3*) from serine (which is converted to glycine). 5-10-Methylene THF can (a) be reduced to 5-methyl THF (reaction *2*); (b) be oxidized in two steps to 10-formyl THF (reactions *6, 7*); (c) transfer its -CH2 and an H to deoxyuridylate to form thymidylate (reaction *4*). This step is important in DNA synthesis and may be the limiting reaction in causing megaloblastic anemia. Dihydrofolate is generated by this reaction and requires dihydrofolate reductase (reaction *5*) to regenerate THF. 10-Formyl (CHO) THF contributes two different carbon atoms of the purine structure (reactions *8* and *9*) and regenerates THF. Reactions *2, 3, 4, 8,* and *9* probably function in vivo only when the folate is in the polyglutamyl form, formed from THF and CHO-THF by folylpolyglutamyl synthetase

viding a possible site of interaction between the folate and pterin metabolic pathways.

A total of 21 cases have been described in the literature [29–37] (Table 1), but others may exist. The clinical syndrome is somewhat variable but includes a variety of different neurological syndromes, ranging from mental retardation through a variety of mild tremors and hypotonia, with single examples of schizophrenic syndrome and typical subacute combined degeneration of the brain and spinal cord. In two cases, pathological lesions in the brain were suggestive of arterial vascular occlusions possibly due to homocystinemia [32, 34]. Serum and erythrocyte folates are decreased in the severe cases, probably because of the lability in the plasma of folates other than 5-methyl THF. Megaloblastic anemia, when it occurs, is associated with low folate levels.

Recognition is based on the presence of homocystinuria and cystathioninuria, with decreased levels of plasma methionine and the presence of a large proportion of folates other than 5-methyl THF in serum and cells, determined by differential microbiological assay [29]. The diagnosis is confirmed by demonstration of decreased or abnormal enzyme activity in extracts of fibroblasts cultured from a skin biopsy [35]. The cultured fibroblasts are usually auxotrophic for methionine. the severity of the clinical syndrome is closely related to residual enzyme activity in cultured fibroblasts and also closely parallels both the decrease in the proportion of total cell folates which are 5-methyl THF and the decreased methionine synthesis within the cell. One case has been diagnosed in utero using

Table 1. Published reports of methylene tetrahydrofolate deficiency

Patient	Sex	Age at diagnosis	Residual enzyme activity (%)	Reference
CP	M	16 years	24	[30]
BM	F	17 years	15	[30]
LM	F	15 years	17	[30]
WM	M	11 years	13	[30]
EC	F	died at 9 years	–	[30]
MC	F	died at 13 years	–	[30]
TC	F	9 years (died at 10 years)	7	[30]
MEC	F	7 years 11 months	12	[30]
GS	M	11 months (died at 2 years)	10	[30]
GP	M	3 years	3	[30]
TK	F	23 days	0	[30]
SS	F	5 months (died at 9 months)	0	[30]
SM	F	7 months	2	[33]
X-1	M	18 years	4	[34]
X-2	F	died at 2 years	–	[34]
X-3	(Fetus)	16 weeks gestation	4	[34]
X-4	F	2 years	1	[36]
KM	–	24 months	–	[37]
MC-2	–	60 months	–	[37]
RJ	–	36 months	–	[37]
LC	–	27 months	–	[37]

X-1, etc. refers to patients without initials known; MC-2 is described as MC by the authors.

cultures of amniocytes obtained early in pregnancy (about 16th week) by amniocentesis, and one fetus at risk has been found unaffected [36].

It has been reported that in some patients with this abnormality, 5-hydroxy indole acetate (HIAA) and homovanillate (HVA) in urine are below the normal range [37]. These are products of the neurotransmitters serotonin and DOPA, the synthesis of which from tryptophane and phenylalanine require tetrahydrobiopterin (BH_4) dependent enzymes. In these children, BH_4 excretion was also decreased. Because a biochemical link between folate and BH_4 metabolism has not been established, such metabolites have not been sought in most patients with this illness. It is possible that some of the neurological manifestations of methylene THF reductase deficiency are related to secondary abnormalities in BH_4 metabolism, thus requiring examination of this pathway in future cases.

All severely affected children treated before 1979 died of complications of this disease. Patient SM (Table 1), who was treated with methionine [33], oral folinic acid, and vitamins B_{12} and B_6, survived. Such children have recently been treated with betaine (3-20 g/day), which appears to increase plasma methionine levels to normal, as well as with folate and vitamin B_{12}. There is some evidence that 5-methyl THF may increase BH_4 levels in tissue extracts by a little understood mechanism [38]. Since oral folinic acid is converted to this folate by the small intestine [39], it is possible that this folate would be the form preferred. Treatment of these children thus requires correction of methionine levels in plasma with betaine and/or with supplements of methionine; correction of BH_4, HIAA, and HVA excretion by appropriate treatment; and supplementation with either folinic acid or folic acid. It is recommended that therapy be begun with daily doses of 3 mg/kg betaine, 15 mg folinic acid, and 900 mg methionine, all given per os if possible. Patients should also receive 1 mg/vitamin B_{12} subcutaneously weekly; vitamin B_6 may also be added. If metabolites are corrected with this treatment, doses may gradually be decreased, with care to maintain normal levels of these metabolites. Vitamin B_{12} is recommended because of the observation of subacute combined degeneration of the spinal cord in one patient dying at age 3 years [37] following 6 months of treatment with large doses of 5-methyl THF. In many patients, betaine supplementation probably suffices, and methionine supplements may not be required, but insufficient data are available at this time to predict this. Very large doses of methionine have provoked acute psychosis in patients with schizophrenia, and hypermethioninemia should thus probably be avoided.

Defects in Glutamate Formiminotransferase and Formimino-THF Cyclodeaminase
Table 2 and Fig. 1 [40-45]. During the catabolism of histidine, a formimino group is transferred to THF, followed by release of ammonia and generation of 5-10-methenyl THF. The two enzyme activities (glutamate formiminotransferase and formimino-THF cyclodeaminase) share a single octameric enzyme which channels polyglutamate folate molecules from one reaction to the next [46]. The pathway represents a minor source of single carbons, and may exist only in liver and kidney. The enzyme appear to be absent from fibroblasts and blood cells, but other tissues have not been examined. Defects in the pathway cause excre-

tion of formiminoglutamic acid (FIGLU). FIGLU is also excreted in folate deficiency and in patients deficient in vitamin B_{12}.

It has been reported that normal subjects fed 15 g (100 mmol) 1-histidine excrete a mean of about 9 mg (.05 mmol) FIGLU [47], whereas in folate deficiency this increases to 200–2000 mg (1–10 mmol). Based on excretion of metabolites of histidine in two children with severe deficiency of glutamate formiminotransferase, basal metabolism of histidine appeared to be equivalent to about 50% of that metabolized after ingestion of about 200 mg (1 mmol) per kilogram 1-histidine, suggesting that in these children about .5 mmol/kg endogenous histidine was metabolized daily.

Eleven patients with inherited deficiency in this pathway have been reported (Table 2). The clinical spectrum of these patients is extremely variable, making definition of a clinical syndrome difficult. The only common feature is excretion of FIGLU without evidence of folate deficiency. Enzyme activity was measured in livers of five patients (numbers 1–5, Table 2) and ranged from 14% to 54% of normal. Abnormalities of mental development were observed in six patients, one of which cases was thought to be secondary to neonatal anoxia (number 5), and in two, delayed speech development was the primary abnormality noted (numbers 7, 8). Four patients may have had hematological abnormalities, but cobalamin defects such as deficiency of transcobalamin II were not excluded in these. Serum folate was elevated in at least three patients and may also have been elevated in two others (numbers 10, 11). Erythrocyte folate levels were reported only in patients with normal serum folate, and distribution of folate coenzyme forms in liver or other cells or in plasma have not been described.

The relationship between the FIGLU excretion and mental retardation or the hypersegmentation of neutrophils is unclear. In all cases, the patients were investigated because of abnormalities in mental development, an unusual appearance,

Table 2. Patients reported with probable defects in glutamate formiminotransferase

Patient number	Age	Sex	Mental development	Megalo-blastic	Serum folate (minimum μg/l)	FIGLU (urine) (maximum μmol)	Reference
1	8 months	F	Retarded	Hyp	10	0.066[a]	[40]
2	19 months	F	Retarded	0	72	0.859[a]	[40]
3	3 months	M	Retarded	Meg	53	0.175[a]	[40]
4	6 months	M	Normal	0	86	0.087[a]	[40]
5	21 years	M	Retarded	0	9	0.267[a]	[41]
6	5 years	F	Normal	0	3	5.185/g creat	[42]
7	7 years	F	Speech	0	5	5.619/g creat	[42]
8	3 years	M	Hypotonia	0	7	3.486/24 h	[43]
9	8 years	F	Normal	0	5	2.834/24 h	[43]
10	42 years	F	Normal	Meg	>10	0.150/l	[44]
11	2 years	M	Normal	Hyp	27	1.700/24 h	[45]

[a] 8-h collection after 0.33 g/kg histidine (24-h collection for patient number 2)
Hyp, hypersegmentation of neutrophils; Speech, delayed speech development as only abnormality. Meg: megaloblastic; creat: creatinine

or other medical condition. Since FIGLU excretion is not routinely measured in subjects without illness, it is possible that the clinical syndromes were unrelated to the enzyme deficiency. Patient numbers 6, 7, 8, 9, and 11 excreted large quantities of FIGLU, suggesting that a large proportion of their daily histidine catabolism was ineffective. Clinical abnormalities did not correlate with the degree of block of histidine metabolism, again suggesting only a coincidental relationship between the enzyme defect and the clinical abnormalities. It has been suggested that the disparity between degree of block and clinical manifestation may be due either to single enzyme defects permitting accumulation of toxic metabolites or to depletion of required intermediates, or a combined block of the two associated enzyme systems. At this time it is impossible to assign any of the clinical manifestations to abnormalities in the folate pathway.

In some of the patients tested, FIGLU excretion decreased but did not disappear during administration of folate or folinate (numbers 3, 9, 10) and did not change in patient numbers 6, 7, 8, and 11. There were three sibling pairs among the patients, suggesting recessive transmission, and FIGLU excretion in excess of normal was observed in parents of some of the patients (numbers 3, 4, 6, 7) but not in the parents of patient numbers 8 and 9. The anemia reported in patient number 3 improved during therapy with folate and pyridoxine, but the effect of therapy on hypersegmentation of neutrophils in the other affected patients has not been described. Therapy thus remains empirical. In future patients, other abnormalities should be carefully excluded, and further attempts are required to evaluate the effect on clinical manifestations of folate and folinate therapy and of correction of metabolic abnormalities which may be detected. Deoxyuridine suppression tests should be carried out on bone marrow or phytohemagglutinin (PHA) stimulated lymphocytes to determine abnormality and the effect of methyl THF, folic and folinic acid, methyl cobalamin, pyridoxine, and excess FIGLU.

Methionine Synthase. Fig. 1 As indicated above, although most of the methionine in the body is ingested as such, it is actively demethylated in the form of SAM to form S-adenosylhomocysteine (SAH), which is hydrolyzed to homocysteine. Failure to regenerate methionine causes depletion of methionine and SAM, and exposes the organism to the toxicity of homocystine, the oxidized form of homocysteine. Homocystine is deactivated by cystathionine synthase and subsequent reactions, followed by excretion. Accumulation of intermediates of this reaction (homocystine, cystine, SAH) is not tolerated by the body without illness. There is evidence that the genes for SAH hydrolase and adenosine deaminase are closely associated on chromosome 20 [48] and may have equal importance within the cell to prevent accumulation of SAH and adenosine.

Methionine synthase requires a reduced cobalamin as prosthetic group, the cobalamin accepting a methyl group from SAM or from 5-methyl THF to convert homocysteine to methionine [49]. In severe deficiency of vitamin B_{12} or when the cobalamin is irreversibly oxidized by nitrous oxide [50], methionine synthesis decreases, with apparent fall in methionine concentration and the appearance of megaloblastic anemia and probably of subacute combined degeneration of the brain and spinal cord. As indicated above, SAM deficiency shunts folates into the 5-methyl THF form, which requires the methionine synthase reaction for regeneration of THF. Patients deficient in methionine synthase are expected to have homocystinuria, deficient levels of methionine in plasma, megaloblastic anemia, and possibly neurological defects caused either by vascular obstruction due to the homocystinemia or to methionine deficiency in the brain. The level of intracellular methionine synthase enzyme decreases if cells are deficient in vitamin B_{12}, probably because the enzyme molecule is stabilized when it is bound to cobalamin. Relative decrease of methionine synthase enzyme is observed in cells deficient in vitamin B_{12}.

One patient has been reported with methionine synthase that is active in cell extracts but has minimal activity within the cell [51]. This defect appears to be due to defective reductase activity which permits oxidation of vitamin B_{12} on the methionine synthase enzyme [52]. The clinical syndrome in this child included failure to thrive at age 5 weeks, with megaloblastic anemia, homocystinuria, homocystinemia, and decreased plasma levels of methionine. One expects children with other defects in methionine synthase to have a similar clinical syndrome.

One patient has been described with possible partial deficiency in this enzyme [40]. This 6-month-old girl had megaloblastic anemia, seizures, dilated cerebral ventricles, and an elevated level of serum folate. Despite low levels of serum vitamin B_{12}, the anemia did not respond to this vitamin, but reticulocytosis without correction of the megaloblastic anemia was observed during therapy with folate or pyridoxal phosphate. Urinary amino acids were normal, and homocystinuria was absent. Methionine synthase activity in an extract of liver was 36% of normal. Methionine synthase was normal in fibroblasts grown from a skin biopsy [5], suggesting either a very mild and localized decrease of methionine synthase of liver only, or that the decreased hepatic methionine synthase was secondary to another illness. Three cases of methionine synthase deficiency have been reported investigated in Montreal [53], one in France [54] and one described by Hallam et al. [55]. In all of these, megaloblastic anemia was observed and homocystinuria in two of them; methionine synthase activity in cultured cells was below normal and was not corrected by culture of the cells in cobalamin. In the Montreal case, the distribution of intracellular cobalamins was abnormal, with marked deficiency of intracellular methyl cobalamin.

Dihydrofolate Reductase. Fig. 1. The enzyme dihydrofolate reductase (DHFR) converts dihydrofolate (DHF) to THF after generation of the former by the thymidylate synthase reaction. One mole DHF is generated per mole thymidylate formed from deoxyuridylate. Because intracellular thymidylate synthesis utilizes methylene THF polyglutamate, the DHF formed is in the polyglutamate form. DHFR uses either nicotinamide adenine dinucleotide phosphate (NADPH) or NADH as a source of electrons: the former with much greater affinity for the enzyme, but the latter present in much larger quantities in the cell. Methotrexate, pyrimethamine, trimethoprim, and most other antifolates are designed to inhibit DHFR, although they also may have other effects. DHFR reduces both monoglutamate and polyglutamate folate, as well as both DHF and folic acid, but the reduction of folic acid to DHF is slow, so that folic acid acts as a partial inhibitor of the enzyme. The level of enzyme in the cell is regulated by the number of copies of the gene. At least three cases have been described which have been presumed to be deficient in DHFR [56–59]. These children were 0–6 weeks old and had megaloblastic anemia with better response to folinic than to folic acid. Deoxyuridine suppression test on one of these was better corrected with folinic than with folic acid. In all three, DHFR activity in liver biopsies was decreased, although in two, the activity became normal when assayed in 0.6 M KCl, which also increases the activity of normal DHFR. One of these patients was studied subsequently and found to be deficient in transcobalamin II and to respond to large doses of vitamin B_{12} [5]. DHFR assayed in the fibroblasts of another patient was normal when assayed subsequently [5]. It is possible that all of these children had defects in cobalamin metabolism which were not identified. The low levels of DHFR in hepatic biopsies are difficult to explain but may have been due to technical problems.

The clinical manifestations of DHFR deficiency are difficult to predict. Accumulation of intracellular DHF might interfere with other folate-dependent enzymes and produce unexpected manifestations. Neonatal megaloblastic anemia, intestinal malfunction, brain damage, and mouth ulcerations such as observed with methotrexate therapy would be expected. These should respond transiently to folinic but not to folic acid therapy. Because DHFR levels increase in response to treatment with antifolates, one would expect that partial defects of DHFR might be spontaneously corrected by gene duplication in many affected cells.

Trifunctional Peptide. Fig. 1 The enzyme activities for methylene THF dehydrogenase, methenyl THF cyclohydrolase, and 10-formyl THF synthetase reside on the same peptide in mammalian cells, but not in some bacteria [58]. They mediate the interconversion of 5-10-methylene THF, 5-10-methenyl THF, and 10-formyl THF and permit either release of the single carbon from folate as formate, or, more probably, the scavenging of potentially toxic formate. This interconversion of folates links the major source of single-carbon units (from serine via serine hydroxymethylase to form methylene THF) with synthesis of thymidylate (thymidylate synthase), or purine (Glycinamide Ribonucleotide Transformylase, GAR and Amino Imidazole Carboxamide Ribonucleotide, AICAR transformylase), and in liver, with the enzyme system which discards excess single carbons.

No documented cases of deficiency of any of these three enzymes has been described. One tentative report of possible methenyl THF cyclohydrolase deficiency in three mentally deficient children with 44% of activity of this enzyme [59] was not confirmed by a later report from the same laboratory [40]. Because of the many other reactions which interact with these pathways, the clinical syndrome caused by reduced activity of one of these is difficult to predict. One would expect defects in purine and thymidylate synthesis to be similar to toxic effects of antimetabolic agents such as 5-fluorouracil and 6-mercaptopurine, which produce decreased cell proliferation, cytopenias, and intestinal pathology and may produce megaloblastic erythropoiesis.

Serine Hydroxymethylase and the Glycine Cleavage Pathway. Fig. 1 The folate-dependent portion of glycine cleavage probably uses mitochondrial serine hydroxymethylase; these will therefore be considered together. As indicated above, the major source of single-carbon fragments available for folate utilization appears to be serine, entering the serine hydroxymethylase reaction in the cytoplasm to form glycine, and glycine cleavage, which utilizes a four-enzyme sequence including mitochondrial serine hydroxymethylase. In megaloblastic anemia, serine in plasma is normal, but plasma glycine levels may be elevated, and these return to normal when the nutritional deficiency is corrected. The direction of the glycine–serine interconversion appears determined by the supply of each amino acid [59]. Serine hydroxymethylase requires pyridoxal phosphate as prosthetic group to bind glycine or serine [58]. A mutant Chinese hamster ovary cell deficient in mitochondial serine hydroxymethylase was auxotrophic for glycine [56], indicating that cytoplasmic enzyme cannot take over all the functions of the mitochondrial enzyme.

Children with defects in the glycine cleavage pathway have nonketotic hyperglycinemia [60] which causes acidosis, ketosis, and failure to thrive. Accumulation of glycine in the CNS [62] has been postulated to cause delayed cerebral development. No cases of this disease have been described which respond to folate, or which are due to defective folate-dependent reactions.

Folylpolyglutamate Synthetase [63]. Human cells require a critical concentration of intracellular folate to permit folate-dependent activities. The critical concentration required to maintain an optimal rate of growth varies from about 50 nM folate in human fibroblasts to about 1 μM in human lymphocytes and certain tumor cells [9]. The K_m for monoglutamate folate of many of the folate-dependent coenzymes is greater than 1 μM, so that the enzyme reactions do not progress within the cell.

A cytoplasmic enzyme is present in all normal cells which adds glutamate residues to selected folate molecules. These glutamates form a peptide bond between the γ-carboxyglutamate already on the molecule and the α-amino group of the additional glutamate to be added. This γ-glutamyl chain is resistant to digestion by the common proteolytic enzymes and is hydrolyzed by specific "conjugase" enzymes, many of which are in lysosomes. Folylpolyglutamate synthetase adds glutamate residues one at a time, requires ATP for its reaction, utilizes THF and certain other folates and antifolates as substrates with different affinities, and reacts poorly with folic acid and 5-methyl THF. Polyglutamate folates of appropriate chain length have much lower K_m values for some of these folate-dependent reactions, which permits folate-dependent metabolism to progress at the concentrations of these folates present in the cell. K_m values of DHFR and methionine synthase for folic acid and 5-methyl THF are in the same general range as for polyglutamate forms of these, so that these reactions can (and do) utilize monoglutamate folate.

No defects in this enzyme have been reported in intact organisms, but several cell lines defective in folate polyglutamate formation have been reported. Cells unable to make polyglutamate folate are auxotrophic for purines, thymidine, and glycine, apparently because reactions generating these within the cell require polyglutamate folates. One cell line of human breast carcinoma cells which makes few polyglutamate forms of methotrexate is more resistant to methotrexate than is the wild type. As indicated above, the rate of folate polyglutamate synthesis appears to be decreased in cells deficient in vitamin B_{12} because 5-methyl THF is a poor substrate for the synthetase enzyme.

Formyl THF Dehydrogenase [64]. This appears to release excess active single – carbon fragments from folate by generating CO_2. It appears restricted to liver [5], and probably serves to maintain sufficient THF to permit acceptance of single carbons in folate-dependent reactions. It is possible that, unlike most other cells in which the need for folate is for synthetic reactions, in liver, THF must be available to accept formimino groups, formate, and other sources of single carbons.

No cases of deficiency have been described. One might expect deficient subjects to excrete FIGLU, be sensitive to the toxic effects of methanol, and perhaps show disturbances of glycine–serine equilibrium. These should respond to administration of folate.

Formyl THF Cyclodehydrase [65]. The activity of formyl THF cyclodehydrase has only recently been described. It is an ATP-dependent enzyme which converts 5-formyl THF to 5-10-methenyl THF. It is the pathway for utilization of 5-formyl THF (folinic acid), and its K_m is appropriate for its function within the cell. 5-Formyl THF appears to be a normal constitient of cells, perhaps formed nonenzymatically from unbound 10-formyl THF and methenyl THF.

No patients with inherited defects of this enzyme have been reported. They would be expected to be relatively unresponsive to folinic acid compared with folic acid and perhaps to have increased folate requirements. It is possible that other folate-dependent pathways would be inhibited by accumulation of excess intracellular 5-formyl THF.

Defects in Tetrahydrobiopterin Metabolism. Human cells synthesize BH_4 by a complicated and labile pathway from guanosine triphosphate (GTP) [66, 67]. BH_4 acts as a redox reagent, donating electrons in selected reductase reactions, which produce the quinonoid form of dihydrobiopterin (BH_2). the latter is reduced to BH_4 by dihydropteridine reductase(DHPR) but not by DHFR. If BH_4 oxidizes nonenzymatically to 7,8-BH_2, this form can be reduced to BH_4 by DHFR. Before the demonstration of this selectivity of DHPR and DHFR, it was believed that the BH_4 and folate systems were linked through competition for DHFR. This was also based on reports indicating absence of DHFR from brain and suggestions that DHPR was responsible for reduction of DHF in brain. Subsequent demonstration of DHFR in brain [5] and the failure of methotrexate to affect endogenous BH_4 metabolism [68] has clarified the separate pathways for reduction of pterins and folates. Because BH_4 is required as coenzyme for conversion (hydroxylation) of phenylalanine and tyrosine to DOPA, and of tryptophane to serotonin, BH_4 deficiency produces mental defects related to neurotransmitter deficiency, and this can be treated with BH_4.

Many BH_4-deficient patients have had low folate levels [5] and have improved clinically when treated with folinic acid. As indicated above, defects in BH_4-

dependent reactions may occur in methylene THF reductase deficiency. Several hypotheses have been proposed to explain the interaction between the two systems, but definitive explanations await further studies.

Because of the empirical observation that children with abnormalities in BH_4 metabolism (DHPR or BH_4 synthesis) are often folate deficient, and that children with methylene THF reductase defects may have defects in BH_4-dependent reactions, in patients with defects in either system such associations should be sought, investigated, and corrected.

Defective Cofactors Affecting Folate-Dependent Reactions. Since reduced nucleotides, flavins, pyridoxal phosphate, etc. are required in different reactions, one might suspect that some folate-dependent reactions might be affected if supply of these became limiting. The cause of erythroblast loss in riboflavin deficiency is unknown, but an effect on methylene THF reductase as well as on other flavin-dependent reactions should be considered.

Defects with Secondary Effects on Folate Metabolism

A number of factors can have secondary effects on folate metabolism through the increased folate requirements that are associated with them. These include:

Pregnancy
Antifolate drugs
Increased need for folate-dependent products
Intrauterine growth
Fragile chromosome sites
Amino acid imbalance
Homocystinuria due to cystathionine synthase deficiency
Methionine adenosyl transferase deficiency

These conditions are dealt with in articles elsewhere in this volume. Attention is called to the response of some children with cystathionine synthase deficiency to treatment with pyridoxine and folate, which maximizes the function of their deficient enzyme, and to the single patient [69] reported with defective generation of SAM from methionine (methionine adenosyltransferase at 6%–12% of control). No clinical manifestations could be detected consequent to this deficiency.

References

1. Roschau J, Date J, Kristoffersen K (1979) Folic acid supplements and intrauterine growth. Acta Obstet Gynecol Scand 58:343–346
2. Laurence KM, James N, Miller MH, Tennant GB, Campbell H (1981) Double blind randomized controlled trial of folate treatment before conception to prevent recurrence of neural-tube defects. Br Med J 282:11509–11511

3. Hecht F, Jacky PB, Sutherland GR (1982) The fragile X syndrome: current methods. Am J Hum Genet 11:489–495
4. Cooper BA (1976) Megaloblastic anaemia and disorders affecting utilisation of vitamin B12 and folate in childhood. Clin Haematol 5:631–659
5. Erbe RW (1979) Genetic aspects of folate metabolism. Adv Hum Genet 9:293–354
6. Erbe RW (1984) In: Whitehead VM, Blakley R (ed), Benkovic SJ (eds) Folates and pterins. New York: John Wiley
7. Hibbard BM (1975) Folates and the fetus. S Afr Med J 49:1223–1226
8. Henderson GB (1986) Transport of folate compounds by hemopoietic cells. Chapter 19 of this book Folate and Cobalamin
9. Watkins D, Cooper BA (1983) A critical intracellular concentration of fully reduced non methylated folate polyglutamate prevents macrocytosis and diminished growth rate in human cell line K 562 in culture. Biochem J 214:456–470
10. Smith GK, Benkovic PA, Benkovic SJ (1981) L(-)10-formyltetrahydrofolate is the cofactor for glycinamide ribonucleotide transformylase from chicken liver. Biochemistry 20:4034–4036
11. Shojania AM, Gross S (1964) Folic acid deficiency and prematurity. J Pediatr 64:323–329
12. Burland WL, Simpson K, Lord J (1971) Response of low birth weight infants to treatment with folic acid. Arch Dis Child 46:189–194
13. Luhby AL, Eagle FJ, Roth E, Cooperman JM (1961) Relapsing megaloblastic anemia in an infant due to a specific defect in gastrointestinal absorption of folic acid. Am J Dis Child 102:482–483
14. Lanzkowsy P, Erlandson ME, Bezan AI (1969) Isolated defect of folic acid absorption associated with mental retardation and cerebral calcification. Blood 34:452–464
15. Santiago-Borrera PJ, Santini R, Perez-Santiago E, Maldonado N, Millan S, Coll-Camalez G (1973) Congenital isolated defect in folic acid absorption. J Pediatr 82:450–455
16. Su PC (1976) Congenital folate deficiency. N Engl J Med 294:1128
17. Poncz M, Colman N, Herbert V, Schwartz E, Cohen AR (1981) Therapy of congenital folate malabsorption. J Pediatr 98:75–79
18. Corbeel L, Van den Berghe G, Jaekane J, Van Tornout J, Eeckels R (1985) Congenital folate malabsorption. Eur J Pediatr 143:284–290
19. Branda RF, Moldow CF, MacArthur JR, Wintrobe MM, Anthony BK, Jacob HS (1978) Folate-induced remission in aplastic anemia with familial defect of cellular folate. N Engl J Med 298:469–475
20. Copenhauer JH, O'Brien KL (1969) Separation and identification of pteroyl polyglutamates by ion exchange thin layer chromatography. Anal Biochem 31:454–462
21. Nakazawa Y, Chiba K, Imatoh N, Kotorii T, Sakamoto T, Ishizaki T (1983) Serum folic acid levels and antipyrine clearance rates in smokers and non-smokers. Drug Alcohol Depend 11:201–207
22. Senti F, Pilch S (1984) Assessment of the folate nutritional status of the US population based on data collected in the second national health and nutrition examination survey 1976–1980. Life Science Research Office, Federation of American Societies for Experimental Biology (FASEB), pp 1–86
23. Zittoun J (1989) Folate abnormalities caused by drugs, and alcohol. Chapter 10, this volume
24. Steinberg SE, Campbell CL, Hillman RS (1979) Kinetics of the normal folate enterohepatic cycle. J Clin Invest 64:83–88
25. Mathews RG (1982) Are the redox properties of tetrahydrofolate cofactors utilized in folate-dependent reactions? Fed Proc 41:2600–2604
26. Lewis GP, Rowe PB (1983) Methylene tetrahydrofolate reductase: studies in a human mutant and mammalian liver. In: Blair J (ed) Chemistry and biology of pteridines. de Gruyter, Berlin, pp 229–233
27. Kutzbach C, Stokstad ELR (1971) Feedback inhibition of methylene tetrahydrofolate reductase in rat liver by S-adenosylmethionine. Biochim Biophys Acta 139:217–220
28. Smith I, Hyland K, Kendall B (1985) Clinical role of pteridine therapy in tetrahydrobiopterin deficiency. J Inherited Metab Dis 8 (Suppl 1):39–45
29. Rosenblatt DS, Cooper BA, Lue-Shing S, Wong PWK, Berlow S, Narisawa K, Baumgartner

R (1979) Folate distribution in cultured cells: studies on 5, 10-CH2-H4PteGlu reductase deficiency. J Clin Invest 63:1019–1025

30. Rosenblatt DS, Cooper BA (1979) Methylene tetrahydrofolate reductase deficiency. Clinical and biochemical correlations. In: Botez IM, Reynolds EH (ed) Folic acid in neurology, psychiatry and internal medicine. Raven, pp 385–390

31. Boss GR, Erbe RW (1981) Decreased rates of methionine synthesis by methylene tetrahydrofolate reductase – deficient fibroblasts and lymphocytes. J Clin Invest 67:1659–1664

32. Kanwar YS, Manaligold JR, Wonk P, WK (1976) Morphologic studies in a patient with homocystinuria due to 5,10-methylene tetrahydro folate reductase deficiency. Pediatr Res 10:598–609

33. Harpey JP, Rosenblatt DS, Cooper BA, Le Moel G, Roy C, Lafourcade J (1981) Homocystinuria caused by 5,10 methylene tetrahydro folate reductase deficiency: a case in an infant responding to methionine, folinic acid, pyridoxine, and vitamin B12 therapy. J Pediatr 98:275–278

34. Christensen E, Brandt NJ (1985) Prenatal diagnosis of 5,10 methylene tetrahydro folate reductase deficiency. N Engl J Med 313:50–51

35. Cooper BA (1983) The assay of cobalamin-dependent enzymes: their utilization for evaluation of cobalamin metabolism. In: Hall C (ed) Methods in haematology, vol 10. Churchill Livingstone, London, pp 165–180

36. Wendel U, Claussen U, Dieckmann E (1983) Prenatal diagnosis for methylene tetrahydro folate reductase deficiency. J Pediatr 102:938–940

37. Hyland K, Smith I, Howells DL, Clayton PT, Leonard JV (1985) The determination of pterins, biogenic amine metabolites, and aromatic amino acids in cerebrospinal fluid using isocratic reverse phase liquid chomatography with in series dual cell coulometric electrochemical and fluorescence detection: use in the study of inborn errors of dihydropteridine reductase and 5,10 methylene tetrahydrofolate reductase. In: Wachter H, Pfleiderer W, Curtius HC (eds) Biochemical and clinical aspects of pteridines, vol 4. de Gruyter, Berlin, pp 85–99

38. Blair JA (1985) Tetrahydrobiopterin metabolism in the central nervous system. In: Wachter H, Pfleiderer W, Curtius HC (eds) Biochemical and clinical aspects of pteridines, vol 4. de Gruyter, Berlin, pp 113–118

39. Whitehead VM, Pratt R, Viallet A, Cooper BA (1972) Intestinal conversion of folinic acid to 5-methyl tetrahydrofolate in man. Br J Haematol 22:63–72

40. Arakawa T (1970) Congenital defects in folate utilization. Am J Med 48:594–598

41. Arakawa T, Yoshida T, Ohara K, Narisawa K, Tanno K, Hondo Y, Higashi O (1972) Defect of incorporation of glycine-1-14C into urinary uic acid in formiminotransferase deficiency syndrome. Tohoku J Exp Med 106:213–218

42. Niederwieser A, Matasovic A, Steinmann B, Baerlocher K, Kemken B (1976) Hydantoin-5-propionic aciduria in folic acid non-dependent formiminoglutamic aciduria observed in two siblings. Pediatr Res 10:215–219

43. Perry TL, Applegarth DA, Evans ME, Hansen S, Jellum E (1975) Metabolic studies of a family with massive formiminoglutamic aciduria. Pediatr Res 9:117–122

44. Herman RH, Rosensweig NS, Stifel FB, Herman YF (1969) Adult formiminotransferase deficiency: a new entity. Clin 17:304

45. Beck B, Chistensen E, Brandt NJ (1981) Formiminoglutamic aciduria in a slightly retarded boy with chronic obstructive lung disease. J Inherited Metab Dis 4:225–228

46. Beaudet R, MacKenzie RE (1976) Formiminotransferase-cyclohydrolase from porcine liver: an octameric enzyme containing bifunctional polypeptides. Biochim Biphys Acta 453:151–161

47. Chanarin I (1969) The megaloblastic anaemias. Blackwell, London, pp 393–396

48. Hershfield MS, Francke U (1982) The human genes for S-adenosylhomocysteine hydrolase and adenosyl deaminase are syntenic on chromosome 20. Science 216:739–740

49. Taylor RT (1982) B12-dependent methionine biosynthesis. In: Dolphin D (ed) Vitamin B12. Wiley, New York, pp 307–355 (Biochemistry and medicine, vol 2)

50. Chanarin I (1989) Cobalamin folate interrelationships. Haematol This book "Folate and Cobalamins"

51. Schuh S, Rosenblatt DS, Cooper BA, Schroeder M, Bishop A, Sargeant LE, Haworth JC

(1984) Homocystinuria and megaloblastic anemia responsive to vitamin B12 therapy: a defect in cobalamin metabolism. N Engl J Med 310:686–690

52. Rosenblatt DS, Cooper BA, Pottier A, Lue-Shing H, Matziaszuk N, Grauer K (1984) Altered vitamin B12 metabolism in fibroblasts of a patient with megaloblastic anemia and homocystinuria due to a new defect in methionine biosynthesis. J Clin Invest 74:2149–2156

53. Rosenblatt DS, Thomas IT, Watkins D, Cooper BA, Erbe RW (1987) Vitamin B_{12} responsive homocystinuria and megaloblastic anemia: heterogeneity in methylcobalamin deficiency. Am J Med Genet 26:377–383

54. Zittoun J, Fischer A, Marquet J, Perignon JL, Lagrue A, Griscelli C (1987) Megaloblastic anemia and immune abnormalities in a patient with methionine synthase deficiency. Acta Paediatr Scand 76:991–998

55. Hallam L, Sawyer M, Clark AL, Van Der Weyden MB (1987) Vitamin B_{12}-responsive neonatal megaloblastic anemia and homocystinuria associated with reduced levels of methionine synthetase. Blood 69:1128–1133

56. Walters T (1967) Congenital megaloblastic anemia responsive to N5-formyl tetrahydrofolic acid administration. J Pediatr 70:686–687

57. Rowe PB (1978) Inherited disorders of folate metabolism. In: Stanbury JB, Wyngaarden JB, Fredrickson DS (eds) The metabolic basis of inherited diseases. McGraw-Hill, New York, pp 430–457

58. Tan LUL, Drury EJ, MacKenzie RE (1977) Methylene tetrahydrofolate dehydrogenase – methenyl tetrahydrofolate cyclohydrolase – formyl tetrahydrofolate synthetase: a multifunctional protein from porcine liver. J Biol Chem 234:1830–1846

59. Schirch L (1984) Folates in serine and glycine metabolism. In: Blakley R, Benkovic SJ (eds) "Folates and pterins, vol 1. Chemistry and biochemistry of folates. Wiley Interscience, New York, pp 135–190

60. Nyhan WL (1978) Non ketotic hyperglycinemia. In: Stanbury JB, Wyngaarden JB, Frederickson DS (eds) the metabolic basis of inherited disease. McGraw Hill, New York, pp 518–527

61. McBurney MW, Whitmore GF (1974) Isolation and biochemical characterization of folate deficient mutants of Chinese hamster cell. Cell 2:173–182

62. Perry TL, Urquhart N, MacLean J, Evans ME, Hansen S, Davidson AGF, Applegarth DA, MacLeod PJ, Lock JE (1975) Non-ketotic hyperglycinemia. Glycine accumulation due to absence of glycine cleavage in brain. N Engl J Med 292:1269–1273

63. McGuire JJ, Coward JK (1984) Pteroylpolyglutamates: biosynthesis, degradation and function. In: Blackley K, Benkovic J (eds) Folates and pterins, vol 1. Chemistry and biochemistry of folates. Wiley Interscience, New York, pp 135–190

64. Krebs HA, Hems R, Tyler B (1976) The regulation of folate and methionine metabolism. Biochem J 158:341–353

65. Hopkins S, Schirch L (1984) 5,10-methenyltetrahydrofolate synthetase. J Biol Chem 259:5618–5621

66. Brown GM, Switchenko AC, Primus JP (1985) Enzymatic formation of H4-biopterin in Drosophila melanogaster. In: Wachter H, Curtius HC, Pfleiderer W (eds) biochemical and clinical aspects of pteridines, vol 4. de Gruyter, Berlin, pp 119–141

67. Hentil D, Leimbacher W, Redweik U, Blau N, Niederwieser A, Curtius H (1985) Tetrahydrobiopterin biosynthesis in man. In: Wachter, H. Curtius HC, Pfleiderer W (eds) Biochemical and clinical aspects of pteridines, vol 4. de Gruyter, Berlin, pp 119–141

68. Nichol CA, Viveros OH, Duch DH, Abou-Donia M, Smith GK (1983) Metabolism of pteridine cofactors in neurochemistry. In: Blair JA (ed) Chemistry and biology of pteridines; pteridines and folic acid derivatives. de Gruyter, Berlin, pp 131–151

69. Gaull GE, Tallan HH (1974) Methionine adenosyltransferase deficiency: new enzymatic defect associated with hypermethioninemia. Science 186:59–60

Chapter 17

The Relationship Between Biopterin and Folate Metabolism

J. P. HARPEY

Phenylketonuria (PKU) is a metabolic disorder that is inherited in an autosomal recessive pattern and is caused by a defect in the hepatic phenylalanine (PA) hydroxylating system. This system consists of two enzymes, phenylalanine 4-hydroxylase (PAH) (EC 1.14.16.1) and dihydropteridine reductase (DHPR) (EC 1.6.99.7), and a coenzyme, tetrahydrobiopterin (BH_4), which acts not only on PAH but also on tyrosine 3-hydroxylase (EC 1.14.16.2) and tryptophan 5-hydroxylase (EC 1.14.16.4) – the former on the dopamine and norepinephrine synthesis pathway, the latter on the serotonin synthesis pathway (Fig. 1). During each hydroxylation cycle of these three enzymes, BH_4 is oxidized to quininoid dihydrobiopterin (q-BH_2), the latter being transformed partly into L-erythro-7,8-dihydrobiopterin (7,8-BH_2). They are reduced to BH_4 by DHPR and by dihydrofolate reductase (DHFR) (EC 1.5.1.3), respectively (Fig. 1).

The classic form of PKU is caused by a lack of PAH activity, which normally transforms PA into tyrosine. This defect produces hyperphenylalaninemia, which untreated results in severe mental retardation. Hyperphenylalaninemia impedes the transfer of L-tryprophan and L-tyrosine at the blood–brain barrier level. This results in a decreased synthesis of biogenic amines. Furthermore, increased levels of phenylalanine in brain seems directly to inhibit tyrosine 3-

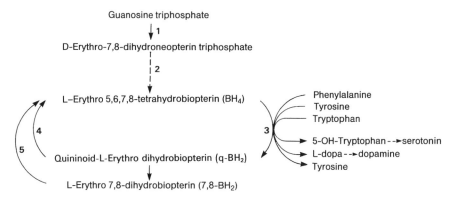

Fig. 1. The biopterin system, coenzyme of aromatic amino acid hydroxylases: *1*, Guanosine triphosphate cyclohydrolase; *2*, "dihydrobiopterin-synthetase;" *3*, phenylalanine 4-hydroxylase, tyrosine 3-hydroxylase, tryptophan 5-hydroxylase; *4*, dihydropteridine reductase; *5*, dihydrofolate reductase

hydroxylase and tryptophan 5-hydroxylase activities. The correction of hyper-phenylalaninemia with a low-PA diet leads to a normal biogenic amine synthesis. Neonatal screening for hyperphenylalaninemia and early low-PA diet allow a normal development in children with PKU.

In France, the prevalence of classic PKU is $1:13000$ newborn infants. Some 1%–3% of PKU cases are not caused by a defect in PAH but by a defect in BH_4 synthesis or in the regeneration of BH_4 (DHPR deficiency; Fig. 1). In either case, the result is the development of malignant hyperphenylalaninemia, i.e., PKU which does not respond to a good dietary control of blood-PA levels and leads to an irreversible neurologic deterioration which becomes apparent between 3 and 5 months of age. This is characterized by myoclonic seizures, axial hypotonia, extrapyramidal hypertonia with lead-pipe rigidity of upper limbs, and distal tremor; oculo motor dysfunction with episodes of upper and sideward gaze deviation; expressionless facies, buccofacial dyskinesia and drooling, and swallowing difficulties. Death usually occurs by age 5–7 years [1, 2]. This deterioration which occurs in spite of satisfactory blood-PA levels is due to biogenic amine deficiency (dopamine, norepinephrine, and serotonin).

Of paramount importance is early recognition of these cases. DHPR and biopterin assay on dried blood spots and biopterin assay on filter paper spots of urine should be carried out in every infant with a positive neonatal screening for hyperphenylalaninemia. Positive results require dietary control of blood-PA levels, a supplementation with neurotransmitter precursors, L-dopa and 5-hydroxytryptophan (with a peripheral inhibitor of aromatic amino acid decarboxylase, carbidopa) [2], and in cases of BH_4 synthesis deficiency, a replacement therapy with BH_4.

However, it has recently become evident that the treatment of DHPR deficiency is not fully satisfactory if folinic acid (5-formyl-tetrahydrofolate, (THF)) supplementation is not started early [3, 4, 5]. Study of DHPR-deficient patients has led us to a better understanding of the relationship between folate and biopterin metabolism. In such patients, folic acid deficiency in blood, and brain seems usual [6], with development of basal ganglia calcifications [3], as in cases of inherited defect of folate transfer into brain. Furthermore, some cases of DHPR deficiency with good dietary control of blood-PA levels, sufficient neurotransmitter precursor supplementation, and normal neurologic status have developed at age 2–3 years an unexpected neurologic deterioration with signs and symptoms similar to early manifestations of untreated cases and associated with decreased levels of CSF amine metabolites. This late neurologic deterioration is secondary to brain and CSF 5-methyl-THF (5-CH_3-THF) deficiency, and has been cured by 5-formyl-THF in two patients [3, 4].

DHPR is well represented in brain, even in the area without tyrosine or tryptophan hydroxylase activity [7]. DHPR has been shown to catalyze the reduction of quininoid dihydrofolate to THF (Fig. 2). DHPR deficiency results in brain and CSF in reduced folate deficiency [6]. DHFR, well represented in liver but not in brain, can reduce 7,8-BH_2 to BH_4 [8] (Fig. 2). In DHPR-deficient patients, a competition between 7,8-BH_2 (derived from q-BH_2 accumulated above the metabolic block) and dihydrofolate could lead to THF deficiency peripherally and secondarily in brain CSF. 5,10-Methylene THF reductase (MTHFR) (EC 1.1.1.68),

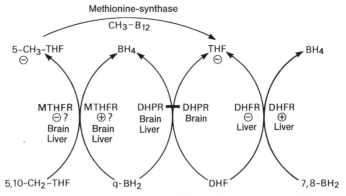

Fig. 2. Salvage pathways in dihydropteridine reductase (DHPR) deficiency and possible secondary effects on folate metabolism. *q-BH₂*, quininoid dihydrobiopterin; *BH₄*, tetrahydrobiopterin; *7,8-BH₂*, L-erythro-7,8-dihydrobiopterin; *DHF*, dihydrofolate; *THF*, tetrahydrofolate; *5,10-CH₂-THF*, 5,10-methylene tetrahydrofolate; *5-CH₃-THF*, 5-methyltetrahydrofolate; *MTHFR*, 5,10-methylene tetrahydrofolate reductase; *DHFR*, dihydrofolate reductase; *CH₃-B₁₂*, methyl-cobalamin

which normally catalyzes the reduction of 5,10-methylene THF to 5-CH₃-THF (the essential folate derivative), present in liver and brain, could, in DHPR-deficient patients, help to reduce q-BH₂ to BH₄ [9]. A competition between q-BH₂ and 5,10-methylene THF, could decrease the production of 5-CH₃ THF (Fig. 2).

Patients with MTHFR deficiency, who also have a severe deficiency in brain CSF 5-CH₃-THF, have low levels of amine metabolites in CSF [11, 12]. It has been shown in a DHPR-deficient patient that some salvage of BH₄ exists, likely by means of MTHFR and DHFR [10, 12]. In DHPR-deficient patients, the decreased level of reduced folate in brain seems to impede the control of biogenic amine release rather than the amine synthesis [3]. It has been demonstrated on homogeneized fresh rat brain that the addition of 5-CH₃-THF, hydroxocobalamin, and ascorbate increased the in vitro synthesis of biopterin [13].

Folinic acid supplementation, by providing increased amounts of reduced folate for MTHFR, peripherally and in the brain, can correct 5-CH₃-THF deficiency, and thus prevent basal ganglia calcifications and delayed neurologic deterioration in DHPR-deficient patients. 5-CH₃-THF seems too unstable to be used as a therapeutic agent. Not only folic acid is ineffective [2, 4], but by competitive inhibition with 5-CH₃-THF at the blood–brain barrier level, it can aggravate the neurologic status of such children [3].

References

1. Smith I (1974) Atypical phenylketonuria accompanied by a severe progressive neurological illness unresponsive to dietary treatment. Arch Dis Child 49:242
2. Harpey J-P (1983) Les défauts de synthèse des bioptérines: les déficits complets (réductase et synthétase). Arch Fr Pédiatr 40:231–235

3. Smith I, Hyland K, Kendall B. Leeming RJ (1985) Clinical role of pteridine therapy in tetrahydrobiopterin deficiency. J Inherited Metab Dis 8 (Suppl 1):39–45

4. Harpey J-P, Rey F, Leeming RJ (1984) Seven year follow-up in a child with early-treated dihydropteridine-reductase deficiency. 22nd Annual symposium, Society for the Study of Inborn Errors of Metabolism. Newcastle upon Tyne, 1984

5. Irons M, Levy HL, O'Flynn ME, Stack CV, Langlais PJ, Butler IJ, Milstien S, Kaufman S (1987) Folinic acid therapy in treatment of dihydropteridine reductase deficiency. J Pediatr 110:61–67

6. Pollock RJ, Kaufman S (1978) Dihydropteridine reductase may function in tetrahydrofolate metabolism. J Neurochem 31:115–123

7. Turner AJ, Ponzio F, Algeri S (1974) Dihydropteridine reductase in rat brain: regional distribution and the effect of catecholamine-depleting drugs. Brain Res 70:550–558

8. Nichol CA, Lee CL, Edelstein MP, Chao JY, Duch DS (1983) Biosynthesis of tetrahydrobiopterin by de novo and salvage pathways in adrenal medulla extracts, mamalian cell cultures and rat brain in vivo. Proc Natl Acad Sci USA 80:1546–1550

9. Matthews RG, Kaufman S (1980) Characterization of the dihydropteridine reductase activity of the pig liver methylenetetrahydrofolate reductase. J Biol Chem 255:6014–6017

10. Howells D, Smith I, Leonard J, Hyland K (1986) Tetrahydrobiopterin in dihydropteridine reductase deficiency. N Engl J Med 314:520–521

11. Harpey J-P, Le Moël G, Zittoun J (1983) Follow-up in a child with 5,10-methylenetetrahydrofolate reductase deficiency. J Pediatr 103(6):1007

12. Clayton PT, Smith I, Harding B, Hyland K, Leonard K, Leeming RJ (1986) Subacute combined degeneration of the cord, dementia and parkinsonism due to an inborn error of folate metabolism. J Neurol Neurosurg Psychiatry 49:920–927

13. Leeming RJ, Harpey J-P, Brown SM, Blair JA (1983) Tetrahydrofolate and hydroxocobalamin in the management of dihydropteridine reductase deficiency. J Ment Defic Res 26:21–25

Chapter 18

Inherited Disorders of Cobalamin Metabolism

J. ZITTOUN and J. MARQUET

Congenital defects of cobalamin (Cbl) metabolism have been identified at each of its steps: absorption transport, and intracellular utilization. These can be summarized as follows:

Malabsorption
    Defect of synthesis or secretion of intrinsic factor
    Functional abnormality of intrinsic factor
    Juvenile pernicious anemia
    Specific malabsorption due to miscellaneous diseases
Defective transport
    Congenital absence of transcobalamin II
    Functional impairment of transcobalamin II
    Congenital absence of R proteins
    Combined deficiency of R proteins and intrinsic factor
Abnormalities of intracellular cobalamin metabolism
    Defect of synthesis (or retention) of adenosyl-Cbl
    Deficiency of methylmalonyl-CoA mutase
    Defect of synthesis of adenosyl-Cbl and methyl-Cbl
    Defect of synthesis (or retention) of methyl-Cbl
    Deficiency of methionine synthase

The only source of Cbl in the body is food; Cbl bound to dietary proteins is released in the stomach by gastric acidity and pepsin. Cbl then becomes bound in gastric juice to two specific binders, intrinsic factor (IF) and R proteins; the latter has a higher affinity for Cbl than does the former. Both are glycoproteins, but only IF is necessary for normal absorption. Specific receptors for the complex Cbl-IF are located on microvillus membranes in the distal gut. The steps inside the enterocyte are not well understood, but Cbl enters the circulation bound to transcobalamin II (TC II), a polypeptide required for the cellular uptake of Cbl, while transcobalamin I (TC I), very close to R binders of gastric juice, binds endogenous Cbl in plasma but is not involved in the delivery of Cbl to cells.

Inside the cells, the complex Cbl-TC II is internalized after attachment to specific receptors. A process of lysosomal proteolysis is then essential to the release of Cbl, and TC II is degraded in this process. Released Cbl becomes available for the formation of active forms, methyl-Cbl and adenosyl-Cbl. The synthesis of these two coenzymes requires that trivalent Co (III) be reduced to divalent Co (II) and then to monovalent Co (I) by two apparently distinct reductase systems. Methyl-Cbl, the coenzyme of methionine synthase, is involved in methionine synthesis, while adenosyl-Cbl, the coenzyme of methylmalonyl-CoA mutase leads to the conversion of methylmalonate to succinate (Fig. 1).

Absorption and transport of Cbl are extensively detailed in Chaps. 4, 5 (this volume).

Malabsorption

Defects of absorption are the most frequent congenital abnormalities of Cbl metabolism. Malabsorption usually induces, besides megaloblastic anemia, a delay in physical and even mental development.

Congenital Intrinsic Factor Deficiency. Selective failure to synthesize or secrete intrinsic factor (IF) [6, 10] is a common occurrence; more than 50 cases with this disorder have been reported. The usual feature at presentation is megaloblastic anemia due to vitamin B_{12} deficiency. This disorder is characterized by: (a) impaired vitamin B_{12} absorption corrected by exogenous IF; (b) normal acid secretion and normal gastric fundic histology; and (c) absence of IF in the gastric secretion and absence of autoantibodies against IF and parietal cells in serum (Table 1). A retrospective immunohistochemical study performed on gastric biopsies from patients with such a disorder, however, has shown that in some, immunogenic IF was detected in parietal cells [24]. This disease is thus heterogeneous, with different defects of either synthesis or secretion but with the same clinical and hematological expression.

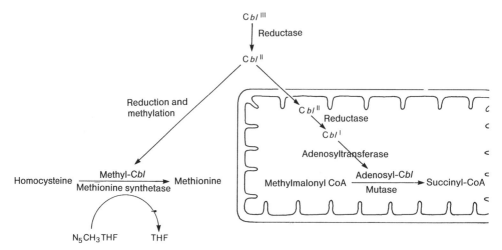

Fig. 1. Synthesis and functions of active cobalamins (*Cbl*): methylcobalamin and adenosyl-cobalamin. *THF*, Tetrahydrofolate; N_5CH_3THF: N_5 methyltetrahydrofolate

Table 1. Discriminating features of congenital cobalamin malabsorptions

	Stomach gastric secretion			Schilling test			Associated abnormalities
	IF immuno-logical assay	HCl	Histology	−IF	+IF	Anti-IF antibodies	
Defect of synthesis or secretion of IF	−	+	N	↓	N	−	−
Functional impairment of IF	+	+	N	↓	N	−	−
Juvenile pernicious anemia	−	−	Atrophy	↓	N	+	Endocrinopathies, hypothyroidism, hypoparathyroidism, adrenal atrophy, moniliasis, hypo- or agamma-globulinemia, deficiency of IgA
Imerslund's disease	+	+	N	↓	↓	−	Proteinuria without renal insufficiency

IF, Intrinsic factor. Schilling test: −IF, without intrinsic factor; +IF, with intrinsic factor.

There is great age variability at presentation [6, 22]. Megaloblastic anemia usually occurs during the first 2 years of life, often associated with neurological dysfunction, such as ataxia and other features of subacute combined degeneration of the spinal cord. The occurrence of megaloblastic anemia in the first few years of life corresponds to the size of prenatal Cbl stores in these children. Some cases, however, have been diagnosed at a later age, such as that of a boy first diagnosed at the age of 17 years who was healthy and without treatment, and who developed a severe and sudden megaloblastic anemia [22]. This great age variability at presentation suggests differences in body stores of Cbl or turnover and, in some cases, the possibility of inadvertent therapy which may have delayed the diagnosis.

Structural and Functional Defects of Intrinsic Factor. Abnormal IF may alter its function and thereby induce the same hematological abnormalities as those due to congenital absence of IF. A case of megaloblastic anemia was reported in a 13-year-old boy in whom IF was present immunologically but was inert biologically. This structurally abnormal IF was unable to bind to the specific ileal receptor of the Cbl-IF complex. Gastric juice from this boy given to patients who had undergone gastrectomy did not correct the Cbl malabsorption [21].

Other variants of abnormal IF have recently been reported. In some of these cases, synthesis and secretion of IF in gastric juice was normal, but this IF was rapidly degraded by acidity and pepsin in the gastric lumen into smaller fragments unable to bind Cbl. Cbl malabsorption was corrected by neutralization of gastric secretion with sodium bicarbonate [43]. In another case, [36] binding to the ileal receptor was very low because the IF-Cbl complex dissociated rapidly at 37°C and reassociated at 4°C. In a preliminary report, it was shown that the abnormal IF formed dimers.

Juvenile Pernicious Anemia. The picture presented by juvenile pernicious anemia mimics that of adult pernicious anemia: atrophic gastritis with achlorhydria and failure of IF secretion (Table 1). Pernicious anemia in adults is extensively detailed in chap. 6 (this volume). However, the frequency of autoantibodies against human IF is far higher in children than in adults. In addition, juvenile pernicious anemia is often associated with other endocrinopathies, such as myxedema, hypoparathyroidism, or adrenal atrophy; some patients also suffer from moniliasis [32]. These children have a genetically determined tendency to develop organ-specific antibodies. In the family of these children, siblings may also show endocrinopathies with a higher frequency. Some cases of hypogammaglobulinemia [11] or even of total and selective IgA deficiency [42] have been reported, associated with juvenile pernicious anemia. The relationship between these two abnormalities is not clear.

Specific Cobalamin Malabsorption. The disorder of congenital and specific malabsorption of Cbl due to a defect of membrane transport in the ileum was first reported by Imerslund [20] and Gräsbeck et al. [14]. It is usually associated with proteinuria but with no other evidence of renal pathology [27]. This disorder is transmitted as a recessive autosomal trait; the largest series of patients are from

Israel [3] among families of Libyan and Tunisian origin and from Scandinavia in some rural areas. The signs usually occur in the first 3 years of life. The clinical pictures at presentation are similar: anorexia, irritability, pallor, vomiting, and weakness. Neurological dysfunctions are rare and have been reported only in patients given folic acid. Children present with megaloblastic anemia. Investigations show vitamin B_{12} deficiency with normal gastric acid and IF secretion. The Schilling test is abnormal and not corrected by exogenous IF (Table 1). Anemia is vitamin B_{12}-responsive, and patients remain well on maintenance vitamin B_{12} therapy.

The precise mechanism of this intestinal and selective malabsorption of Cbl is not yet well defined. Biopsies of the ileal mucosa are normal, and the ileal membrane receptors studied in one patient [26] were not altered. However, a recent study [5] has shown a defect in the attachment in vivo of the Cbl-IF complex to ileal receptors. Renal biopsies are normal or show mild abnormalities.

Miscellaneous Causes of Cobalamin Malabsorption. Children may develop Cbl malabsorption resulting from intestinal diseases. Small-bowel bacterial overgrowth (following intestinal anatomic abnormalities of the gut, including diverticula, anastomoses of the small bowel, blind loops) may induce Cbl malabsorption; Cbl may be converted by bacteria to inactive analogues which are not absorbed [4]. Ileal resections performed for congenital stenosis or volvulus may also induce Cbl malabsorption. Crohn's disease or tuberculosis of the ileum are inflammatory disorders which may affect the ileal mucosa, as may neoplastic disorders such as lymphoma of the small bowel. Impaired pancreatic function, usual in mucoviscidosis may be responsible of Cbl malabsorption by a failure to split R proteins to which Cbl is bound (see chap. 7, this volume). However, in this latter condition, vitamin B_{12} deficiency and megaloblastic anemia are very uncommon.

Defective Transport

Hereditary Transcobalamin II Deficiency. TC II is a protein essential for cellular uptake of Cbl [16]. Congenital absence of TC II thus produces a life-threatening disease with severe clinical and hematological signs that usually occur during the first few months of life [15, 19]. Hematological features consist of pancytopenia with megaloblastic bone marrow changes. Macrocytosis and hypersegmentation of neutrophils are not constant. Cases presenting with a decrease of erythroid precursors and megacaryocytes and some with excess of blasts of the myeloid series have been reported [31, 33]. Clinical features usually consist of a failure to thrive, vomiting, and diarrhea, with malabsorption of Cbl. Neurological dysfunctions are rare except in cases treated with folic acid for a long time before the correct diagnosis was made. Infections are frequent and often severe, especially in upper respiratory and intestinal tract as well as mouth ulcers. These infections may be partly related to immunological abnormalities which are sometimes associated to this disorder [40]: hypogammaglobulinemia, impaired functions of lymphocytes and granulocytes, and especially impaired bactericidal

Table 2. Main abnormalities observed in defective transport of cobalamin

	Plasma vitamin B_{12} level	Schilling test		Hematological abnormalities	Other abnormalities
		−IF	+IF		
Deficiency of TC II	N or ↓	↓	↓	Megaloblastic anemia + pancytopenia	Hypo- or agammaglobulinemia Cellular and humoral immunity impairment Functional impairment of leukocytes
Functional impairment of TC II	N	N	N	Megaloblastic anemia	–
Deficiency of R protein	↓↑	N	N	0	–
Combined deficiency of IF and R proteins	↓↑	↓	N	Megaloblastic anemia	–

TC II, Transcobalamin II; IF, intrinsic factor. Schilling test: −IF, without intrinsic factor; +IF, with intrinsic factor.

activity, normalized by plasma infusion or by treatment with pharmacological doses of Cbl. The diagnosis of congenital absence of TC II must be considered in all cases of neonatal megaloblastic anemia. Serum vitamin B_{12} is usually normal (Table 2), but two cases have been recently reported [8, 30] with low levels of serum Cbl.

Functional Impairment of Transcobalamin II. TC II may be immunologically detected but functionally abnormal and incapable of promoting the cellular uptake of Cbl. Megaloblastic anemia reported in two women 32- and 34-years-old [18, 41], with episodes of Cbl-responsive anemia in childhood, has been ascribed to impairment of TC II function. No infections or immunological problems were present. In one of these two cases, TC II was able to bind Cbl in vitro but could not facilitate the uptake of the vitamin, probably due to an incapacity to attach to the specific cell receptors. In the second case, TC II demonstrated decreased ability to bind Cbl in vitro.

Congenital Absence of R Protein. Carmel and Herbert [7] have reported two cases of R protein deficiency in two brothers: this absence was detected in adult life. No hematological abnormalities were present except a low serum vitamin B_{12} (since R proteins bind the major part of circulating vitamin B_{12}). The absence of these proteins was also demonstrated in other secretions (saliva, gastric juice) and in leukocytes. Recently, a combined deficiency in IF and R proteins [45] has been identified in a young Algerian boy with a megaloblastic anemia and who presented with neurological dysfunctions consistent with vitamin B_{12} deficiency. Gastric juice analysis showed a complete deficiency of IF and R proteins while acid secretion and gastric histology were normal; R proteins were also absent from saliva, serum, and polymorphonuclear leukocytes.

Defects in Intracellular Cobalamin Metabolism

Cbl enters cells bound to TC II and is then converted into two active forms: methyl-Cbl, (the coenzyme of methionine synthase involved in methionine synthesis; this step occurs in the cytosol) and adenosyl-Cbl (the coenzyme of methylmalonyl-CoA mutase which catalyzes the conversion of L-methylmalonyl-CoA into succinyl-CoA; this step is intramitochondrial). Active Cbl formation involves the stepwise reduction (by reductases) of cobalt from an oxidation trivalent state to a monovalent state (Fig. 1). Several inborn errors of intracellular Cbl metabolism have been identified; they have been found to be due to abnormalities in synthesis or retention of one of the two coenzymes or to a defect of one of the corresponding enzymes.

Defect of Synthesis (or Retention) of Adenosyl-Cbl. Inability of cells to synthesize or retain normal levels of adenosyl-Cbl results in a urinary methylmalonic acid excretion in spite of normal plasma levels of vitamin B_{12}. Clinically, children present with significant development delay, somnolence, and vomiting, with severe metabolic abnormalities which appear in the first months of life: ketoacidosis, accompanied by hyperglycinemia, hyperglycinuria, and hyperammoniema. Hematological investigations have failed to show either anemia or megaloblastic bone marrow changes. Serum Cbl is normal. Such patients are usually responsive to massive doses of Cbl (1000-2000 µg hydroxo-Cbl per 24 h) which reduce methylmalonic aciduria and induce clinical improvement [34].

This congenital defect consists of two distinct types of mutant (Table 3): Cbl A mutant may have a deficiency of one of the two mitochondrial reductases transforming cobalt from an oxidation trivalent state to a monovalent state, while Cbl B mutant may have a specific defect in adenosyltransferase [29].

Methylmalonyl-CoA Mutase Deficiency. In some patients methylmalonic aciduria results from methylmalonyl-CoA mutase deficiency. The clinical picture in patients with this disorder resembles that in patients with specific inability to form adenosyl-Cbl, but the disease may vary greatly in severity according to the degree of enzyme deficiency [28]. These patients usually fail to respond to treatment with Cbl, but treatment which restricts dietary intake of some amino acids (isoleucine, valine, methionine) may reduce the urinary excretion of methylmalonic acid.

Failure of Synthesis (or Retention) of both Adenosyl-Cbl and Methyl-Cbl. The abnormality in which there is a failure in the synthesis or retention both of adenosyl-Cbl and of methyl-Cbl is related to a cellular abnormality formation (or possibly retention within the cells) of the two active forms, adenosyl-Cbl and methyl-Cbl (2, 9, 12, 25) probably due to a defect in reduction of hydroxo-Cbl (Fig. 1). The block seems to be located in the early steps of this common metabolic pathway and produces methylmalonic aciduria and homocystinuria, hypomethioninemia, and in some cases cystathioninuria. Serum vitamin B_{12} is normal (Table 3). Cells of these patients have been classified by complementation analysis into two distinct groups, called Cbl C and D mutants. Cbl C mutants are more severely affected than Cbl D mutants, and both sexes may be affected. The clinical features reported in patients with Cbl C mutant consist of development and mental delay, more or less severe in different cases and problems with poor feeding. Among cases extensively studied, (Table 3) most have presented with severe megaloblastic anemia; hypersegmentation of neutrophils and thrombopenia have been reported in some of them. Metabolic disorders, especially ketoaci-

Table 3. Main abnormalities identified in A, B, C, D, E, F, Cbl mutants

	A	B	C	D	E	F
Adenosyl-Cbl	↓	↓	↓	↓	N	↓
Methyl-Cbl	N	N	↓	↓	↓	↓
Methylmalonylaciduria	+ + +	+ + +	+ +	+ +	↓	+ +
Homocystinuria	−	−	+ + +	+ + +	+ + +	−
Megaloblastic anemia	−	−	unconstant	−	+	−
Thrombopenia	−	−	unconstant	−	+	−
Hypersegmentation of neutrophils	−	−	unconstant	−	+	−

dosis, may be less severe than in patients with defects of adenosyl-Cbl or methyl-malonyl-CoA mutase deficiency. Four of these patients died at 7 weeks, 3 months, 4 months, 7 years of age, and the children alive are mentally retarded.

Hemolytic anemia with renal insufficiency is a frequent complication, often responsible for death. Autopsy showed severe vascular lesions with renal changes characteristic of thrombotic microangiopathy, supporting a diagnosis of hemolytic uremic syndrome. Clinical response to treatment with hydroxo-Cbl is variable. Massive doses of hydroxo-Cbl may induce a net decrease in methylmalonic aciduria, but only a slight decrease in homocystinuria. The two brothers presenting with Cbl D mutant were much less severely affected. In one of them, the diagnosis was made at age 4 because of mental retardation and recurrent episodes of thrombophlebitis. The other brother was asymptomatic except for the presence of biochemical abnormalities [13].

More recently, a new mutant (Cbl F) has been reported in an infant with vitamin B_{12}-responsive methylmalonic aciduria but no homocystinuria or megaloblastic anemia [36]. This patient presented with convulsions and developmental delay. Cultured fibroblasts showed defective synthesis of both adenosyl-Cbl and methyl-Cbl, caused by an impairement of Cbl efflux from lysosomes.

Defect of Synthesis (or Retention) of Methyl-Cbl. The congenital error of defective synthesis or retention of methyl-Cbl was first reported in an infant who presented early in life with megaloblastic anemia, homocystinuria, developmental delay, and neurological manifestations [39]. The anemia and homocystinuria were vitamin B_{12}-responsive. This disorder (Cbl E disease) was related to an abnormality of methionine biosynthesis; fibroblasts cultured from the patient with Cbl E were auxotrophic for methionine and contained low amounts of methyl-Cbl. Methionine synthase activity in extracts was decreased below normal at suboptimal concentrations of thiol [35]. Fibroblasts from several other patients with methyl-Cbl deficiency had low levels of methionine synthase activity even under reducing conditions, suggesting a heterogeneity in this disorder [37].

Methionine Synthase Deficiency. Methionine synthase deficiency is a rare disorder. The common feature is megaloblastic anemia with or without homocystinuria [1, 17, 23, 44]. This anemia was quite corrected by Cbl and folate in one case [17] but only partially and transiently by folic or folinic acid in other cases [1, 44]. In two of these cases, methionine synthase deficiency was suspected due to abnormal deoxyuridine (dU) suppression test, corrected in vitro only by folinic acid and not by 5-methyltetrahydrofolate or by methyl-Cbl. On one occasion, methionine was tested in vitro and normalized the dU test [23]. The first case, reported by Arakawa et al. [1] in a 6-month-old child, presented with megaloblastic anemia, mental retardation, and ventricular dilatation. The enzyme activity, normal in fibroblasts, was decreased in liver. Another case was identified in a 14-year-old boy who presented with megaloblastic anemia, unresponsive to Cbl and folic acid, and with developmental and mental delay in the first months of life. The anemia was associated with cellular immune deficiencies, possibly related to the enzyme deficiency. Enzyme activity was undetectable in bone marrow cells and very low in lymphocytes and liver [44].

It appears that among numerous congenital abnormalities of Cbl metabolism, some are quite responsive to Cbl and without sequelae if the diagnosis is made early enough, and if the treatment is correct, e.g., congenital malabsorptions or TC II deficiency. Intracellular abnormalities of Cbl metabolism, however, are more complex and very heterogeneous, and many of them are poorly responsive to Cbl.

These different abnormalities have been useful in clarifying many of physiological steps of Cbl metabolism.

References

1. Arakawa T, Narisawa K, Tanno K et al. (1967) Megaloblastic anemia and mental retardation associated with hyperfolic-acidemia: probably due to N5-methyl tetrahydrofolate deficiency. Tohoku J Exp Med 93:1–22
2. Baumgartner ER, Wick H, Maurer K, Egli N, Steinmann B (1979) Congenital defect in intracellular cobalamin metabolism resulting in homocystinuria and methymalonyl aciduria. Helv Paediatr Acta 34:465–482
3. Ben-Bassat I, Feinstein A, Ramot B (1969) Selective vitamin B_{12} malabsorption with proteinuria in Israel. Clinical and genetic aspects. Isr J Med Sci 5:62–68
4. Brandt LJ, Bernstein LH, Wagle A (1977) Production of vitamin B_{12} analogues in patients with small bowel bacterial overgrowth. Ann Intern Med 87:546–551
5. Burman JF, Jenkins WJ, Walker-Smith JA et al. (1985) Absent ileal uptake of IF-bound vitamin B_{12} in vivo in the Imerslund-Grasbeck syndrome (familial vitamin B_{12} malabsorption with proteinuria). Gut 26:311–314
6. Carmel R (1983) Gastric juice in congenital pernicious anemia contains no immune active intrinsic factor molecule: study of three kindreds with variable ages at presentation, including a patient first diagnosed in adulthood. Am J Hum Genet 35:67–77
7. Carmel R, Herbert V (1969) Deficiency of vitamin B_{12} binding alpha globulin in two brothers. Blood 33:1–12
8. Carmel A, Ravindranath Y (1984) Congenital transcobalamin II. deficiency presenting atypically with a low serum cobalamin level: studies demonstrating the coexistence of a circulating transcobalamin I (R binder) complex. Blood 63:598–605
9. Carmel R, Bedros AA, Mace JW, Goodman SI (1980) Congenital methyl malonic aciduria-homocystinuria with megaloblastic anemia: observations on response to hydroxocobalamin and on the effect of homocysteine and methionine on the deoxyuridine suppression test. Blood 55:570–579
10. Chanarin I (1979) The megaloblastic anaemias, 2nd edn. Blackwell Scientific, Oxford, 430
11. Clark R, Tornyos K, Herbert V, Twomey JJ (1967) Studies on two patients with concomitant pernicious anemia and immunoglobulin deficiency. Ann Intern Med 67:403–410
12. Dillon MJ, England JM, Gompertz D et al. (1974) Mental retardation, megaloblastic anemia, methylmalonyl-aciduria and abnormal homocysteine metabolism due to an error in vitamin B_{12} metabolism. Clin Sci Mol Med 47:43–61
13. Goodman SI, Moe PG, Hammond KG, Mudo SH, Uhlendorf BW (1970) Homocystinuria with methylmalonic aciduria: two cases in sibship. Biochem Med 4:500–515
14. Gräsbeck R, Gordin R, Kantero I (1960) Selective vitamin B_{12} malabsorption and proteinuria in young people a syndrome. Acta Med Scand 167:289–296
15. Hakamii N, Nieman PE, Canellos GP, Lazerson J (1971) Neonatal megaloblastic anaemia due to inherited transcobalamin II deficiency in two siblings. N Engl J Med 285:1163–1170
16. Hall CA (1981) Congenital disorders of vitamin B_{12} transport and their contributions to concepts. Yale J Biol Med 54:485–495
17. Hallam L, Clark AL, Van der Weyden MB (1987) Neonatal megaloblastic anemia and homocystinuria with reduced levels of methionine synthetase. Blood 69:1128–1133

18. Haurani FI, Hall CA, Rubin R (1979) Megaloblastic anemia as a result of an abnormal transcobalamin II (Cardeza). J Clin Invest 64:1253–1259

19. Hitzig WH, Dohmann U, Pluss HJ, Vischer D (1974) Hereditary transcobalamin II deficiency: clinical findings in a new family. J Pediatr 85:622–628

20. Imerslund O (1960) Idiopathic chronic megaloblastic anaemia in children. Acta Paediatr 49 (Suppl 119):1–115

21. Katz M, Lee SK, Cooper BA (1972) Vitamin B_{12} malabsorption due to biologically inert intrinsic factor. N Engl J Med 287:425–429

22. Le Prise PY, Zittoun J, Roudier G, Richier JL (1974) Anémie pernicieuse chez un enfant de 17 ans. Déficience congénitale en facteur intrinséque type A de Hoffbrand. Nouv Presse Med 3:2633–2634

23. Leverge G, Zittoun J, Marquet J, Bancillon A, Schaison S (1984) Anémie mégaloblastique par déficit en methionine synthetase. Actual Hematol 18:142–148

24. Levine JS, Allen RH (1985) Intrinsic factor within parietal cells of patients with juvenile pernicious anemia. A retrospective immunohistochemical study. Gastroenterology 88:1132–1136

25. Levy HL, Mudd SH, Schulman JD, Dreyfus MP, Abeles RH (1970) A derangement in B_{12} metabolism associated with homocystinemia, cystathioninemia, hypomethioninemia and methyl malonyl aciduria. Am J Med 48:390–397

26. Mackenzie H, Donaldson RM, Trier JS, Mathan VI (1972) Ileal mucosa in familial selective vitamin B_{12} malabsorption. N Engl J Med 286:1021–1025

27. Marsden KA, Newman NM, Marden DE (1979) Imerslund's syndrom. A case from Australia and review of the literature. Aust Paediatr J 15:49–52

28. Matsui SM, Mahoney MJ, Rosenberg LE (1983) The natural history of the inherited methyl malonic acidemias. N Engl J Med 308:857–861

29. Mohoney MJ, Hart AC, Steen VD, Rosenberg LE (1975) Methylmalonic acidemia: biochemical heterogeneity in defects of 5-deoxyadenosyl-cobalamin synthesis. Proc Natl Acad Sci USA 72:2799–2803

30. Myers PA, Carmel R (1984) Hereditary transcobalamin II deficiency with subnormal serum cobalamin levels. Pediatrics 74:866–871

31. Niebrugge DJ, Benjamin DR, Christie D, Scott CR (1982) Hereditary transcobalamin II deficiency presenting as red cell hypoplasia. J Pediatr 101:732–735

32. Quinto MG, Leikin SL, Hung W (1964) Pernicious anemia in a young girl associated with idiopathic hypoparathyroidism familial Addison's disease and moniliasis. J Pediatr 64:241–247

33. Rana SR, Colman N, Goh KO, Herbert V, Klemperer MR (1983) Transcobalamin II deficiency associated with unusual bone marrow findings and chromosomal abnormalities. Am J Hematol 14:89–96

34. Rosenberg LE (1983) Disorders of propionate and methylmalonate metabolism. In: Stanbury JB, Wyngaarden JB, Frederickson DS (eds) The metabolic basis of Inherited Disease, 5th edn. Mc Graw-Hill. New York, pp 474–497

35. Rosenblatt DS, Cooper BA, Pottier A, Lue-Shing H, Matiazuk N, Graner K (1984) Altered vitamin B_{12} metabolism in fibroblasts from a patient with megaloblastic anemia and homocystinuria due to a new defect in methionine biosynthesis. J Clin Invest 74:2149–2156

36. Rosenblatt DS, Laframboise R, Pichette J, Langevin P, Cooper BA, Costa T (1986) New disorder of vitamin B_{12} metabolism (cobalamin F) presenting as methyl-malonic aciduria. Pediatrics 78:51–54

37. Rosenblatt DS, Thomas IT, Watkins D, Cooper BA, Erbe RW (1987) Vitamin B_{12} responsive homocystinuria and megaloblastic anemia: heterogeneity in methylcobalamin deficiency. Am. J. Med. Genet 26:377–383

38. Rothenberg SP, Quadros EV, Straus EW, Kaperner S (1984) An abnormal intrinsic factor (IF) molecule: a new cause of «pernicious anemia» (PA). 26th annual meeting of the ASH, 74:419

39. Schuh S, Rosenblatt D, Cooper BA et al. (1984) Homocystinuria and megaloblastic anemia responsive to vitamin B_{12} therapy. N Engl J Med 310:686–690

40. Seger R, Galle J, Wildfouer A, Frater-Schroeder M, Linnell J, Hitzig W (1980) Impaired functions of lymphocytes and granulocytes in transcobalamin II deficiency and their re-

sponse to treatment in primary immunodeficiencies. In: Seligman M, Hitzig WH (eds) Inserm symposium no 16. Elsevier North Holland: Biomedical Press. New York, pp 353–362

41. Seligman PA, Steiner LL, Allen RH (1980) Studies of a patient with megaloblastic anaemia and an abnormal transcobalamin II. N Engl J Med 303:1209–1212
42. Spector JI (1974) Juvenile achlorydric pernicious anemia with IgA deficiency. A family study. Jama 228:334
43. Yang Y, Ducos R, Rosenberg AJ et al. (1985) Cobalamin malabsorption in 3 siblings due to an abnormal intrinsic factor that is markedly susceptible to acid and proteolysis. J Clin Invest 76 (6):2057–2065
44. Zittoun J, Fischer A, Marquet J, Perignon JL, Lagrue A, Griscelli C (1987) Megaloblastic anemia and immune abnormalities in a patient with methionine synthase deficiency. Acta Paediatr. Scand 76:991–998
45. Zittoun J, Leger J, Marquet J, Carmel R (1988) Combined congenital deficiencies of intrinsic factor and R binder. Blood 72:940–943

Chapter 19

Transport of Folate Compounds by Hematopoietic Cells

G. B. HENDERSON

Introduction

Hematopoietic cells are unable to synthesize folate de novo and thus have developed efficient transport systems that can retrieve this vitamin from the extracellular environment. The principal source of folate for hematopoietic cells is 5-methyl-H_4folate, the major circulating form of folate in mammals. Utilization of 5-methyl-H_4folate depends on efficient uptake by membrane-associated transport systems and also on its intracellular conversion to tetrahydrofolate via methionine synthase. Internalized 5-methyl-H_4folate is then trapped within the cell by polyglutamylation. Transport system for folate compounds also serve as the primary route by which methotrexate and other antifolate drugs enter hematopoietic cells [12, 41, 50]. In addition, the extent of drug uptake has been correlated in several cases to the overall sensivity of tumor cells to antifolates [44, 53], and resistance to these compounds can occur via alteration in membrane transport [41]. The toxicity of antifolate drugs is also affected by the presence of folate transport systems in normal proliferating cells (e.g., in bone marrow, liver, and the intestinal mucosa) that mediate the efficient uptake of antifolate drugs.

General Properties of Folate Transport Systems in Tumor Cells of Hematopoietic Origin

Various studies have shown that murine and human leukemia cells have the ability to transport a variety of folate compounds, including methotrexate, 5-methyl-H_4folate, 5-formyl-H_4folate, and folate, and that uptake is carrier mediated [12, 15, 36, 41, 43, 46, 52]. K_t values for half-maximal influx (in buffered-saline solutions) range from 1 to 10 μM for reduced folates and methotrexate but are much higher for folate (Table 1). V_{max} values are less variable and are in the general range of 1–4 nmol min^{-1} g^{-1} dry weight (Table 1). The transport system of L1210 mouse lymphoblasts has a broad specificity in the binding of various folate analogs [56]. The more effective compounds include 3', 5'-dichloromethotrexate, aminopterin, 10-ethyl-aminopterin, 10-deaza-aminopterin, 10-oxa-aminopterin, and 5-chloro-5, 10-dideaza-aminopterin. Conversely, a relatively poor affinity is observed for folate derivatives with substitutions at either the α- or γ-carboxyl position [55, 56]. The transport system of L1210 cells also shows unu-

Table 1. Kinetic parameters for the transport of folate compounds by various mammalian cells

Cell line	Origin	Substrate	K_t (μM)	V_{max} (nmol min^{-1} g^{-1} dry weight)	Reference
L1210	Mouse	Methotrexate	3.1	3.4	[52]
		5-Methyl-H$_4$folate	1.0	3.0	[12]
		5-Formyl-H$_4$folate	2.5	–	[46]
		Folate	150	1.7	[36]
Ehrlich	Mouse	Methotrexate	5.8	2.4	[52]
P388	Mouse	Methotrexate	5.2	2.6	[52]
P288	Mouse	Methotrexate	3.9	3.3	[52]
Sarcoma 180	Mouse	Methotrexate	11.2	3.5	[52]
L-cells	Mouse	Methotrexate	0.7	–	[17]
CCRF-CEM	Human	Methotrexate	4.5	1.0	[37]
WI-L2	Human	Methotrexate	6.7	1.5[a]	[47]
MCF-7	Human	Methotrexate	8.2	2.1[a]	[49]

Measurements were performed at pH 7.4 in various buffered saline solutions.
[a] Values in nmol min^{-1} g^{-1} dry weight were estimated from published values reported in other units.

sual stereochemical specificity. Both the natural (6S) and unnatural (6R) diastereoisomers of 5-methyl-H$_4$folate are transported with equal facility [5, 62], while the system accommodates only the 6S form of 5-formyl-H$_4$folate [5, 54]. Other compounds, including glucose, xanthine derivatives, and prostaglandins, are noncompetitive inhibitors of methotrexate transport [30]. Since these compounds also increase intracellular levels of cAMP, it has been suggested that the transport system in L1210 cells may be regulated by cyclic nucleotides [30].

A single high-affinity transport system appears to be the only route by which folate compounds enter L1210 cells [26, 36]. This conclusion is supported by observations that methotrexate influx reaches a maximum value when measured at substrate concentrations up to 50 μM, and that high concentrations of competitive inhibitors reduce [³H]methotrexate influx in a monophasic, hyperbolic fashion and to an extrapolated maximum of 100%. Complete inhibition of influx is also achieved with irreversible inhibitors of this system. Previous results suggesting the presence of a minor low-affinity route [14, 15] may have been influenced by trace amounts of labeled impurities that occur in commercial preparations of [³H]methotrexate, and that also arise during short-term storage of the labeled substrate. Recent studies in CCRF-CEM cells also provide evidence for a single influx route for methotrexate [37].

Trace impurities have a pronounced effect on folate transport measurements and have led to incorrect conclusions regarding the number and identity of transport routes for folate. The effects of impurities are accentuated with folate since its relatively high K_t value for transport has required the use of much higher concentrations of the ³H-labeled substrate. [³H]Folate is also less stable than [³H]methotrexate, and interfering impurities can arise rapidly during storage at $-20°C$ [36]. The principal breakdown product which interferes with folate transport appears to be 6-hydroxymethylpterin [36]. The latter compound can enter L1210 cells via a separate transport system with a lower K_t and a higher V_{max} than folate [60] and thus can obscure the true uptake properties of folate. Interference occurs at impurity levels as low as 1%.

Highly purified [^{3}H]folate enters L1210 cells via the same system that mediates the uptake of reduced folates and methotrexate [36]. This conclusion is supported by the observations that folate influx is competively inhibited by methotrexate, and this inhibition is complete with excess unlabeled methotrexate. Folate also inhibits the influx of methotrexate, and the K_i is comparable to the K_t for folate influx, and folate and methotrexate influx both exhibit the same high sensitivity to several irreversible inhibitors and to buffer composition [36]. These results contradict previous suggestions that folate enters mouse leukemia cells by other unidentified routes [13, 14, 16, 38, 42, 48] or by a system shared with adenine [59].

Methotrexate transport in L1210 cells can be irreversibly inhibited by a variety of compounds, including sulfhydryl reagents [15, 21, 48]. The most effective inhibitor of this group is p-chloromercuriphenylsulfonate, which inactivates transport by 50% at the relatively low concentration of 3 μM [21]. Methotrexate fails to protect the transport system against this inactivation, indicating that the critical sulfhydryl residue which reacts with p-chloromercuriphenylsulfonate is not located at the methotrexate-binding site. Irreversible inhibition of methotrexate transport can also be achieved by irradiating L1210 cells in the presence of 8-azido-AMP [31]. Inactivation in this case is half-maximal at a reagent concentration of 65 μM, and protection against inactivation is afforded by the addition of unlabeled methotrexate during the irradiation step. Complete inhibition is achieved only when several cycles of azido-AMP treatment are performed. Inactivation of transport has also been demonstrated with DIDS (4,4'-diisothiocyanostilbene-2,2'-disulfonate) [22], a potent covalent inhibitor of the bicarbonate/chloride exchange carrier of these cells. Inhibition with this compound occurs at a relatively slow rate and is half-maximal at a reagent concentration of 10 μM.

Carboxyl-activated derivatives of methotrexate have been the most effective irreversible inhibitors of methotrexate transport in L1210 cells. These compounds include 1-ethyl-3 (3-dimethylaminopropyl)-carbodiimide (EDC) activated methotrexate, which can be synthesized by direct reaction of EDC with methotrexate [32]. Its structure has not been established but appears to be the α, γ-anhydride of methotrexate. Inhibition of methotrexate transport by this agent occurs rapidly, is temperature-dependent, and is half-maximal at a concentration of 1.0 μM. Specific reaction at the substrate binding site was indicated by various findings, including protection against inactivation by substrates, lack of inhibition of other transport systems, and a direct correlation between the ability of various EDC-activated folate compounds to inhibit methotrexate influx irreversibly and the affinity of the system for the corresponding nonactivated compounds. At 4°C, the inhibitor forms a reversible complex with the binding site prior to inactivation, and the K_i (0.8 μM) for binding is comparable to the concentration of reagent required for 50% inactivation (1.0 μM at 37°C).

The most potent irreversible inhibitor of methotrexate transport is an *N*-

hydroxysuccinimide ester of methotrexate (NHS-methotrexate) [35]. In cells exposed to NHS-methotrexate at 37°C, inactivation occurs optimally at pH 6.8 and is half-maximal at a reagent concentration (20 nM) which was nearly stoichiometric with the level of the binding protein. At higher levels of the reagent inhibition is complete. The loss in transport activity also occurs at an unusually rapid rate. Over a range of reagent concentrations from 5 to 200 nM, inhibition reaches a maximum after pretreatment for 1 min at 37°C. NHS-Methotrexate appears to react with the substrate-binding site on the transport protein since protection against inactivation is afforded by the presence of unlabeled methotrexate. Irreversible inhibition is also observed in cells exposed to NHS-methotrexate at 4°C. Inactivation in this case is qualitatively similar to the corresponding process at 37°C, except that the extent of inhibition reaches a maximum of only 75%. The proposed explanation for these results is that the transport protein exists in two forms at 4°C, one (75% of the total) containing binding sites which are accessible to NHS-methotrexate and the other (25% of the total) with inaccessible sites. Since this transport system is unable to mediate substrate translocation at 4°C [33], it was suggested that these sites represent transport proteins which have outward and inward orientations, respectively. NHS-Methotrexate is also an irreversible inhibitor of methotrexate influx in CCRF-CEM cells [37]. Inhibition in this case (at 37°C) is half-maximal at a concentration of 80 nM and is complete at higher levels of the inhibitor.

Effect of Anions on Methotrexate Influx and Efflux

An inhibitory effect of anions on methotrexate transport was first observed by Kessel and Hall [43], who reported that the uptake of methotrexate by L1210 cells is lower in a phosphate-containing buffer than in a buffer-containing bicarbonate. Goldman [12] subsequently extended the list of inhibitory anions to in-

Table 2. Effect of buffer composition on methotrexate transport parameters in L1210 cells

Buffer	K_t influx (μM)	Influx V_{max} (pmol min^{-1} mg^{-1} protein)	$t_{1/2}$ Efflux (min)	Steady state (pmol mg^{-1} protein)	Concentration ratio ($[C_{in}]/[C_{out}]$)
PBS	10.0	15.6	18	20	1.6
HBS	4.3	17.7	24	57	7.6
Na-HEPES	1.6	17.6	38	140	23
K-HEPES	1.4	17.1	40	177	30
Mg-HEPES-sucrose	0.7	19.4	100	432	107

Buffer compositions: PBS (phosphate-buffered saline), 138 mM NaCl, 2.7 mM KCl, 8.1 mM Na$_2$HPO$_4$, 1.5 mM KH$_2$PO$_4$, 1.0 mM CaCl$_2$, and 0.5 mM MgCl$_2$, pH 7.4; HBS (HEPES-buffered saline), 140 mM NaCl, 10 mM KCl, 20 mM HEPES, and 1 mM MgCl$_2$, pH 7.4 with NaOH; Na-HEPES, 160 mM HEPES and 2 mM MgCl$_2$, pH 7.4 with NaOH; K-HEPES, 160 mM HEPES and 2 mM MgCl$_2$, pH 7.4 with KOH; and Mg-HEPES-sucrose, 20 mM HEPES and 225 mM sucrose, pH 7.4 with MgO.

clude nitrate, sulfate, adenine nucleotides, nicotinamide adenine dinucleotide (NAD), and various sugar phosphates, while other studies on the effect of buffer composition [18–20, 25] showed that the anionic components of standard saline buffers were also inhibitory. The latter effect is illustrated by the improvement in K_t value for methotrexate that occurs as phosphate and chloride ions are removed from saline buffers and replaced isotonically with a combination of sucrose and 4-2(2-hydroxyethyl)-1-piperazine-ethanesulfonate (HEPES) (Table 2). The V_{max} for methotrexate remains constant under these conditions, indicating that phosphate and chloride are competitive inhibitors of methotrexate influx, and that, in addition, the non-physiological buffer conditions employed during these measurements do not adversely affect either the transport system or cell integrity [25]. K_t values for methotrexate range from 10 μM is a phosphate-buffered saline to a minimum of 0.7 μM in an anion-deficient buffer system (Mg-HEPES-sucrose) (Table 2).

A wide variety of anions competitively inhibit methotrexate influx in L1210 cells [22, 24, 31, 33]. K_i values for some of these anions are listed in Table 3. The highest affinity is observed with folate compounds, although other large organic anions of diverse structure are also effective in the same concentration range. Moderate inhibition occurs with small divalent anions, such as phosphate, while small monovalent anions (e.g., chloride) are the least effective. The transport system exhibits a strong preference for divalent, rather than monovalent anions, as illustrated by a 50-fold higher affinity for 1,2-dicarboxybenzene (o-phthalate) than monocarboxybenzene (benzoate).

Buffer composition has even greater effects on the ability of L1210 cells to accumulate methotrexate [18, 19, 25]. Concentration gradients at the steady state are in the range of 1.6-fold when measurements are performed in phosphate-buffered saline, but this increases progessively as various anions are removed from the medium and replaced with sucrose and/or HEPES (Table 2). The highest uptake is observed with the Mg-HEPES-sucrose buffer in which steady-state concentration gradients of methotrexate exceed 100-fold.

The anionic composition of the external medium also has pronounced effects on methotrexate efflux [20, 24–26]. The removal of various anions and their replacement isotonically with sucrose and HEPES leads to a dramatic reduction in efflux. The extent of this reduction is substantial and exceeds 5-fold when cells are transferred from phosphate-buffered salaline to Mg-HEPES-sucrose (Table 2). Restoration of efflux to levels observed in saline buffers occurs upon the re-addition of selected anions to the external medium [24]. Effective compounds include folate compounds, various small monovalent anions, small divalent anions, AMP, and thiamine pyrophosphate (see Table 3). The extent of stimulation is 3- to 5-fold, and this enhancement is directly related to an interaction with the methotrexate influx carrier by a close correlation between K_i values for inhibition of influx and anion concentrations required for 50% stimulation of efflux [20]. Competitive inhibitors of methotrexate influx that do not enhance methotrexate efflux include trivalent anions, NAD, most nucleotides, a fluorescein derivative of methotrexate [34], and other large organic anions. Treatment of cells with NHS-methotrexate has no effect on methotrexate efflux in anion-deficient buffers, but it eliminates the ability of folate compounds and other anions to

Table 3. Effect of various anions on the influx and efflux of methotrexate in L1210 cells

Anion	K_i for influx (μM)	Stimulation of efflux
Methotrexate	0.7	+ + +
5-Methyl-H$_4$folate	0.3	+ + +
5-Formyl-H$_4$folate	0.5	+ + +
Folate	14	+ +
p-Aminobenzoylglutamate	50	+ +
F-MTX	0.3	0
Chloride	30 000	+ + +
Bicarbonate	10 000	+ +
L-Lactate	7 800	+ + +
Acetate	8 700	+ + +
Succinate	510	+ +
Tricarboxypropane	190	0
Citrate	410	0
Phosphate	400	+ + +
Sulfate	540	+ +
Thiamine pyrophosphate	3	+ + +
NAD	530	+
NADP	500	0
AMP	50	+ + +
ADP	55	+ +
ATP	70	0
IMP	100	+ +
GMP	140	+
UMP	38	+
CMP	430	0
Monocarboxybenzene	1 100	+ + +
1,2-Dicarboxybenzene	20	+ + +
1,2,4-Tricarboxybenzene	50	0
1,2,4,5-Tetracarboxybenzene	460	0
BSP	2	0
ANS	7	+
NAP-Taurine	20	0
Probenecid	50	0

K_i values were determined in Mg-HEPES-sucrose buffer [24] from a Dixon plot in the inverse of methotrexate influx versus anion concentration. (Note: corresponding K_i values are approximately six fold higher when measured in buffered saline solutions.) Efflux determinations were performed [24] in the absence and presence of the indicated anions at concentrations (where feasible) which were ten fold higher than their corresponding K_i values for inhibition of influx. Ability to promote methotrexate efflux: + + +, strong; + +, moderate; +, weak; 0, none. Efflux buffer, Mg-HEPES-sucrose.

F-MTX, fluorescein-aminopentyl-methotrexate [34]; BSP, bromosulfophthalein; ANS, 8-anilino-1-naphthalenesulfonate; NAP-taurine, N-4-azido-2-nitrophenyl-2-aminoethylsulfonate; NAD, nicotinamide adenine dinucleotide; NADP, nicotinamide adenine dinucleotide phosphate

stimulate efflux [24, 26], indicating that the methotrexate influx carrier mediates the anion-sensitive portion of methotrexate efflux.

The effect of anions on methotrexate influx and efflux have also been examined in membrane vesicles derived from L1210 cells [63]. Influx and efflux are both stimulated when high concentrations of sulfate, phosphate, or folate compounds are present in the opposite compartment, while stimulation is not observed when the same equivalence of two different anions are present on opposing sides of the membrane. The extent of stimulation by various anions was 2- to 4-fold.

Energetics of Methotrexate Transport

Early studies on energy requirements had established that L1210 cells can accululate methotrexate to intracellular levels which are greater than 10-fold higher than predicted by a passive uptake process [8, 11, 12, 15]. Uptake does not require a Na^+ and is insensitive to ouabain, indicating that the energy source is not a Na^+ gradient. It had also been established that methotrexate uptake at the steady state is enhanced by various metabolic inhibitors and, likewise, is diminished by glucose and pyruvate [11, 12]. The basis for these latter effects was traced to the inhibition and stimulation of efflux by these agents, respectively. It was concluded that the influx of methotrexate is dependent upon an energy source which is not immediately affected by the energy levels of the cells, while efflux is under the direct influence of ATP or some other high-energy phosphate compound. Steady-state levels of methotrexate would then be determined by the net contribution of these separate influx and efflux components.

Clarification of these varied observations on the energetics of transport has been obtained from the recent demonstration that methotrexate efflux occurs via at least three distinct routes [26]. The methotrexate influx carrier, whose contribution can be quantitated by its sensitivity to NHS-methotrexate, was found to represent the major route for methotrexate efflux when the cells are suspended in saline buffers without glucose. Under these conditions, the contribution by this system is approximately 71% of total efflux (Table 4). A second route, which accounted for 6% of the total, was identified by its sensitivity to bromosulfophthalein, while the remaining efflux component accounted for 23% of the total and could be inhibited by probenecid [28]. Initial efflux and contributions to total efflux by each of these routes are summarized in Table 4.

Glucose enhances methotrexate efflux in L1210 cells by approximately three fold, and it also dramatically changes the proportion of efflux that proceeds via each route (Table 4) [26]. The predominant route is now the bromosulfophthalein-sensitive component which, by increasing 25-fold in rate, accounted for nealy half (46%) of total efflux. In the reciprocal experiment, oligomycin was shown to reduce total efflux, and since this occured mostly at the expense of the bromosulfophtalein- and probenecid-sensitive components, the proportion of efflux proceeding via the methotrexate influx carrier increased to greater than 80% (Table 4).

The individual routes which mediate the efflux of methotrexate in CCRF-

Table 4. Effect of buffer composition on the relative activity of individual efflux routes for methotrexate in L1210 and CCRF-CEM cells

Cells	Buffer/additions	Efflux route	Methotrexate efflux (pmol min^{-1} mg^{-1} protein)	Relative contribution (%)
L1210	HBS	A	2.5	71
		B	0.2	6
		C	0.8	23
L1210	HBS + glucose	A	3.1	29
		B	4.9	46
		C	2.7	25
L1210	HBS + oligomycin	A	2.3	83
		B	0.1	4
		C	0.4	13
CCRF-CEM	HBS	A	0.8	50
		B	0.19	12
		C	0.60	38
CCRF-CEM	HBS + glucose	A	1.18	46
		B	0.16	6
		C	1.22	48
CCRF-CEM	HBS + oligomycin	A	0.73	87
		B	0.04	5
		C	0.07	8

Separation of the individual efflux routes was performed as described in [26]. HBS, HEPES-buffered saline. Concentration of additions: glucose, 5 mM; oligomycin, 1.0 μM. Load of [³H]methotrexate for L1210 cells, 124 pmol/mg protein; load of [³H]methotrexate for CCRF-CEM cells, 32 pmol/mg protein. Efflux routes: A, methotrexate influx carrier; B, bromosulfophthalein-sensitive route; C, probenecid-sensitive route.

CEM cells have also been analyzed [37]. Separate efflux routes that are sensitive to NHS-methotrexate, bromosulfophthalein, and probenecid were identified in these cells, although their contribution to total methotrexate efflux differed from L1210 cells (Table 4). The primary difference in CCRF-CEM cells was the substantially lower activity of the component sensitive to bromosulfophthalein. This route contributed a relatively small portion of total efflux under various conditions and was not substantially enhanced by glucose (Table 4) or by glucose plus bicarbonate. The major portion of methotrexate efflux proceeded via the methotrexate influx carrier.

The identification of at least three routes which contribute to methotrexate efflux and the demonstration that the relative contribution of these routes to total efflux is dramatically influenced by the energetic state of the cells helps to explain various observations regarding the transport process. Two of these efflux routes do not contribute to methotrexate influx and are enhanced by glucose and thus appear to represent the ATP-dependent efflux pump proposed by Goldman [11, 12]. The latter routes also appear to represent the alternate efflux system for methotrexate that was observed kinetically by Sirotnak and his coworkers [8, 50, 51, 57]. Other studies [20, 21] in which it was suggested that methotrexate influx and efflux proceed via the same route could likewise be explained, since the alternate efflux routes for methotrexate had been suppressed by the absence of extracellular exchange anions and/or glucose.

Mechanism of Transport

Various findings support the hypothesis that the 5-methyl-H_4folate/methotrexate transport system of L1210 cells mediates substrate translocation via an anion-exchange mechanism [19, 20, 22-26, 29, 63]. The proposed model is that the uptake of folate compounds is coupled to the exit of intracellular anions of equal negative charge, and that gradients of these exchange anions serve as the energy source for the active accumulation of methotrexate. This mechanism would allow methotrexate transport across the membrane to be electroneutral and not directly under the influence of the membrane potential. In L1210 cells, the latter may be as high as -70 mV [25] and thus could act as a substantial outward driving force for methotrexate and similar divalent anions. It was proposed further that the carrier protein contains only a single binding site which can accommodate both folate compounds and exchange anions, and that the anion exchange mediated by this protein is obligatory, i.e., unloaded binding sites cannot traverse the cell membrane.

A crucial observation supporting the anion-exchange model is that the portion of methotrexate efflux which occurs via the carrier is blocked by the removal of extracellular anions, and that the addition of certain anions back to the medium restores efflux to its original level [20, 24, 26, 37]. The promotion of methotrexate efflux is associated with a variety of folate and nonfolate anions (see Table 3), and these anions enhance efflux via an interaction with the binding site on the methotrexate carrier since anion concentrations required to stimulate efflux by 50% correspond directly to their K_i values for inhibition of influx [20, 37]. Several proposed exchange anions, including phosphate [23], sulfate [22] and phtalate [29], are also transported directly into L1210 cells via the methotrexate carrier system, and phosphate and sulfate mediate the trans-stimulation of methotrexate influx and efflux in membrane vesicles [63]. The transport system exhibits specificity since not all anions can serve as exchange substrates [20, 24].

Intracellular anions that exchange for extracellular folate compounds have not been identified, although the most likely candidates are inorganic phosphate and AMP [23, 24, 37]. The latter anions are equivalent to methotrexate as transport substrates, occur within cells at relatively high concentrations, are retained against a concentration gradient, and have a moderate affinity for the carrier protein. Efficient means are also available for the recovery of these anions from the medium. The reuptake of phosphate could occur directly via the Na^+-dependent phsophate transport system of these cells [23], while extruded AMP could be recovered following extracellular hydrolysis by 5'-nucleotidase and subsequent uptake via separate transport systems for adenosine and phosphate.

Other anions which may participate to a lesser degree in the exchange cycle include chloride, bicarbonate, and lactate [24].

The unusually high uptake of methotrexate that is observed in cells suspended in anion-deficient buffers (see Table 2) can also be explained by the anion-exchange model. In the absence of extracellular anions, large concentration gradients of various intracellular anions would be generated and these could act as the energy source for accumulating methotrexate. The driving force generated under these conditions is substantial since cells in the Mg-HEPES-sucrose buffer can concentrate methotrexate in excess of 100-fold [25]. Cells in anion-deficient buffers also have relatively low levels of ATP [25]. This further increases intracellular pools of potential exchange anions (e.g., phosphate and AMP), and it also suppresses the activity of the energy-dependent efflux routes for methotrexate, which allows uptake to proceed almost entirely by the methotrexate exchange system.

Properties of the Binding Protein

Low-temperature binding studies have identified and quantitated a binding protein for folate compounds that resides on the external surface of L1210 cells [33]. Binding to this protein (in anion-deficient buffers) occurs rapidly and exhibits saturation kinetics for 5-methyl-H_4folate ($K_D = 0.11$ μM) and methotrexate ($K_D = 0.35$ μM). A role for this protein in the transport process has been indicated by similarities in the magnitude of K_D values (at 4°C) for the binding of 5-methyl-H_4folate and methotrexate and corresponding K_t values for the transport of these compounds into the cell (at 37°C), and, moreover, by the ability of various folate compounds and other anions to inhibit 5-methyl-H_4folate binding and transport in a parallel fashion [33]. Protein-bound 5-[^{14}C]methyl-H_4folate is displaced rapidly (at 4°C) by the addition of excess unlabeled methotrexate, indicating that the substrate/binding-site complex remains at the external membrane surface and is not internalized at low temperature. The amount of binding protein (1 pmol/mg protein) corresponds to approximately 80000 copies per cell. The corresponding binding protein in CCRF-CEM human lymphoblasts has a K_D for 5-methyl-H_4folate of 0.10 μM and is present at the slightly lower level of 0.3 pmol/mg protein [37]. Chinese hamster ovary (CHO) cells also contain a similar binding protein for folate compounds [9]. The K_D of this protein for methotrexate is 2 μM, and the amount of binder is 1.3 pmol/mg protein. Mutant CHO cells with the inability to transport methotrexate are unable to bind the drug, indicating that this binding protein is a component of the transport system.

The folate transport protein of L1210 cells has been covalently labeled with NHS-[^{3}H]methotrexate, and the incorporated radioactivity has been utilized as a guide for extraction and partial purification [27]. Radioactivity was incorporated into a single membrane component which was present in the plasma membrane fraction of the cell, could be extracted by Triton X-100 and various other detergents, appeared as a single symmetrical peak on Sephacryl S-300, and has a molecular weight of 36000 as determined by SDS-gel electrophoresis and autoradiography. The protein fraction obtained after Sephacryl chromatography had been enriched by approximately 40-fold relative to whole cells but was less than 5% pure. Labeling of this membrane component was not observed in cells exposed to NHS-[^{3}H]methotrexate in the presence of an excess of unlabeled methotrexate or thiamine pyrophosphate or in a mutant cell line that is defective in methotrexate transport.

The likelihood of obtaining homogenous folate transport protein has been improved by the isolation of mutant sublines of L1210 cells that over-produce the binding protein [58]. Up-regulated mutants were selected by repeated passage of the cells in medium containing limiting concentrations of folinate (5 nM) or limiting folinate plus metoprine, a nonclassical inhibitor of dihydrofolate reductase. The amount of binding protein and the rate of methotrexate influx increased coordinately in these cells and was elevated by as much as 14-fold.

Transport of Folate Compounds in Normal Hematopoietic Cells

Human lymphocytes in vitro do not express detectable levels of a transport system for methotrexate [45]. Even at low concentrations of the drug (0.2 μM), uptake is temperature-insensitive and nonsaturable, and the drug does not accumulate within the cells. An increase in the transport capacity could be achieved, however, by exposure of lymphocytes to phytohemagglutinin (PHA) and subsequent culturing in vitro for 2–3 days [39]. Methotrexate transport in PHA-stimulated lymphocytes is inhibited (ca. 40%) by actinomycin D and also by modulators of intracellular cAMP levels, suggesting that cyclic nucleotides may regulate the transport process. The transport of folate and 5-methyl-H$_4$folate is also enhanced by PHA in cultured human lymphocytes, and saturation kinetics and inhibition by methotrexate were noted for both processes [7]. 5-Methyl-H$_4$folate uptake is not inhibited by folate, suggesting that these folate compounds enter the cells by alternative pathways.

Human granulocytes have a higher capacity for transporting methotrexate than human lymphocytes, and uptake appears to be carrier-mediated since the process is temperature-dependent and saturable [45]. An energy-coupling mechanism may not be operative, however, since steady-state concentration ratios for methotrexate in granulocytes do not exceed 0.2, even at low levels (0.2 μM) of the substrate.

Human erythrocytes transport 5-methyl-H$_4$folate by a carrier-mediated process which is similar to that in other normal and neoplastic cells of hematopoietic origin [2–4]. Influx is saturable ($K_t = 0.6$ μM), temperature-dependent, enhanced by preincubation of the cells with 5-formyl-H$_4$folate, and inhibited by unlabeled 5-methyl-H$_4$folate, 5-formyl-H$_4$folate, and methotrexate, but not by folate. Total uptake at the steady state is also increased by metabolic inhibitors, and extracellular folate compounds enhance the efflux of intracellular 5-methyl-H$_4$folate. However, metabolic inhibitors did not reduce substrate efflux, suggesting that an energy-dependent efflux pump for methotrexate is not present in erythrocytes. Transport of 5-methyl-H$_4$folate by erythrocytes is inhibited by various compounds which also interact with the general anion exchange carrier of these cells

[2], suggesting that mechanistic or structural similarities exist between these two transport systems. pH studies suggested further that substrate permeation may occur via a process which does not translocate electric change, possibly by a mechanism involving co-permeation with protons or anion exchange [4].

In human bone marrow cells, methotrexate, folate, and 5-methyl-H$_4$folate are each transported via a process which is temperature-dependent and saturable, indicating that membrane carrier mechanisms are involved [6, 39]. Methotrexate inhibits the uptake of folate, although it was unclear whether this involved a block in folate metabolism (by inhibiting dihydrofolate reductase) or resulted from competition at a common receptor site on a carrier protein. Puromycin, dibutyl-cAMP, and theophylline each inhibit uptake of methotrexate and 5-methyl-H$_4$folate, but not folate [39], indicating that methotrexate and reduced folates may enter bone marrow cells by a route which is separate from that of folate. Bone marrow cells are essentially impermeable to the triglutamate of folate, as judged by its inability to act as a coenzyme in intracellular DNA synthesis [40].

Human platelets also transport folate [10], although the kinetic parameters which characterize this process have not been evaluated. It was shown that folate readily enters these cells and is accumulated to intracellular concentrations which can exceed extracellular substrate levels by 20-fold. However, since metabolic inhibitors have no effect on total uptake, it was suggested that apparent concentration gradients of folate may have resulted from passive uptake followed by the binding of folate to intracellular proteins. Methotrexate and other inhibitors of dihydrofolate reductase decrease the ability of the cells to accumulate folate.

Rabbit reticulocytes isolated from phenylhydrazine-treated rabbits transport methotrexate by a carrier mechanism which is quantitatively similar to that in human and murine leukemia cells [1]. Both processes are Na$^+$-independent, sensitive to the anionic composition of the medium, strongly inhibited by reduced folate compounds, but not by folic acid, stimulated by preloading cells with substrate analogs, and enhanced by metabolic inhibitors. Michaelis-Menten analysis revealed further that the K_t for methotrexate influx in reticulocytes (1.4 μM) is similar to values obtained in various mouse and human tumor cells (see Table 1), although the V_{max} is about 50-fold lower. The above results contrast with rabbit erythrocytes which are essentially devoid of methotrexate transport activity [1].

References

1. Bobzien WF, Goldman ID (1972) The mechanism of folate transport in rabbit reticulocytes. J Clin Invest 51:1688–1696
2. Branda RF, Anthony BK (1979) Evidence for transfer of folate compounds by a specialized erythrocyte membrane system. J Lab Clin Med 94:354–360
3. Branda RF, Anthony BK, Jacob HS (1978) The mechanism of 5-methyltetrahydrofolate transport in human erythrocytes. J Clin Invest 61:1270–1275
4. Branda RF, Nelson NL (1982) Effect of pH on 5-methyltetrahydrofolic acid transport in human erythrocytes. Biochem Pharmacol 31:2300–2302
5. Chello PL, Sirotnak E, Wong E, Kisliuk RL, Gaumont Y, Combepine G (1971) Further

studies of stereospecificity at carbon 6 for membrane transport of tetrahydrofolates. Br J Haematol 20:503-509

6. Corcino JJ, Waxman S, Herbert V (1971) Br J Haematol 20:503-509

7. Das KC, Hoffbrand AV (1971) Studies of folate uptake by phytohemagglutininstimulated lymphocytes. Br J Haematol 19:203-221

8. Dembo M, Sirotnak FM (1976) Antifolate transport in L1210 cells: kinetic evidence for the non-identity of carriers for influx and efflux. Biochim Biophys Acta 448:505-516

9. Flintoff WF, Nagainis CR (1983) Transport of methotrexate in Chinese hamster kidney cells: a mutant defective in methotrexate uptake and cell binding. Arch Biochem Biophys 233:433-440

10. Gaut ZN, Solomin HM, Baugh CM (1970) Accumulation of [3',5'-3H] folic acid by human blood platelets: effect of dihydrofolic acid reductase inhibitors. Biochim Biophys Acta 215:194-197

11. Goldman ID (1969) Transport energetics of the folic acid analogue, methotrexate, in L1210 leukemia cells. J Biol Chem 244:3779-3785

12. Goldman ID (1971) The characteristics of the membrane transport of amethopterin and the naturally occurring folates. Ann NY Acad Sci 186:400-422

13. Goldman ID (1975) Membrane transport of Methotrexate (NSC-740) and other folate compounds: Relevance to rescue protocols. Cancer Chemother Rep 6:63-72

14. Goldman ID (1977) Membrane transport of antifolates as a critical determinant of drug toxicity. Adv Exp Med Biol 84:85-111

15. Goldman ID, Lichtenstein NS, Oliverio VT (1968) Carrier-mediated transport of the folic acid analogue, methotrexate, in the L1210 leukemia cell. J Biol Chem 243:5007-5017

16. Goldman ID, Snow R, White JC (1985) The effect of Methotrexate on the cellular accumulation of 3H after incubation of Ehrlich ascites tumor with 3H-folic acid. Fed Proc 34:807

17. Grzelakowska-Sztabert B (1976) Uptake of folate and its analogue amethopterin by mouse L cells. Acta Biochim Polonica 23:309-320

18. Henderson GB, Grzelakowska-Sztabert B, Zevely EM (1980) Binding Properties of the 5-Methyltetrahydrofolate/Methotrexate Transport System in L1210 Cells. Arch Biochem Biophys 202:144-149

19. Henderson GB, Montague-Wilkie B (1983) Irreversible inhibitors of methotrexate transport in L1210 cells: Characteristics of inhibition by an N-hydroxysuccinimide ester of methotrexate. Biochem Biophys Acta 735:123-130

20. Henderson GB, Russel A, Whitely JM (1980) A fluorescent derivative of methotrexate as an intracellular marker for dihydrofolate reductase in L1210 cells. Arch Biochem Biophys 202:29-34

21. Henderson GB, Suresh MR, Vitols KS, Huennekens FM (1986) Transport of folate compounds in L1210 cells: Kinetic evidence that folate influx proceeds via the high-affinity transport system for 5-methyltetrahydrofolate and methotrexate. Cancer Res 46:1639-1643

22. Henderson GB, Tsuji JM, Kumar HP (1986) Characterization of the individual transport routes that mediate the influx and efflux of methotrexate in CCRF-CEM human lymphoblastic cells. Cancer Res 46:1633-1638

23. Henderson GB, Zevely EM (1979) Energetics of methotrexate transport in L1210 mouse leukemia cells. In: Kisliuk R, Brown G, eds. Chemistry and Biology of Pteridines. New York: Elsevier/North Holland 549-554

24. Henderson GB, Zevely EM (1980) Transport of methotrexate in L1210 cells: effect of ions on the rate and extent of uptake. Arch Biochem Biophys 200:149-155

25. Henderson GB, Zevely EM (1981) Anion exchange mechanism for transport of methotrexate in L1210 cells. Biochem Biophys Res Commun 99:163-169

26. Henderson GB, Zevely EM (1981) Transport of methotrexate in L1210 cells: mechanism for inhibition by p-chloromercuriphenylsulphonate and N-ethylmaleimide. Biochim Biophys Acta 640:549-556

27. Henderson GB, Zevely EM (1982) Functional correlations between the methotrexate and general anion transport systems of L1210 cells. Biochem Internat 4:493-502.

28. Henderson GB, Zevely EM (1982) Intracellular phosphate and its possible role as an ex-

change anion for active transport of methotrexate in L1210 cells. Biochem Biophys Res Commun 104:474–482

29. Henderson GB, Zevely EM (1983) Structural requirements for anion substrates of the methotrexate transport system of L1210 cells. Arch Biochem Biophys 221:438–446

30. Henderson GB, Zevely EM (1983) Use of non-physiological buffer systems in the analysis of methotrexate transport in L1210 cells. Biochem Internat 6:507–515

31. Henderson GB, Zevely EM (1984) Transport routes utilized by L1210 cells for the influx and efflux of methotrexate. J Biol Chem 259:1526–1531

32. Henderson GB, Zevely EM (1984) Affinity labeling of the 5-methyltetrahydrofolate/methotrexate transport protein of L1210 cells by treatment with an N-hydroxysuccinimide ester of [3H] methotrexate. J Biol Chem 259:4558–4562

33. Henderson GB, Zevely EM (1985) Inhibitory effects of probenecid on the individual transport routes which mediate the influx and efflux of methotrexate in L1210 cells. Biochem Pharmacol 34:1725–1729

34. Henderson GB, Zevely EM (1985) Characterization of the multiple transport routes for methotrexate in L1210 cells using phthalate as a model anion substrate. J Membr Biol 85:263–268

35. Henderson GB, Zevely EM, Huennekens FM (1978) Cyclic adenosine 3′, 5′-monophosphate and methotrexane transport in L1210 cells. Cancer Res 38:859–861

36. Henderson GB, Zevely EM (1979) Photoinactivation of the methotrexate transport system of L1210 cells by 8-azido-adenosine 5′-monophosphate. J Biol Chem 254:9973–9975

37. Henderson GFB, Zevely EM, Huennekens FM (1980) Irreversible inhibition of the metrotrexate transport system of L1210 cells by carbodiimide-activated stubstrates. J Biol Chem 255:4829–4833

38. Hill BT, Bailey BD, White JC, Goldman ID (1979) Characteristics of transport of 4-amino antifolates and folate compounds by two lines of L5178Y lymphoblasts, one with impaired transport of methotrexate. Cancer Res 39:2440–2446

39. Hoffbrand AV, Tripp E, Catovsky D, Das KC (1973) Transport of methotrexate into normal haemopoietic cells and into leukemic cells and its effect on DNA synthesis. Br J Haematol 25:497–511

40. Hoffbrand AV, Tripp E, Houlihan CM, Scott JM (1973) Studies on the uptake of synthetic conjugated folates by human bone marrow cells. Blood 42:141–146

41. Huennekens FM, Vitols KS, Henderson G (1978) Transport of folate compounds in bacterial and mammalian cells. Adv Enzymol 47:313–346

42. Jackson RC, Niethammer D, Huennekens FM (1975) Enzymic and transport mechanisms of amethopterin resistance in L1210 mouse leukemia cells. Cancer Biochim Biophys 1:151–155

43. Kessel D, Hall TC (1967) Amethopterin transport in Ehrlich ascites carcinoma and L1210 cells. Cancer Res 27:1539–1543

44. Kessel D, Hall TC, Roberts D, Wodinsky I (1965) Uptake as a determinant of methotrexate response in mouse leukemias. Science 150:752–754

45. Kessel D, Thomas TC, DeWayne R (1968) Modes of uptake of methotrexate by normal and leukemic human leukocytes in vitro and their relation to drug response. Cancer Res 28:564–570

46. Nahas A, Nixon PF, Bertino JR (1972) Uptake and metabolism of 5-formyltetrahydrofolate by L1210 leukemia cells. Cancer Res 32:1416–1421

47. Niethammer D, Jackson RC (1976) Transport of folate compounds through the membrane of human lymphoblastoid cells. In: Pfleiderer W, ed. Chemistry and Biology of Pteridines: Walter de Gruyter, 197–202

48. Rader JI, Niethammer D, Huennekens FM (1974) Effects of sulfhydryl inhibitors upon transport of folate compounds in L1210 cells. Biochem Pharmacol 23:2057-2059

49. Schilsky RL, Bailey GD, Chabner BA (1981) Characteristics of membrane transport of methotrexate by cultured human breast cancer cells. Biochem Pharmacol 30:1537–1542

50. Sirotnak FM (1980) Correlates of folate analog transport, pharmacokinetics, and selective antitumor action. Pharmacol Ther 8:71–103

51. Sirotnak FM, Chello PL, Brockman RW (1979) Potential for exploitation of transport systems in anticancer design. Methods Cancer Res 16:381–447

52. Sirotnak FM, Chello PL, Moccio DM et al. (1979) Stereospecificity at carbon 6 of formyl-tetrahydrofolate as a competitive inhibitor of transport and cytotoxicity of methotrexate in vitro. Biochem Pharmacol 28:2993–2997
53. Sirotnak FM, Chello PL, Piper JR, Montgomery JA (1978) Growth inhibitory, transport, and biochemical properties of the γ-glutamyl and γ-aspartyl peptides of methotrexate in L1210 leukemia cells in vitro. Biochem Pharmacol 27:1821–1825
54. Sirotnak FM, Chello PL, Piper JR, Montgomery JA, DeGraw JI (1979) Structural specificity of folate analog transport. In: Kisliuk R, Brown G, eds. Chemistry and Biology of Pteridines. New York: Elsevier/North Holland 597–602
55. Sirotnak FM, Donsbach RC (1974) Stereochemical characteristics of the folate-antifolate transport mechanism in L1210 leukemia cells. Cancer Res 34:371–377
56. Sirotnak FM, Donsbach RC (1976) Kinetic correlates of methotrexate transport and therapeutic responsiveness in murine tumors. Cancer Res 36:1151–1158
57. Sirotnak FM, Moccio DM, Yang CH (1984) A novel class of genetic variants of the L1210 cells up-regulated for folate analogue transport inward. J Biol Chem 259:1339–1344
58. Sirotnak FM, Moccio DM, Young CW (1981) Increased accumulation of methotrexate by murine tumor cells in vitro in the presence of probenecid which is mediated by a preferential inhibition of efflux. Cancer Res 41:996–970
59. Suresh MR, Henderson GB, Huennekens FM (1979) Folate uptake by L1210 cells: mediation by an adenine transport system. Biochem Biophys Res Commun 87:135–139
60. Suresh MR, Huenneken FM (1982) Transport of 6-hydroxymethylpterin by L1210 mouse leukemia cells. Biochem Internat 4:533–541
61. Warren RD, Nichols AP, Bender RA (1978) Membrane transport of methotrexate in human lymphoblastoid cells. Cancer Res 38:668–671
62. White JC, Bailey BD, Goldman ID (1978) Lack of stereospecificity at carbon 6 of methyltetrahydrofolate transport in Ehrlich ascites tumor cells. J Biol Chem 253:242–245
63. Yang C-H, Sirotnak FM, Dembo M (1984) Interaction between anions and the reduced folate/methotrexate transport system in L1210 plasma membrane vesicles: Directional symmetry and anion specificity for differential mobility of loaded and unloaded carrier. J Membr Biol 79:285–292

Chapter 20

Methotrexate and 5-Fluorouracil: Cellular Interactions with Folates

J. JOLIVET

The anticancer effect of methotrexate (MTX) and of 5-fluorouracil (5-FU) are intimately related to their effects on intracellular folates. MTX is used alone or in combination with other cytotoxic agents in the treatment of many cancers, including acute leukemia in children, trophoblastic tumors, non-Hodgkins lymphoma, cancers of the breast, lung, head and neck, and in osteogenic tumors. 5-FU is also used in many situations, most often as palliative therapy for gastrointestinal cancer. MTX is sometimes administered in high doses followed by rescue with 5-formyl tetrahydrofolate (THF; leucovorin, LV; folinic acid) with the expectation that the high plasma concentrations so achieved will kill cancer cells resistant to conventional doses [31]. 5-FU has recently been administered with LV to overcome resistance in tumors containing insufficient levels of intracellular folates to permit its maximum effect [36]. Understanding of the mechanism of interaction of these drugs with intracellular folate metabolism is thus important in developing the most effective treatment regimens. We therefore review the mechanisms of intracellular resistance to MTX and to 5-FU prior to describing the rationale of their use with LV.

Methotrexate

Mechanisms of Action and of Tumor Refractoriness. MTX is the 4-NH$_2$-10 CH$_3$-analogue of pteroyl-glutamic acid (folic acid; PteGlu) and acts to disrupt normal intracellular folate metabolism. MTX enters the cytoplasm of cells via a specific transmembrane protein which is the normal carrier of reduced (tetrahydro-) folates. It has classically been considered that MTX exerts its cytotoxic effect by blocking dihydrofolate reductase (DHFR): during synthesis of thymidylate, 5-10-methylene THF (CH$_2$-H$_4$PteGlu) is metabolized to dihydrofolate (H$_2$PteGlu), which is inactive in metabolism unless reduced to THF (H$_4$PteGlu). It would thus be expected that MTX should cause accumulation of H$_2$PteGlu and depletion of other intracellular folaltes, with consequent interference in the synthesis of thymidylate and with AICAR and GAR transformylases which function in the de novo synthesis of purines. The effect of MTX on the distribution of intracellular folates remains unclear because of difficulties in analysis.

By this mechanism, MTX must for effective cytotoxicity block all available DHFR sites and must accumulate to such a high level in the cytoplasm as to prevent its displacement from the enzyme by the high levels of H$_2$PteGlu generated consequent to its action. By this classical schema, resistance to MTX could arise through four distinct mechanisms: (a) altered transmembrane transport which prevents accumulation of a toxic level of drug in the cytoplasm [24], (b) increased DHFR which would prevent blocking of all DHFR sites by MTX [4], (c) decreased affinity of DHFR for MTX [21], and (d) reduced de novo thymidylate synthesis with consequent reduction in the need for DHFR [41].

Recent demonstration of intracellular metabolism of MTX has permitted fresh perspectives on its mechanism of action. MTX is a substrate for the enzyme folypolyglutamate synthetase (FPGS), as are most biologically active folates [39]. This enzyme adds glutamyl residues to intracellular folates (including MTX). MTX plyglutamates (MTXPGs) have been indentified in liver and erythrocytes of patients treated with MTX [5, 27] and in a variety of tissues of treated patients and experimental animals [5, 11, 17, 18, 27, 30, 33, 35, 40, 42, 43, 48]. In studies with mammary tumor cell lines in vitro we have found [30] that (a) the quantity of MTXPGs formed depends both on the extracellular concentration of MTX and on the duration of incubation, (b) the capacity of different tumor cell lines to generate MTXPGs varies, and (c) this intracellular metabolism permits generation of higher intracellular MTX concentrations than would be possible if the drug were not so metabolized. MTXPGs thus accumulate to concentrations far greater than required to saturate DHFR sites [31].

MTXPGs replace MTX on DHFR as soon as they appear in the cytoplasm [29]. Their affinity for DHFR appears to be greater than that of the monoglutamyl, unmetabolized MTX in vitro [9], and within the cell, MTXPGs appear to dissociate less effectively than MTX from DHFR [29].

As with folylpolyglutamates, MTXPGs leave the cell less readily than do MTX. We have found that in human breast carcinoma cells, MTXPGs with glutamate chain lengths of two or three residues leave the cell when the cell is placed in MTX-free medium, whereas those with four or more residues are almost completely retained [29, 30]. Such retention permits continued MTXPG action with prolonged inhibition of cell growth after removal of extracellular drug [11, 29, 30].

Additional glutamyl residues attached by γ linkage to the γ-carboxyl of the glutamate on MTX increases the affinity of MTX for many enzymes, as it does for many of the folate-requiring enzymes [38]. Unlike MTX, which has little affinity for thymidylate synthase (K_i $1.3 \cdot 10^{-5}$ M) and for several other enzymes, K_i values for MTXPGs (1–3) for human thymidylate synthase ($6.1 \cdot 10^{-8}$ M), AICAR transformylase ($7.1 \cdot 10^{-6}$ M) and methylene THF reductase ($1.5 \cdot 10^{-7}$ M) correspond to readily achieved intracellular concentrations [30]. It is thus probable that MTXPGs may interfere with de novo synthesis of both thymidylate and purines.

Despite the general toxicity of MTX when used clinically, its apparently greater toxicity in certain tumors compared with normal tissues probably explains cure of choriocarcinoma with this agent. This selectivity is probably at least partly due to the different rate of MTXPG formation in normal and neoplastic tissues. MTXPG formation is much less in murine intestinal mucosa than in Ehrlich ascites carcinoma cells [18], and in man, normal myeloblasts and promyelocytes generate much less MTXPG than do promyelocytic leukemia cells [35].

The evidence that MTXPGs formation is probably required for MTX-mediated cytotoxicity includes the following observations:

1. Correlation between MTXPG formation and cytotoxicity in certain murine tumors [15, 33] and in human cell lines derived from leukemia [40], cancer of the breast [30], and small-cell carcinoma of the lung [11].
2. A mammary cancer cell line induced to resistance to MTX by prolonged culture in graded concentrations of MTX was found to have lost its capacity to generate MTXPGs [10].
3. γ-Fluoro-MTX, which cannot form polyglutamates, is considerably less toxic than is MTX [20]. This analogue inhibits DHFR as effectively as does MTX.

Any mechanism which increases MTXPG synthesis would be expected to increase the toxicity of MTX for that cell. The mechanisms which have been defined include cellular growth rate and intracellar folate pools. Tumor cells in logarithmic growth synthesize ten times more MTXPGs than do cells in which growth rate has reached a plateau [42]. Other maneuvers which decrease the growth rate of cells also reduce MTXPG synthesis. The antagonism between the cytotoxicity of L-asparaginase and MTX in certain leukemia cells appears due to this phenomenon [33]. L5187Y cells exposed during active growth to both drugs decreased growth rate but showed less cytotoxicity than when exposed to MTX alone. This mechanism may explain the resistance of noncycling cells to MTX. Intracellular folate affects the cytotoxicity of MTX by competing with MTX for enzymes. Cells depleted of intracellular folates by prolonged growth in folate-deficient media produced twice as much MTXPG, while cells grown in excess folate to increase intracellular folate concentration produced less MTXPG than did folate replete cells [42]. The clinical implications of these observations are discussed below.

Methotrexate–Leucovorin Interactions. The above considerations permit some understanding of the basis of using high doses of MTX (HDMTX) with LV rescue to overcome MTX resistance in tumors without producing unacceptable levels of toxicity [31].

HDMTX therapy was initially designed to overcome resistance due to decreased transmembrane transport of MTX by tumor cells [16]. At high extracellular MTX concentrations enough MTX may diffuse passively across the cell membrane to bind the active sites on DHFR [24]. The other mechanisms of resistance to MTX enumerated above are not equally overcome by high concentrations of MTX. (a) Increased levels of intracellular DHFR may not be saturated by the levels of MTX achieved by diffusion, since MTX itself may induce an acute increase in DHFR level by unknown mechanisms [32]. (b) The affinity for MTX of the DHFR described in human acute myeloblastic leukemia cells is so small that it is not readily inhibited even by very high concentrations of MTX [12]. (c) Human cancer cell lines with decreased thymidylate synthase activity are relatively insensitive to MTX [11]. (d) Increased concentrations of MTX in the cell do not always correct inadequate synthesis of MTXPGs [10, 11].

HDMTX is utilized with LV to reduce toxicity but with the expectation that the therapeutic ratio will increase because of greater rescue to normal than to malignant cells. Doses and timing of LV have been developed in models using mice [47]. These studies revealed that rescue with either too much LV or LV given too soon after MTX could reduce the anti-tumor effect of the treatment. The mechanisms by which LV rescues cells from MTX toxicity are described below.

Since MTX and LV share the same cell membrane transport system, they affect each other's accumulation by the cell. In human fibroblasts studied in vitro, LV could completely block entry of MTX into the cells [44]. By analogy, very large doses of LV given in vivo could block further entry of MTX into neoplastic cells. If tumor cells do not effectively transport MTX, they have a similar defect in LV transport [16].

If inhibition of DHFR by MTX reduces the intracellular concentration of some biologically active folates, LV therapy replenishes these by conversion of the LV to methenyl THF by the enzyme 5-10-methenyl THF synthetase [25], while the product of LV in plasma (5-methyl THF) enters the cell to be converted to THF by methionine synthase [13]. In MTX resistance by mechanisms other than transmembrane transport, this decreased activity of the two enzymes may induce selective rescue of normal compared with tumor cells by LV. Some authors suggest that methionine synthase may be important in generating such an effect [23], but other studies have not confirmed this [22].

As indicated above, LV has important effects on MTX cytotoxicity. When administered *simultaneously* with the MTX, LV prevents entry of MTX into the cell [44]; when administered after the MTX, it blocks its conversion to MTXPGs and abolishes its cytotoxicity. The latter effect probably is mediated by competition between intracellular folates and MTX for FPGS [19, 45]. We have shown that treatment of human mammary carcinoma cells with LV before MTX increased intracellular folates, with consequent decreased metabolism of MTX [28]. In clinical treatment regimens using repeated treatments with HDMTX with LV rescue, the LV may induce refractoriness to subsequent treatments with HDMTX, thus making scheduling of treatments important.

5-Fluorouracil

5-FU is a prodrug which must be metabolized within the cell to form its three active agents: (a) 5-fluorodeoxyuridine monophosphate (5-FdUMP) which inhibits thymidylate synthase by forming a covalent bond with the enzyme and its coenzyme folate, 5-10-methylene THF; (b) 5-fluorouridine triphosphate (5-FUTP) which is incorporated into RNA; and (c) 5-fluorodeoxyuridine triphosphate (5-FdUTP) incorporated into DNA [46]. The mechanism of toxicity differs in different cells [7]. Because of the various reactions involved in forming the active metabolites, several mechanisms of resistance to the drug have been described, including decreased activity of some of the enzymes and increased intracellular levels of thymidylate synthase [7].

In some cells, resistance was found to be due to insufficient levels of 5-10-methylene THF to permit binding of all thymidylate synthase molecules by

FdUMP [26]. Addition of LV to the culture medium of such cells has been shown to increase the cytotoxicity of 5-FU [14], but in other model systems such an effect could not be demonstrated [34]. These preliminary observations were rapidly introduced into clinical situations in which trials of 5-FU with and without LV were tested in patients with cancer of the bowel [36, 37]. Results have been encouraging, but definitive evaluation awaits prospective randomized trials in such patients.

Summary

The antineoplastic activities of MTX and 5-FU reflect on their effect on intracellular folates. MTX affects folate metabolism by blocking DHFR and subsequent reduction of oxidized folates generated during thymidylate synthesis. Its cytotoxicity appears determined in most cells by formation of polyglutamyl forms of MTX, which increase cellular persistence of MTX with consequent prolongation of DHFR inhibition, and direct inhibition by the MTXPGs of other folate-dependent enzymes. Reduced MTXPG accumulation caused by decreased growth rate of cells or exposure to folate or LV may reduce the cytotoxicity of MTX. Intracellular folates also mediate resistance to 5-FU if the concentration of 5–10 methylene THF is insufficient to permit irreversible inhibition of thymidylate synthase. The latter type of resistance might be overcome by LV administration.

References

1. Allegra CJ, Chabner BA, Drake JC, Lutz R, Rodbard D, Jolivet J (1985) Enhanced inhibition of thymidylate synthase by methotrexate-polyglutamates. J Biol Chem 260:9720–9726
2. Allegra CJ, Drake JC, Jolivet J, Chabner BA (1985) Inhibition of AICAR transformylase by methotrexate and dihydrofolic acid polyglutamates. Proc Natl Acad Sci USA 82:4881–4885
3. Allegra CJ, Drake JC, Jolivet J, Chabner BA (1985) Inhibition of folate-dependent enzymes by methotrexate polyglutamates. In: Goldman ID (ed) Proceedings of the second workshop on folyl and antifolyl polyglutamates. Praeger, New York
4. Alt FW, Kellems RE, Schimke RT (1976) Synthesis and degradation of folate reductase in sensitive and methotrexate-resistant lines of S-180 cells. J Biol Chem 25:3063–3074
5. Baugh CM, Krumdieck CL, Nair MG (1973) Polygammaglutamyl metabolites of methotrexate. Biochem Biophys Res Commun 52:27–34
6. Bruckner HW, Rustum YM (1984) The current status of 5-fluorouacil-leucovorin calcium combination. In: Bruckner HW, Rustum YM (eds) Advances in Cancer Chemotherapy. Wiley, New York
7. Chabner BA (1982) Pyrimidine antagonists. In: Chabner BA (ed) Pharmacologic principles of cancer treatment. Saunders, Philadelphia, pp 183–212
8. Chabner BA, Allegra CJ, Curt GA (1985) Polyglutamation of methotrexate. Is methotrexate a prodrug? J Clin Invest 76:907–912
9. Clendeninn NJ, Drake JC, Allegra CJ, Welch AD, Chabner BA (1985) Methotrexate polyglutamates have a greater affinity and more rapid on-rate for purified human dihydrofolate reductase than MTX (Abstract 915). Proc Am Assoc Cancer Res, vol 232:915
10. Cowan KH, Jolivet J (1984) A methotrexate-resistant human breast cancer cell line with

multiple defects, including diminished formation of methotrexate polyglutamates. J Biol Chem 259:10793–10800

11. Curt GA, Jolivet J, Carney DN (1985) Determinants of the sensitivity of human small cell lung cancer cell lines to methotrexate. J Clin Invest 76:1323–1329

12. Dedhar S, Hartley D, Fitz-Gibbons D, Phillips G, Goldie JH (1985) Heterogeneity in the specific activity and methotrexate sensitivity of dihydrofolate reductase from blast cells of acute myelogenous leukemia patients. J Clin Oncol 3:1545–1552

13. Dudman NPB, Slowiaczek P, Tattersall MHN (1982) Methotrexate rescue by 5-methyltetra-hydrofolate or 5-formyltetrahydrofolate in lymphoblast cell lines. Cancer Res 42:502–507

14. Evans RM, Laskin JD, Kakala MT (1981) Effect of excess folates and deoxyinosine on the activity and site of action of 5-fluorouracil. Cancer Res 41:3288–3295

15. Fabre I, Fabre G, Goldman ID (1984) Polyglutamylation, an important element in methotrexate cytotoxicity and selectivity in tumor versus murine granulocytic progenitor cells in vitro. Cancer Res 44:3190–3195

16. Frei III E, Jaffe N, Tattersall MHN, Pitman S, Parker L (1975) New approaches to cancer chemotherapy with methotrexate. N Engl J Med 292:846–851

17. Frei III E, Rosowsky A, Wright JE et al. (1984) Development of methotrexate resistance in a human squamous cell carcinoma of the head and neck in culture. Proc Natl Acad Sci USA 81:2873–2877

18. Fry DW, Anderson LA, Borst M, Goldman ID (1983) Analysis of the role of membrane transport and polyglutamation of methotrexate in gut and the Ehrlich tumor in vivo as factors in drug sensitivity and selectivity. Cancer Res 43:1087–1092

19. Galivan J, Nimec Z (1983) Efects of folinic acid on hepatoma cells containing methotrexate polyglutamates. Cancer Res 43:551–555

20. Galivan J, Inglese J, McGuire JJ, Nimec Z, Coward JK (1985) γ-fluoromethotrexate: synthesis and biological activity of a potent inhibitor of dihydrofolate reductase with greatly diminished ability to form poly-γ-glutamates. Proc Natl Acad Sci USA 82:2598–2602

21. Goldie JH, Krystal G, Hartley D, Gudauskas G, Dehar S (1980) A methotrexate insensitive variant of folate reductase present in two lines of methotrexate-resistant L5178Y cells. Eur J Cancer 16:1539–1546

22. Groff JP, Blakley RL (1978) Rescue of human lymphoid cells from the effects of methotrexate in vitro. Cancer Res 38:3847–3853

23. Halpern RM, Halpern BC, Clark BR (1975) New approach to antifolate therapy of certain cancers as demonstrated in tussue culture. Proc Natl Acad Sci USA 72:4018–4022

24. Hill BT, Dedhar S, Goldie JH (1982) Evidence that at "high" extracellular methotrexate concentrations the transport barrier is unlikely to be an important mechanism of drug resistance. Biochem Pharmacol 31:263–266

25. Hopkins S, Schirch V (1984) 5–10 methenyltetrahydrofolate synthetase. Purifications and properties of the enzyme from rabbit liver. J Biol Chem 259:5618–5622

26. Houghton JA, Maroda SJ Jr, Phillips JD, Houghton PJ (1981) Biochemical determinants of responsiveness to 5-fluorouracil and its derivatives in xenografts of human colorectal adenocarcinoma in mice. Cancer Res 41:144–149

27. Jacob SA, Derr CJ, Johns DG (1977) Accumulation of methotrexate diglutamate in human liver during methotrexate therapy. Biochem Pharmacol 26:2310–2313

28. Jolivet J (1985) The influence of intracellular folate pools on methotrexate metabolism and cytotoxicity (Abstract 919). Proc Am Assoc Cancer Res 26:233

29. Jolivet J, Chabner BA (1983) Intracellular pharmacokinetics of methotrexate polyglutamates in human breast cancer cells: selective retention and less dissociable binding of 4-NH_2-10-CH_3PteGlu$_4$and$_5$ to dihydrofolate reductase. J Clin Invest 72:773–778

30. Jolivet J, Schilsky RL, Bailey BD, Drake JC, Chabner BA (1982) The synthesis, retention and biological activity of methotrexate polyglutamates in cultured human breast cancer cells. J Clin Invest 70:351–360

31. Jolivet J, Cowan KH, Curt GA, Clendeninn NH, Chabner BA (1983) The pharmacology and clinical use of methotrexate. N Engl J Med 309:1094–1104

32. Jolivet J, Levine RM, Cowan KH (1983) The regulation of dihydrofolate reductase in gene-amplified methotrexate-resistant human breast cancer cells. Clin Res 31:509 A

33. Jolivet J, Cole DE, Holcenberg JS, Poplack DG (1985) Prevention of methotrexate cytotox-

icity by asparaginase inhibition of methotrexate polyglutamate formation. Cancer Res 45:217-220

34. Klubes P, Cerna I, Meldon MA (1981) Effect of concurrent calcium leucovorin infusion on 5-fluorouracil cytotoxicity against murine L1210 leukemia. Cancer Chemother Pharmacol 6:121-125

35. Koizumi S, Curt GA, Fine RL, Griffin JD, Chabner BA (1985) Formation of methotrexate polyglutamates in purified myeloid precursor cells from normal human bone marrow. J Clin Invest 75:1008-1014

36. Machover D, Schwarzenberg L, Goldschmidt E et al. (1982) Treatment of advanced colorectal and gastric adenocarcinoma with 5-FU combined with high-dose folinic acide: a pilot study. Cancer Treat Rev 66:1803-1807

37. Madajewicz S, Petrelli N, Rustum YM et al. (1984) Phase I-II trial of high-dose calcium leucovorin and 5-fluorouracil in advanced colorectal cancer. Cancer Res 44:4667-4669

38. McGuire JJ, Bertino JR (1981) Enzymatic synthesis and function of folylpolyglutamates. Mol Cell Biochem 38:19-48

39. McGuire JJ, Hsieh P, Coward JK, Bertino JR (1980) Enzymatic synthesis of folylpolyglutamates: characterization of the reaction and its products. J Biol Chem 255:5776-5778

40. McGuire JJ, Mini E, Hsieh P, Bertino JR (1985) Role of methotrexate polyglutamates in methotrexate- and sequential methotrexate-5-fluorouracil-mediated cell kill. Cancer Res 45:6395-6400

41. Moran RG, Mulkins M, Heidelberg C (1979) Role of thymidylate synthase activity in development of methotrexate cytotoxicity. Proc Natl Acad Sci USA 76:5924-5928

42. Nimec Z, Galivan J (1983) Characteristics of methotrexate polyglutamate formation in cultured hepatic cells. Arch Biochem Biophys 226:671-680

43. Rosenblatt DS, Whitehead VM, Dupont MM, Vuchich MJ, Vera N (1978) Synthesis of methotexate polyglutamates in cultured human cells. Mol Pharmacol 14:210-214

44. Rosenblatt DS, Whitehead VM, Vuchich MJ (1981) Inhibition of methotrexate polyglutamate accumulation in cultured human cells. Mol Pharmacol 19:87-97

45. Schoo MMJ, Pristura ZB, Vickers PJ, Scrimgeour KB (1985) Folate analogues as substrates for mammalian folylpolyglutamate synthetase. Cancer Res 45:3034-3041

46. Schuetz JD, Wallace HJ, Diasio RB (1984) 5-Fluorouracil incorporation into DNA of CF-1 mouse bone marrow cells as a possible mechanism of toxicity. Cancer Res 44:1358-1363

47. Sirotnak FM, Moccio DM, Dorick DM (1978) Optimization of high-dose methotrexate with leucovorin rescue therapy in the L1210 leukemia and sarcoma 180 murine tumor models. Cancer Res 38:345-353

48. Whitehead VM (1977) Synthesis of methotrexate polyglutamates in L1210 murine leukemia cells. Cancer Res 37:408-412

Chapter 21

Methotrexate Pharmacokinetic Studies: Their Clinical Use

Y. NAJEAN and O. POIRIER

Asking those who performed analyses of drug levels to explain their utility is a difficult proposition. The present article is therefore based more upon a review of the literature and the opinions of clinical collegues than on our own experience. These sources are, however, essentially in agreement. There is, for example, abundant literature from the period 1978–1982 describing the development of specific techniques of dosage and the use of new and potentially dangerous protocols for high-doses methotrexate (MTX). The declining number of publications thereafter suggests that the bulk of useful information has been obtained.

The current views of clinicians incorporate this body of information, and the number of requests for serum MTX levels has thus decreased considerably in the past 2 or 3 years. While pharmacokinetic studies were needed previously for the design of high-dose therapeutic protocols, interest in pharmacokinetic studies has now diminished, perhaps excessively. Although it is true that the technical costs of pharmacokinetic studies are high (price of reagents, time for determination of drug levels, time for calculation), and that the results (always demanded in emergency by the clinician) impose a heavy burden on the laboratory, especially in the present period of financial austerity, it may nervertheless be wrong to suppose that such studies have no practical value.

Current Understanding: Methodology

High-dose MTX chemotherapy could not have been developed without techniques for determining serum levels in the range of 10^{-3} to 10^{-8}, 10^{-9} M. Currently utilized methods [6] include radioimmunoassay, competitive binding assays, enzymatic inhibition, microbiological techniques, detection by high-pressure liquid chromatography (HPLC), and fluorescence polarization.

Radioimmunologic values are valid at nanomolar concentrations, but this method suffers from a lack of specificity. The same problem exists for other competitive binding studies. This is especially important for cross-reactions with trimethoprim sulfate, which is occasionally given to some of these patients [13]. Drug levels measured by enzyme inhibition or inhibition of bacterial growth are tedious and have poor reproducibility. These techniques are poorly adapted to clinical practice, where rapid response is required. Serum levels measured by HPLC have good specificity, and various refinements of the technique allow the detection of levels on the order of 10^{-8} M, which is clinically adequate [6]. Two methods may be used for the separate measurement of MTX and its metabolites, notably of 7-OH-MTX [26], a potential nephrotoxic waste, due to its very limited aqueous solubility. However, 95% of MTX is eliminated "as is" [3], and the dosage of 7-OH-MTX is probably not important in clinical practice [26, 29].

Mathematic analysis of plasma MTX kinetics has been the subject of numerous experimental studies. Some authors have devised very sophisticated metabolic models and based the validity of their investigations on mathematically complicated equations [2], which seem to be a mathematical game even if the use of diffusion chambers has demonstrated that kinetic parameters of the drug varies from organ to organ [30]. In addition, a portion of the MTX loss is intestinal and subject to enterohepatic cycling [15], a fact which complicates the kinetic model.

The generally accepted model is a simplified system describing three slopes with three exponential coefficients [23]:

$$\phi(t) = A\,e^{-\lambda_1 t} + B\,e^{-\lambda_2 t} + C\,e^{-\lambda_3 t}$$

where $A + B + C = 1$ and the λ coefficients are a fraction of time t, and ϕ is the concentration.

Exact determination of these six parameters (A, B, C, λ_1, λ_2, and λ_3) depends on the precision of the measurements, the number and adequacy of sampling, and the existence of a steady state during the study. Actually none of these conditions is generally fulfilled perfectly. In particular, it is frequently overlooked that the last of these coefficients can be disturbed during the study because of changes in renal clearance caused by the nephrotoxicity of MTX itself [4]. Some authors have proposed nonlinear models, with constants varying in relation to drug concentration [23, 32]. It is generally stated that the initial slope corresponds to flow into extravascular compartments in equilibrium with the plasma (as the observed concentration in the pleural fluid and peritoneal fluid equilibrates rapidly with the plasma) [15], the second to renal clearance of the equilibrated vascular pool, and the third to the slow return to the plasma of MTX from a cellular compartment that is slowly renewed (MTX accumulates in target cells in the form of slowly exchangeable polyglutamate). The level of transformation of MTX into polyglutamates is a parameter seemingly related to the drug efficiency, which unfortunately is not very accessible clinically [12, 37, 39]. Figure 1 is a schema of a multicompartimental model, where Q_1 is the plasma pool, Q_2 the extravascular pool and Q_3 the cellular pool. The drug loss is assumed to

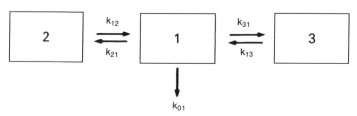

Fig. 1. Representation of the parameters of a three-component model of drug kinetics

occur from the plasma pool (k_{01}). Exchanges between the pools are assumed to be aleatory (exponential).

The k values of flows are not the values of the equation components. Numerous publications assume λ_2 to be k_{01} or λ_3 to be k_{13}, but this reveals a poor understanding of the basic calculation of the kinetics of the compartments [34]. From the six parameters of the equation and the value of the MTX concentration at t_0, it is possible to calculate the parameters of the model, either from classical equations or by simulation. The ultimate aim in measuring plasma kinetics is calculation of the concentration of the drug within the target cells: in malignant cells for evaluating drug efficiency and in normal cells for evaluating its toxicity. Obviously, in the clinical setting there is no direct access to these parameters as it is in the in vitro experiments on cell lines in culture.

Some theoretical and empirical considerations suggest the clinical use of a dose leading to a plasma concentration of 1×10^{-6} M. The model permits calculation of the plasma concentration curve and of the length of time that MTX remains bound in the cellular compartment (Q3). An important factor is indeed the duration of the adequate concentration of the drug. This is dependent on the injected dose and also on the time during which the plasma concentration is adequate [7, 28]. The usual in vitro tests made on normal or cancerous cell lines in culture with brief exposure are thus of limited practical interest [17]. Mathematical analysis, on the other hand, permits the calculation, as a function of time, of the concentration of drug in the compartment Q3, yet it does not indicate the proportion of drug in the target cell present as free drug or as polyglutamates.

Clinical Applications

The use of high-dose MTX to reach high plasma concentrations (10^{-4} or 10^{-5} M) for extended periods (12–36 h) is justified on the basis of pharmacological data [17]: (a) at high concentrations, the passive transport of MTX exceeds the resistance due to diminished active transmembrane transport; (b) the increased intracellular concentration of free MTX can overcome resistance due to an increased level of dihydrofolate reductase; and (c) the extended period of time during which the cellular concentration is elevated exposes more cells in the process of DNA synthesis to the toxic action of the drug.

However, the systemic risk of a high, although uncertain, concentration requires the use of an antidote (folinic acid) after 48 h. The majority of authors have fixed the dose and duration of the rescue as a function of the plasma concentration of MTX at the 48th hour after the start of treatment. The level at which the use of folinic acid is necessary is generally estimated at 5×10^{-5} M [17]; the dose is adjusted for the observed, or expected, plasma concentration of MTX (i.e., 20 mg/m^2 every 6 h for 5×10^{-7} M, 100 mg/m^2 for 1×10^{-6} M, 200 mg/m^2 for 2×10^{-6} M). In such a situation it is therefore not a question of pharmacokinetics but of a serum level, from which, in fact, one can predict neither the intracellular concentration nor the duration of exposure of cells to an excessive concentration, as one does not known the clearance, k_{01}. This simplification of the biological study is, however, often justified in clinical practice [25].

Most authors indicate good reproducibility of kinetic parameters, both in single patients on repeated perfusions [24] and from one patient to another, between low and high doses [2, 27], or according to the route of administration (arterial [14], venous, or intramuscular [5]). Testing these kinetic parameters in a given patients at a moderate dose and then extrapolating the values for high doses in order to develop a therapeutic regimen has been proposed [20, 27].

A recent article [9], however, takes issue with these conclusions, indicating wide variation in MTX clearance in children with acute lymphoblastic leukemia treated with identical doses (1 g/m^2). The difference in clearance leads to serum concentrations varying among individual subjects by a factor of 1 to 3, and the probability of a prolonged remission is linked to the concentration achieved.

Pharmacological studies have also been useful in the analysis of MTX absorption. It has been demonstrated that absorption varies widely from subject to subject [1, 19], and that it decreases progressively in percentage with increasing drug administration [35]. This can explain the unpredictable effects of oral treatment. Thus, plasma level measurements could be of practical use in clinical protocols using oral MTX.

Studies have been performed in ascitic fluids, pericardial fluids, and malignant pleural effusions [11, 18]. Concentrations exceeding the plasma level up to 20 times have been noted. The clearance from these cavities is slow enough to permit a sufficient duration of these very elevated levels. However, few cases have been studied, and follow-up has generally been insufficient. Many studies have been performed on the cerebro-spinal fluid (CSF) because of the frequency of leukemic meningitis and its sensitivity to MTX. This drug diffuses slowly across the blood-brain barrier. The ratio of its concentration in CSF to that observed in plasma is in the range of $1:20–1:100$ [8, 31]. In order to obtain a sufficient concentration of $1 \times 10^{-6} M$, one must obtain a plasma concentration of $1 \times 10^{-4} M$, which is at the limit of physiological tolerance. On the other hand, it is known that MTX injected into CSF diffuses poorly to the ventricles. The ventricular concentration may be only $1/1000$ of that in the CSF [33], hence the importance of MTX injections via the Omaya reservoir (12 mg/m^2) The urinary loss of drug observed after intraventricular introduction is significantly slower than that observed after lumbar injection.

Current Usefulness of Pharmacokinetic Studies and Methotrexate Serum Level Measurements

Pharmacokinetic studies are long, difficult, expansive, and sometimes painful for the patient due to the number of samples needed. Results cannot be delivered rapidly. Some of the calculated parameters are without clinical interest. In our opinion, this type of study is of interest only for clinical research studies. For such investigations, we suggest four indications:

1. In the design of a new protocol, using previously seldom employed dosages of the drug, unusual time intervals, or poorly explored clinical indications, it seems logical to perform preliminary kinetic studies in the phases I and II of such studies.
2. For the study of new routes of administration, for example direct arterial infusion into a tumor, kinetic studies may give the precise concentration and duration of presence of the drug in the tumor itself.
3. For the study of cellular concentration and clearance of MTX, injected as a complex with monoclonal or polyclonal antibodies presumably specifically bound to the malignant cells, kinetic studies must be performed [21, 38].
4. For the study of drug interactions [17, 24, 35], for instance synergy demonstrated between certain drugs (Vinca alkaloids, epipodophyllotoxins) and MTX. The mechanism of such an interaction has been elucidated by pharmacokinetic mechanisms; slowing of the cellular efflux of drug may be demonstrated by a decrease of k_{13}, with an increase in the maximal value of Q3. Another recently noted fact is the alteration of the blood-brain barrier observed after irradiation, which leads to a threefold increase in the cerebral tissue level of MTX at similar plasma levels [36].

What is the clinical interest in MTX plasma measurement in patients treated with high doses? Some find such determinations indispendable [3, 31]. Other feel, on the contrary, that these measurements are without any practical value, and some others that even high-dose chemotherapy itself needs reevaluation [22]. At the present time, caution would seem to dictate that it would be useful, perhaps necessary, to measure MTX plasma levels in some clinically disturbing conditions, for instance, in patients in poor clinical condition, with insufficient renal status, in the event of an unusual drug level, when the nurse is concerned about the dose actually administered; or whenever an unusual interaction between drugs may interfere with MTX excretion. One additional indication is the administration of drug via the Ommaya reservoir [16] to evaluate its diffusion into the spinal fluid.

In summary, from a clinical point of view, pharmacokinetic studies of MTX are possible, but complicated when one wishes to correctly perform and completely interpret them. These studies have provided much interesting and useful data, but now are essentially of research interest.

Measuring the level of a single drug may sometimes be useful to determine precisely the dose and the necessity of folinic acid rescue, but in general practice it is of limited usefulness.

References

1. Balis FM, Savitch JL, Bleyer WA (1983) Pharmacokinetics of oral methotrexate in children. Cancer Res 43:2342–2345
2. Bischoff KB, Dedrick RL, Zaharko DS, Longstreth JA (1971) Methotrexate pharmacokinetics. J Pharm Sci 60(8):1128–1133
3. Breithaupt H, Kuenzlen E (1982) Pharmacokinetics of methotrexate and 7-hydroxymethotrexate following infusion of high-dose methotrexate. Cancer Treat Rep 66(9) 1733–1741

 4. Christophidis N, Louis WJ, Lucas I, Moon W, Vajda FJE (1981) Renal clearance of methotrexate in man during high-dose oral and intravenous infusion therapy. Cancer Chemother Pharmacol 6:59–64

 5. Colls BM, Robson RA, Robinson BA, Tisch GW (1983) Serum profiles and safety of intermediate-dose (500–1000 mg) methotrexate following IV and IM administration. Cancer Chemother Pharmacol 11:188–190

 6. Deen WM, Levy PF, Wei J, Pariridge RD (1981) Assay for methotrexate in nanomolar concentrations with simultaneous infusion of citrovorum factor and vincristine. Anal Biochem 114:355–361

 7. Eichholtz H, Trott KR (1980) Effect of methotrexate concentration and exposure time on mammalian cell survival in vitro. Br J Cancer 41:277–284

 8. Ettinger LJ, Chervinsky DS, Freeman AI, Creaven PJ (1982) Pharmacokinetics of methotrexate following intravenous and intraventricular administration in acute lymphocytic leukemia and non-Hodgkin's lymphoma. Cancer 50:1676–1682

 9. Evans WE, Crom WR, Abramowitch M et al. (1986) Clinical pharmaco-dynamics of high-dose methotrexate in acute lymphocytic leukemia. N Engl JMed 314:471–477

10. Harding NGL, Martelli MF, Huennekens FM (1970) Amethopterin-induced changes in the multiple forms of dihydrofolate-reductase from L 1210 cells. Arch Biochem Biophys 137:295–296

11. Howell SB, Chu BBF, Wung WE, Bipin M (1981) Long duration intracavitary infusion of methotrexate with systemic leucovorin protection in patients with malignant effusions. J Clin Invest 67:1161–1170

12. Koizumi S, Curt GA, Fine RS, Griffin JD, Chabner BA (1985) Formation of methotrexate polyglutamate in purified myeloid precursor cells from normal human bone marrow. J Clin Invest 75:1008–1014

13. Icke GC, Davis RE, Thom J (1983) A microbiological assay for the measurement of methotrexate in biological fluids. J Clin Pathol 36:1116–1119

14. Ignoffo RJ, Oie S, Friedman MA (1981) Pharmacokinetics of methotrexate administered via the hepatic artery. Cancer Chemother Pharmacol 5:217–220

15. Jaffe N, Howell S (1979) Antifolate rescue. Use of high-dose MTX and citrovorum factor. In: Resowsky (ed) Advances in cancer chemotherapy. New York, Basel: M. Dekker, 111:141 111–171

16. Janvier M, Leverger G, Renier D, Poirier O, Tobelem G, Boiron M (1984) Utilisation des réservoirs d'Ommaya dans le traitement des rechutes méningées des leucémies lymphoblastiques. Nouv Rev Fr Hematol 26:295–298

17. Jolivet J, Cowan KH, Curt GA, Clendeninn NJ, Chabner BA (1983) The pharmacology and clinical use of methotrexate. N Engl J Med 309:1094–1104

18. Jones RB, Collins JM, Myers CE, Brooks AE, Hubbard SM Balow JE, Brennan MF, Dedrick RL, Devita VT (1981) High volume intraperitoneal chemotherapy with methotrexate in patients with cancer. Cancer Res 41:55–59

19. Kearney PJ, Light PA, Preece A, Mott MG (1979) Unpredictable serum levels after oral methotrexate in children with acute lymphoblastic leukaemia. Cancer Chemother Pharmacol 3:117–120

20. Kerr IG, Jolivet J, Collins JM, Drake JC, Chabner BA (1983) Test dose for predicting high-dose methotrexate infusions. Clin Pharmacol Ther 33:44–51

21. Kulkarni PN, Blair AH, Ghose T, Mammen M (1985) Conjugation of methotrexate to IgG antibodies and their F (ab) fragments and the effect of conjugated methotrexate on tumor growth in vivo. Cancer Immunol Immunother 19:211–214

22. Lange B, Levine AS (1982) Is it ethical not to conduct a prospectively controlled trial of adjuvant chemotherapy in osteosarcoma? Cancer treatment reports. 66, 1699–1703, 1982.

23. Leme PR, Creaven PJ, Allen LM, Berman M (1975) Kinetic model for the disposition and metabolism of moderate and high-dose methotrexate (NSC-740) in man. Cancer Chemother Rep 59:811–817

24. Lokiec F, Poirier O, Gisselbrecht C, Marty M, Boiron M, Najean Y (1982) Effect of combination chemotherapy, duration of methotrexate administration, and patient's age on methotrexate pharmacokinetics. Cancer Chemother Pharmacol 9:165–168

25. Marty JJ, Shaw J (1979) Methotrexate pharmacokinetics long and short infusions. Clin Exp Pharmacol Physiol [Suppl] 5:29–34
26. Milano G, Thyss A, Renee N, Schneider M, Namer M, Boublil JL Lalanne CM (1983) Plasma levels of 7-hydroxymethotrexate and high-dose methotrexate treatment. Cancer Chemother Pharmacol 11:29–32
27. Monjavel S, Rigault JP, Cano JP, Carcassone Y, Favre R (1976) High-dose methotrexate: preliminary evaluation of a pharmacokinetic approach. Cancer Chemother Pharmacol 3:189–196
28. Pinedo HM, Zahorko DS, Bull J, Chabner BA (1977) The relative contribution of drug concentration and duration of exposure to mouse bone marrow toxicity during continuous methotrexate infusion. Cancer Res 37:445–450
29. Pommier Y, Lokiec F (1981) Le methotrexate. Chimiother Anticancer 31 (43):3083–3089
30. Porpaczy P, Schmidbauer CP, Georgopoulos A, Endler AT (1983) Pharmacokinetics of high-dose methotrexate in dogs. An experimental model with diffusion chambers. Cancer Chemother Pharmacol 11:172–176
31. Rechnitzer C, Scheibel E, Hendel J (1981) Methotrexate in the plasma and cerebrospinal fluid of children treated with intermediate dose methotrexate. Acta Paediatr Scand 70:615–618
32. Reich SD (1979) Mathematical modeling-guide to high-does methotrexate infusion therapy. Cancer Chemother Pharmacol 3:25–31
33. Shapiro WR, Young DF, Mehta BM (1975) Methotrexate: distribution in cerebrospinal fluid after intravenous, ventricular and lumbar injections. N Engl J Med 293:161–166
34. Sheppard CW (1962) Basic principles of the tracer method. Introduction to mathematical tracer kinetics. Wiley, New York
35. Smith DK, Omura GA, Ostroy (1980) Clinical pharmacology of intermediate-dose oral methotrexate. Cancer Chemother Pharmacol 4:117–120
36. Storm AJ, Van der Kogel AJ, Nooter K (1985) Effect of X-irradiation on the pharmacokinetics of methotrexate in rats: alteration of the blood-brain barrier. Eur J Cancer Clin Oncol 21:759–764
37. Tsursawa M, Sasaki K, Matsuoka H, Yammamoto Y, Katano N, Fujimoto T (1986) Flow cytometric analysis of marrow cell kinetics in children treated with high-dose MTX and CF rescue. Cancer Chemother Pharmacol 16:277–281
38. Vadia P, Blair AH, Ghose T, Ferrone S (1985) Uptake of methotrexate linked to polyclonal and monoclonal antimelanoma antibodies by a human melanoma cell line. J Natl Cancer Inst 74:29–35
39. Yalowich JC, Fry DW, Goldman ID (1982) Teniposide (VM-26) and Etoposide (VP-16-213)-induced augmentation of methotrexate transport and polyglutamylation in Ehrlich ascites tumor cells in vitro. Cancer Res 42:3648–3653

Subject Index

A

Absorption
- of cobalamin 29–31, 55, 71
- of folate 137
Adenosylcobalamin 148, 225–226
Alcohol 79, 137, 141–143
Anti-intrinsic factor antibody 29–31, 88

B

Bacterial proliferation and cobalamin absorption 78
Biopterin and folate metabolism 210, 215

C

Carcinoid tumors 95
Cobalamin
- absorption and malabsorption 73–74, 105, 220–223
- analogues 55, 105
- assay 23–25
- binding proteins 53
- metabolism 53, 71
- - interrelations with folate 41
Cobalophilin 54
Corrinoïds 55, 105
Cystic fibrosis and cobalamin absorption 112

D

Deoxyribonucleotide triphosphate 1
Deoxythymidine triphosphate (dTTP) 1, 4, 10, 15–18

Deoxyuridine
- monophosphate (dUMP) 3–4, 10–11
- suppression 5, 10–11, 27–28, 43
Dihydrobiopterin reductase 216
Dihydrofolate reductase 5, 207, 217, 245
DNA
- polymerase 6, 8–9
- replication 1, 5, 8, 15
- synthesis 1, 3, 5
Drugs interfering with cobalamin and folate metabolism 137

F

5-Fluorouracil 138–139, 245
Folate
- assay 26–27
- deficiency
- - in developing nations 171
- - in geriatrics 179
- - in intensive care patients 191
- - in pregnancy 161
- metabolism 137
- requirements 122
- status 119, 134
Folylpolyglutamate synthetase 209
Formate 44–46, 48
Formyltetrahydrofolate 11–13

G

Gastrectomy 74
Gastric cancer and pernicious anemia 92
Gastritis 75, 88

H

Haptocorrin 54, 105, 107
Homocysteine 11, 12, 14
Homocystinuria 203–207, 226–227

I

Inherited defects of cobalamin metabolism
    219
Inherited defects of folate metabolism 199
Intrinsic factor 29–30, 54, 105

L

Lactation and folate deficiency 168

M

Macrocytosis 22
Macro-ovalocytes 23
Megaloblastic anemia 1, 10, 12, 14, 16, 21
Megaloblastosis 1, 48
Methionine 11–13
Methionine synthase 11–14, 207, 227
Methotrexate 5, 12, 245
Methotrexate pharmacokinetics 253
Methylcobalamin 225–227
Methylenetetrahydrofolate 3
Methylenetetrahydrofolate deficiency 202,
    217
Methylmalonylaciduria 226
Methyltetrahydrofolate 11–13
Methyltetrahydrofolate trap 12, 42–43

N

Neuropsychiatric illness in relation with co-
    balamin or folate deficiency 145
Nitrous oxide 44

O

Okasaki fragments 7, 13, 15

P

Pancreatic insufficiency and cobalamin ab-
    sorption 76, 105–108

Pernicious anemia 29–30, 85
Polyglutamates 4–5, 45
Pregnancy and folate deficiency 161

R

R binder, R proteins 53, 219, 225
Replicon 7–8
Replitase 9

S

S-adenosylmethionine (SAM) 13, 47, 206
Schilling test 29–32, 73–74, 90

T

Tetrahydrobiopterin and folate metabolism
    215
Tetrahydrofolate 5
Thymidine 2, 3, 46
Thymidine kinase 4, 9, 11
Thymidine monophosphate (dTMP) 3
Thymidylate synthase 5, 9–12
Transcobalamine 29–32, 54
Transcobalamin I 56–57
Transcobalamin II 60–61, 219, 223
Transcobalamin II receptor 53
Transport of folate 231
Transport of methotrexate 233–238

18324643R10149

Made in the USA
Lexington, KY
27 October 2012